Solid Waste Management in Indian Cities
Status and Emerging Practices

Editors

Darshini Mahadevia
Jeanne M. Wolfe

CONCEPT PUBLISHING COMPANY, NEW DELHI-110059

ISBN: 13-978-81-8069-522-3
ISBN: 10-81-8069-522-0

First Edition 2008

Published and Printed by
Ashok Kumar Mittal
Concept Publishing Company
A/15-16, Commercial Block, Mohan Garden
New Delhi-110059 (INDIA)

Phones : 25351460, 25351794
Fax : 091-11-25357103
Email : publishing@conceptpub.com

Preface

This book is the result of three years of collaborative work between faculty and graduate students of the Schools of Planning of CEPT University in Ahmedabad and McGill University in Montreal. It is the follow up to a research project on solid waste management entitled "Urban Governance and Environmental Management: Action-Oriented Policy Studies on Waste Management in Jabalpur and Raipur" conducted with local NGOs, SAMVAD and LAHAR, between 2003 and 2005. The entire work was funded by the Shastri Indo-Canadian Institute, to whom the researchers are extraordinarily grateful since it set them on paths of exciting investigation.

One NGO partner, Mr. Jayant Verma of SAMVAD (Society for Advocacy and Development), Jabalpur, suggested that one of the problems in undertaking advocacy work in the field of solid waste management is the absence of knowledge about precisely how other urban centres attack solid waste problems: what are the success stories and what are the bottlenecks? While there is an enormous literature on the subject, both Indian and international, there is little detailed work that links the chain of municipal management systems, the actors, the institutional decision-making mechanisms, the budget, the achievements and the concerns. Most case studies focus on the technical mechanics of the process. This work is an attempt to fill that gap.

This book is the result of comparative research work on solid waste management projects undertaken in seven cities in India. Each case study examined the scale of the project, target groups, institutional framework, actors involved, manpower, innovation in practice, technology, finance systems, environmental impacts, conflicts, sustainability and potential for up-scaling. It was found that while some systems work well in small-circumscribed neighbourhoods, they do not work well for the greater part of the city, or that while some activities work smoothly, such as street sweeping, others do not – particularly the management of final disposal sites. Research also shows that some cases that have been the subject of uncritical praise in the literature, show very mixed results.

We would like to thank not only the Shastri Indo-Canadian Institute and Mr. Jayant Verma, but also all the people who freely extended information, opinions and hospitality to our researchers. These include municipal officials and employees of all ranks in the case-study cities, Bangalore, Hyderabad, Kolkata, Mumbai, Ranchi, Suryapet and Vegalpur, along with NGO workers, personnel of various research institutes, colleagues, and residents association members.

Darshini Mahadevia
Jeanne Wolfe

Contents

Preface v

List of Tables ix

List of Coloured Photos xii

List of Figures and Boxes xiii

Acronyms xv

List of Contributors xxi

1. **Introduction** 1
   Darshini Mahadevia and *Jeanne Wolfe*
2. **Solid Waste Management in Indian Cities: An Overview** 24
   Darshini Mahadevia, *Jeanne Wolfe* and *Appeeji Parasher*
3. **Centralized and Decentralized Solid Waste Management Practices in Bangalore** 85
   Natalie Kehle, *Darshini Mahadevia* and *Appeeji Parasher*
4. **Local Government's Initiative: Hyderabad** 133
   Darshini Mahadevia and *Manish Pathak*
5. **Modernizing Municipal Management: A Case Study of Kolkata Municipal Corporation** 170
   Natalie Kehle, *Darshini Mahadevia* and *Jeanne Wolfe*
6. **Multiple Partnerships: Mumbai** 212
   Darshini Mahadevia and *Bela Pharate*

7. **NGO Initiated Solid Waste Management: Ranchi** 257
Darshini Mahadevia and *Appeeji Parasher*

8. **A Litter Free Small Town: Suryapet** 289
Darshini Mahadevia and *Manish Pathak*

9. **Managing Waste in a Peri-urban Locality: Vejalpur, Ahmedabad** 312
Shrawan Kumar Acharya and *Appeeji Parasher*

Bibliography 329

Index 348

List of Tables

1.1 Rationale for the Choice of Case Study Cities 4
1.2 Modes of Governance—Market, Hierarchy and Network 7
1.3 Criteria for the Good City 9
1.4 Normative Goals, Means and Methods for the Inclusive City 10
1.5 Status of Compliance with MSW Rules 2000 in Cities of India, 2004 21
2.1 Research Agenda for SWM Study 26
2.2 Dates of Important New Initiatives 29
2.3 Waste Generated and Composition for Specific Cities, 2005 30
2.4 Waste Generation, Case Study Cities 2005 31
2.5 Recyclable Materials as a Percentage of Total SW, Case Study 32
2.6 Price of Recyclable Materials, Hyderabad 2005 32
2.7 Actors and their Roles 35
2.8 Actors in Residential Solid Waste Processing 38
2.9 Examples of Partnerships and Alliances 48
2.10 Some Examples of Private Sector Firms Active in Solid Waste Disposal 52
2.11 Returns to Actors in the Waste Reycling Chain in Hyderabad 57
2.12 Primary Waste Collection 58
2.13 Percentage Waste Segregation 59
2.14 Secondary Waste Transportation 60
2.15 Waste Disposal Sites in Case Study Cities 61
2.16 Waste Disposal: Composting and Waste to Energy 62
2.17 Waste Composting 63
2.18 SWM Budgets and Wages 66

2.19	Some Cost Recovery Variables	69
2.20	International Agreements on Waste Management	79
3.1	Source and Quantity of Solid Waste in Bangalore	91
3.2	Bangalore's Urban Waste Composition	93
3.3	Quantity of Bio-Medical Waste Generated in Bangalore	93
3.4	Status of Wards Maintained by BMP and Through Outsourcing	95
3.5	Existing Partnerships in SWM	102
3.6	Waste Disposal Practices in Bangalore	105
3.7	KCDC Annual Summaries for 2002-03 and 2003-04	107
3.8	Citizen Anchor Categories in Ward 54	112
3.9	Kalyananagara Waste Committee Monthly Budget	120
3.10	BMP Expenditure Incurred for Collection, Transportation and Disposal	121
3.11	Expenditures on Different SWM Activities	122
3.12	'Swachchha Bangalore' Campaign Budget	123
3.13	Receipts Pertaining to SWM	124
3.14	Proposed Cess on SWM	124
4.1	Solid Waste Composition, MCH	135
4.2	Price of Recyclable Materials, MCH	135
4.3	Change in Quantity of Solid Waste by Sources, 1997-03	136
4.4	Sweeping Logistics, MCH	139
4.5	Costs per Unit for Different Activities	140
4.6	BE of Some Important Departments, MCH (Year 2005-06)	153
4.7	Revenue Expenditure by MCH Over Time	154
4.8	Revenue Receipts by the MCH, Year 2005-06	155
4.9	Comparative Costs	155
5.1	Sources of Funding under the KEIP	175
5.2	Spending for Environmental Improvements under the KEIP	175
5.3	Internal Revenue Changes from 1992-93 and 2005-06	178
5.4	Composition of KMC Waste over Time	179
5.5	KMC: Employees in SWM 2005 (Excluding Private and Informal Sectors)	182
5.6	The Gradual Privatization of SWM Services	184
5.7	Vehicle Report at Dhapa for October 10th 2005	195
5.8	Total Spending for SWM from 2002-03 to 2005-06	199

5.9 Breakdown of SWM Expenses from 2002-03 to 2004-05 200
5.10 Revenue from SWM Receipts compared to SWM Spending 200
6.1 Zones and Wards of MCGM 215
6.2 Wards Committees Details 215
6.3 Garbage Generation by Geographic Locations, Mumbai 218
6.4 Physical Characteristics of Municipal Solid Wastes by Region (% Weight) 218
6.5 Constituents of Recyclable Wastes at Source and Disposal Sites (in %), 1994 219
6.6 Amount of Waste Disposed at Three Dumping Sites 226
6.7 Annual Budget on SWM 227
6.8 Distribution of ALM Groups 231
6.9 Coverage of Slum Adoption Scheme, March 1, 2006 235
6.10 Decentralised Waste Disposal, Mumbai 244
6.11 Stage-wise Partnership with MCGM for Solid Waste Management 248
7.1 Ward-wise Population and Total Waste Generated Per Day 260
7.2 Chronology of Initiative Taken Under the CJP 262
7.3 Number of Capacity Building Sessions Organised under the CJP 269
7.4 User Charge Contribution under CJP 278
7.5 Replication in Ranchi and in other ULBs 282
8.1 Daily Waste Generation (MT) 291
8.2 Basic Figures of ZWM, Suryapet 291
8.3 Work Schedule 298
8.4 Income Details, 2004-05 & 2005-06 304
8.5 Estimated Expenditure 2005-06 305
8.6 Details of Expenditure on Health and Sanitation, 2005-06 306
8.7 Fund Requirement for Upgrading the System 306
9.1 Operating Expenditure 321
9.2 Annual Budget of Vejalpur Nagarpalika 321

List of Coloured Photos

3.1	Doorstep Waste Collection by Pourkarmika—Indira Nagar, Bangalore	131
3.2	Street Sweeping by PMC worker; No Dustbins on Streets and Footpath, Bangalore	131
4.1	Steel Bins Emptied into Trucks at Transfer Stations, Hyderabad	132
4.2	Developed Landfill Site, Hyderabad	132
5.1	Compost heap from the plant at Dhapa, Kolkata	187
5.2	A waste collector depositing the day's collection, Kolkata	187
5.3	Waste truck at Dhapa weighing station, Kolkata	188
5.4	Large concreted transfer station, Kolkata	188
5.5	Informal waste pickers are abundant at the disposal site, Kolkata	188
6.1	Waste Collection Bin at a Park Maintained by Pestom Sagar ALM, Mumbai	241
6.2	Compost Pit in Pestom Sagar ALM, Mumbai	241
6.3	Street side plantations and clean Drains, Joshi Lane, Mumbai	241
6.4	Mittal Apartment ALM, Juhu	242
6.5	Compost Units along Roadside for garden waste, Mumbai	242
6.6	Biogas Unit at BARC Mumbai	242
7.1	PDC meetings in progress, Ranchi	271
7.2	Doorstep collection and special cleanliness during Diwali festival, Ranchi	271
7.3	A Clean Inner Street, Ranchi	272
8.1	A Slum Neighbourhood, Suryapet	293
8.2	Household Waste Collection, Suryapet	293
8.3	A Clean Street, Suryapet	293
8.4	Tractor with Two Partitions, Suryapet	294
8.5	Red & Green Dustbins Distributed, Suryapet	294
9.1	Clean Streets, Vejalpur	315
9.2	Arogya Bhaginis collecting segregated waste, Vejalpur	315
9.3	Dumping Site, Vejalpur	316

List of Figures and Boxes

Figures

2.1 Correlation Between Per Capita Income and Waste Generation 33
3.1 Bangalore District Composition 88
3.2 BMP Health Department Organization Relating to SWM 89
3.3 BMP Engineering Department Organization 90
3.4 Correlation Between Per Capita Income and Waste Generation 92
4.1 Working of Units 138
4.2 Collection System, MCH 141
4.3 Solid Waste Management System, MCH 141
4.4 Logistics of Collection, MCH 142
4.5 Administrative Structure, MCH 149
4.6 Partnerships 151
5.1 Changes in Revenue Sources 178
5.2 SWM Department of the KMC 181
6.1 Municipal Wards of Greater Mumbai 214
6.2 Organisation Chart of SWM Department of MCGM 221
6.3 Organisation Chart of SWM Department of MCGM at Ward Level 222
6.4 Format for ALM Registration 230
7.1 Clean Jharkhand Project (CJP) Coverage by Wards 261
7.2 Organization Structure of RMC 263
7.3 Organization Structure of NBJK 265
8.1 SWM System of Suryapet 296
8.2 Institutional structure of Suryapet Municipality 300

Boxes

1.1 India: Functions of Urban Local Bodies 14

1.2 Laws Relating to Solid Waste 18

5.1 SWM Under the Kolkata Environment Improvement Project (KEIP) 176

Acronyms

ABETS	:	Advanced Bioresidue Energy Technology Society
ACC	:	Assistant Community Coordinator
ADB	:	Asian Development Bank
AHS	:	Assistant Head Supervisor
AIILSG	:	All India Institute for Local Self-Governance
ALM	:	Advance Locality Management
AMC	:	Ahmedabad Municipal Corporation
AP	:	Andhra Pradesh
APIDC	:	Andhra Pradesh Industrial Development Corporation
APM	:	Assistant Project Manager
AUA	:	Ahmedabad Urban Agglomeration
AUDA	:	Ahmedabad Urban Development Authority
BARC	:	Bhabha Atomic Research Centre
BATF	:	Bangalore Agenda Task Force
BDA	:	Bangalore Development Authority
BEST	:	Bombay Electricity Supply and Transport
BOOT	:	Built-Own-Operate-Transfer
BOT	:	Build-Operate-Transfer
BUA	:	Bangalore Urban Agglomeration
BMC	:	Bangalore Municipal Corporation
BMP	:	Bangalore Mahanagar Palika
BMW	:	Bio-medical Waste
CA	:	Constitutional Amendment
CBO	:	Community Based Organisation
CBMWTDF	:	Common Bio-Medical Waste Treatment and Disposal Facility
CBP	:	Capacity Building Programme

CBSWM	:	Community Based Solid Waste Management
CDS	:	Community Development Societies
CIDCO	:	City and Industrial Development Corporation
CIVIC	:	Citizens' Voluntary Initiative for the City
CEE	:	Centre for Environment Education
CEMSAP	:	Calcutta Environmental Management Strategy and Action Plan
CFC	:	Central Finance Commission
CJP	:	Clean Jharkhand Project
CMAG	:	City Manager's Association of Gujarat
CMC	:	Chennai Municipal Corporation
CMFA	:	Controller of Municipal Finance and Accounts
CPCB	:	Central Pollution Control Board
CPHEEO	:	Central Public Health Environment and Engineering Organisation
CPI	:	Communist Party of India
CSI	:	Chief Sanitary Inspector
DFID	:	Department for International Development
DHO	:	Deputy Health Officer
DMC	:	Delhi Municipal Corporation
DWCUA	:	Development of Women and Children in Urban Areas
EMD	:	Earnest Money Deposit
EMS	:	Environmental Management Strategy
ETP	:	Effluent Treatment Plant
EXNORI	:	Excellent Novel and Radical Ideas
FB	:	Forward Bloc
FEDINA	:	Foundation for Education Innovations in Asia
FICCI	:	Federation of Indian Chambers of Commerce and Industry
GIS	:	Geographical Information System
GoI	:	Government of India
GoJ	:	Government of Jharkhand
GoWB	:	Government of West Bengal
HDPE	:	High Density Poly-ethylene
HEC	:	Heavy Engineering Corporation
HH	:	Household
HMC	:	Howrah Municipal Corporation
ICEF	:	Indo-Canadian Environment Facility

IMA	:	Indian Medical Association
IPE	:	Infrastructure Professional Enterprise
ISWM	:	Integrated Solid Waste Management
IT	:	Information Technology
IUEIP	:	Integrated Urban Environment Improvement Project
IWB	:	Itinerant Waste Buyers
JNNURM	:	Jawaharlal Nehru National Urban Renewal Mission
KAIC	:	Karnataka Agro Industries Corporation Limited
KCDC	:	Karnataka Compost Development Corporation
KEIP	:	Kolkata Environmental Improvement Programme
KPCB	:	Karnataka Pollution Control Board
KMA	:	Kolkata Metropolitan Area
KMC	:	Kolkata Municipal Corporation
KMDA	:	Kolkata Metropolitan Development Authority
KSCMF	:	Karnataka State Co-operative Marketing Federation
KUA	:	Kolkata Urban Agglomeration
LACC	:	Local Area Citizens' Committees
LACG	:	Local Area Citizens Group
LIC	:	Life Insurance Corporation
MCH	:	Municipal Corporation of Hyderabad
MCGM	:	Municipal Corporation of Greater Mumbai
MECON	:	Metallurgical and Engineering Consultants
MIC	:	Mayor in Council
MIS	:	Management Information System
MLA	:	Member of Legislative Assembly
MMRDA	:	Mumbai Metropolitan Regional Development Authority
MoEF	:	Ministry of Environment and Forest
MoH	:	Ministry of Health
MoU	:	Memorandum of Understanding
MoUD	:	Ministry of Urban Development
MSW	:	Municipal Solid Waste
MT	:	Metric Tonnes
MUA	:	Mumbai Urban Agglomeration
MW	:	Mega Watt
NORAD	:	Norwegian Agency for Development Cooperation

NBJK	:	Nav Bharat Jagriti Kendra
NC	:	Neighbourhood Committee
NFC	:	National Finance Commission
NGO	:	Non-Governmental Organisation
NHG	:	Neighbourhood Group
NSFD	:	National Slum Dwellers Association
NSS	:	National Sample Survey
O&M	:	Operation and Maintenance
ODA	:	British Government Overseas Development Administration
OSD	:	Officer on Special Duty
PAC	:	Public Affairs Centre
PPF	:	Public Provident Fund
PPP	:	Public-Private Partnership
PSP	:	Private Sector Participation
PDC	:	Pocket Development Committee
PH	:	Public Health
PUBLIC	:	Public United for a Better Calcutta
PVP	:	*Parisar Vikas* Programme
RBI	:	Reserve Bank of India
REDS	:	Rag-pickers Education and Development Scheme
RISE	:	Residents' Initiative for a Safe Environment
RMC	:	Ranchi Municipal Corporation
RRDA	:	Ranchi Regional Development Authority
RTO	:	Road Transport Organization
RSP	:	Revolutionary Socialist Party of India
RUA	:	Ranchi Urban Agglomeration
RWA	:	Resident Welfare Association
SAP or SAS	:	Slum Adoption Programme/Scheme
SANDEC	:	Department of Water and Sanitation in Developing Countries
SBP	:	Swachha Bangalore Programme
SCP	:	Stakeholders' Consultation Process
SEWA	:	Self Employed Women's Association
SHG	:	Self Help Group
SJSRY	:	Swarna Jayanti Shahari Rojgar Yojana
SMS	:	Stree Mukti Sangathan
SNP	:	Slum Networking Programme

SSI	:	Small-Scale Industries
SW	:	Solid Waste
SWM	:	Solid Waste Management
TCS	:	Tata Consultancy Services
TERI	:	The Energy Resources Institute
TIDE	:	Technology Informatics Design Endeavour
TMC	:	Town Municipal Council
TP	:	Town-Planning
TPB	:	Trained Parisar Bhaginies
TS	:	Transfer Station
ULB	:	Urban Local Body
USAID	:	United States Aid for International Development
USEP	:	Urban Self Employment Programme
UWEP	:	Urban Wage Employment Programme
VGDS	:	Voluntary Garbage Disposal Scheme
VMC	:	Vijaywada Municipal Corporation
WBPCB	:	West Bengal Pollution Control Board
WC	:	Ward Committee
WDC	:	Ward Development Committee
WHO	:	World Health Organization
YWVA	:	Youth for Unity and Voluntary Action
ZWM	:	Zero-Waste Management

List of Contributors

Appeeji Parashar, Research Associate, Faculty of Planning and Public Policy, CEPT University and now Executive Analyst with CRISIL Risk and Infrastructure Solutions Limited.
Email: appeejip@yahoo.co.in

Bela Pharate, Research Associate, Faculty of Planning and Public Policy, CEPT University and now Environmental Planner at Maharashtra Pollution Control Board, Mumbai.
Email: bela_p@rediffmail.com

Prof. Darshini Mahadevia, Prof. Darshini Mahadevia is teaches at Faculty of Planning and Public Policy, CEPT University, Ahmedabad. Educated at CEPT, Ahmedabad, School of Planning and Architecture, New Delhi, and PhD from Jawaharlal Nehru University, New Delhi, she has been as visiting scholar at Graduate School of Architecture and Urban Planning, University of California, Los Angeles; and Visiting Scholar at School of Urban Planning, McGill University, and visiting fellow at Tsinghua University, Beijing. Has been recipient of international fellowships such as Canadian Studies Fellowship of Shastri Indo-Canadian Institute for research on Canada and Asian Scholarship Foundation for research on China. Has collaborated on research projects with University of Oslo, St. Mary's University, Halifax, Institute of Social Studies, The Hague and School of Urban Planning, McGill University. Has wide experience of working with NGOs in India. Has number of publications in books and academic journals on urban issues, human and gender development, and environment. Email: darshini@cept.ac.in

Prof. Jeanne Wolfe, Emeritus Professor at the School of Urban Planning, McGill University, Montreal. Educated at London University, the University of Western Ontario and McGill, she worked at the City of Montreal, the Quebec Provincial Government and in the private sector as a planner before beginning her academic career. She was Director of the School of Urban Planning from 1988 to 1999. Her research interests include housing, governance and institutional change, urban policy, and international development studies. She has worked extensively in the Caribbean region, helped in setting up a planning program at the University of the West Indies in Trinidad with CIDA aid, and has participated in projects in China and India. In addition she has often been called upon to be a commissioner in public hearings on planning issues in Montreal. She is widely published in a variety of academic journals. Email: jeanne.wolfe@mcgill.ca

Manish Pathak, Student Intern on the project, at Faculty of Planning and Public Policy, CEPT University and now Senior Associate Analyst, ICRA Limited, India. Email: apmanish@yahoo.co.in

Natalie Kehle, Student Intern on the project, Environmental Studies, McGill University. Email: natalie.kehle@ elf.mcgill.ca

Prof. Shrawan Kumar Acharya, Professor Shrawan Kumar Acharya teaches at the Faculty of Planning and Public Policy, CEPT University, Ahmedabad. After his schooling in Kalimpong he completed his graduation from Santiniketan, post graduation and PhD from the Jawaharlal Nehru University, New Delhi. He also holds M. Tech. in Urban and Regional Planning from CEPT and Diploma in Urban Management from The University of Birmingham, U.K. He was awarded UGC fellowship to undertake his PhD at JNU. He was a SHASTRI Fellow at the McGill University and ASIA Fellow at the Chinese Academy of Social Sciences, Beijing and the Institute of World Economics and Politics, Hanoi. Prof. Acharya has published in international and national journals on urban development, community development and hill areas. Email: shrawan.acharya@gmail.com

1

Introduction

DARSHINI MAHADEVIA and JEANNE WOLFE

Background

Urban residents were shocked by the floods in major metropolitan cities of India during the monsoon of 2005 caused by unusually heavy downpours. The list is formidable: Ahmedabad June 26, Vadodara June 30, Mumbai July 26, Chennai October 10, Kolkata October 21, and Bangalore October 25[1]. Rains persisted in Chennai much longer than usual and so did the flooding, adversely affecting the normal functioning of the city. Ahmedabad, Vadodara and Mumbai were victims of the southwest monsoon, which in any event tends to be heavy, while Chennai, Kolkata and Bangalore were affected by the returning southeast monsoon.

Mumbai, India's financial capital, was paralyzed by an unprecedented rainfall of 944 mm. on July 26, 2005, which coupled with a high-tide, led to unseen levels of flooding. The whole of Mumbai north of the island city and the nearby towns of Thane and Kalyan—Dombivali, were inundated. By the evening of July 26, the entire city had come to a complete halt: one-third of the city's surface area was totally flooded, with water as high as one storey (10 or 12 ft.) in some places, submerging ground floor dwellings. Telecommunication lines were snapped; arterial roads were jammed; the airport was closed and public transport came to a grinding halt. The three suburban train networks, the lifelines of the city, could not operate for close to 36 hours, stranding an estimated 150,000 people. Almost 15,000 children were marooned in schools without food and water. The entire nation's attention was on Mumbai.

According to the state government, 736 people died in the city and 191 in neighbouring Thane[2].

The nightmare does not stop at the deluge. Post-deluge epidemics followed. Concurrent outbreaks of leptospirosis and dengue had occurred in the slums of Mumbai before, in the wake of the monsoon of 2002 (Karande *et al.* 2005). But in 2005 it was worse, outbreaks of dengue fever, not heard of often before 2005, leptospirosis and *Falcipherum malaria* occurred in affected cities. In Mumbai[3], 37 people died within 48 hours of the storm and 250 people were admitted to hospitals for leptospirosis[4]. In Ahmedabad 350[5] people were afflicted by dengue and Kolkata had 945 reported cases[6]. The deluge in Bangalore most severely affected the southeast part of the city, and not only the slum dwellers but the Information Technology companies suffered great losses.

Every monsoon season it is anticipated that on days of heavy rain some low lying urban areas will flood, and that normal life will be disrupted. The slum dwellers of such cities as Mumbai, Surat and Kolkata, which are particularly wet locales, are often forced to sleep with water inside their homes for days on end. Inevitably, epidemics of malaria and water borne diseases follow. The Surat plague of 1995 is remembered for its severity, while the less dramatic are soon forgotten. Life returns to normal as if such hardships are part of urban living, and residents, especially the poor, have to bear them stoically year after year. But two changes have happened in recent times. Twenty-four hour news channels now keep reminding us of current and past disasters and engrave their impacts on our minds. At the same time, the severity of such episodes has increased of late and even normal rains tend to throw life out of gear in many Indian metropolises, including those aspiring to be 'World Class Cities'[7]. Neither the large metros nor the small cities are able to cope with the monsoon downpours and their aftermath.

There are many reasons for the inability to cope even in globalising megacities. The first is very obvious: an urban planning failure. When cities expand into new areas, real estate development often takes place first, and housing is sold and occupied long before infrastructure such as roads, surface-water drainage and sewerage networks are constructed[8]. Then follows confusion about which local authority should build and manage the needed facilities, provide services and deal with waste. Patchy and inadequate infrastructure almost inevitably leads to floods and consequent epidemics in these peri-urban areas.

The third reason is that the lands were acting as natural soak-aways are built up randomly. Storm water drainage systems are either absent or not designed to accommodate monsoon downpours. In addition, many sewers are open canals. In 1998, in 68.8 per cent of cities, sewerage drains were open (Mahadevia and Sarkar 2003). Garbage and solid waste get dumped into them, and scanty maintenance results in blockages. Lack of garbage pick up, and littering by people and animals, results not only in open drains being filled up, but even closed drains becoming clogged. Finally, the increasing use of plastics, and especially plastic bags, is the major contributor to choked drains. This is why ineffective solid waste management is an important cause of the severe flooding of cities and its dreadful consequences.

The management of solid waste is a worldwide concern. Its importance barely needs reiteration: proper management contributes to public health, the protection of the natural environment and the preservation of ecosystems. Improper management leads to decaying filth, litter, infestations by vermin, insects and pathogens, accumulations of toxins in the soil, air and water-bodies, noxious gases and odours, and unsafe food for both people and animals. Mineral, animal and vegetative debris, along with plastics, tyres, metals, hydrocarbons, bio-medical wastes and a wide variety of chemical compounds left behind by industrial transformation, continue to accumulate. The collection, transportation and disposal of solid waste is an ever-growing challenge, especially in the developing world, where galloping urbanization only exacerbates the problem.

While the focus of this book is on solid waste management practices in a number of Indian cities, the approach to the topic originally developed from lively debates on the nature of metropolitan governance. How do cities and their surrounding municipalities manage to provide services to their citizenry? What is the best way to furnish roads, water, sewerage systems, educational and social services, parks playgrounds, garbage removal and other amenities? What level of government, or other agency, should be responsible for what? What are the decision-making processes? What are the financing mechanisms? How can all these be improved? Since this is such a vast subject, it was decided to look at these questions through an examination of one service, namely solid waste management, because this is such an all-encompassing issue in its effect on the health, safety and quality of urban life.

This introductory chapter brings together two lines of enquiry relating to urban policy and urban management. Specifically, this chapter relates urban governance and decentralization to the unfolding story of solid waste management in India and its institutional framework. Chapter two analyses the problems of solid waste management in India through a synthesis of the literature and our own research results from a number of cities, while also looking at the adequacy of information available to properly manage solid waste. The following seven chapters provide detailed case studies of solid waste management practices in the following cities: Bangalore, Hyderabad, Kolkata, Mumbai, Ranchi, Suryapet and Vegalpur—Each of which was chosen in consultation with experts and through literature reviews, because some aspects of their operations have interesting innovations. The reasons for their selection are summarized in Table 1.1 below, and will be treated in detail in chapters three to

Table 1.1: Rationale for the Choice of Case Study Cities

City/ Jurisdiction	*Nature of Innovation*	*Reason for selection*
Bangalore	Institutional-coalition	The Bangalore Agenda Task Force (BATF) formed in 1999 through a state government mandate as a PPP, aimed to be a driving force working with all stakeholders to upgrade the city infrastructure and make it the best city in India by 2005.
Hyderabad	Good waste collection	Winner of the Clean City award given by HUDCO for six consecutive years
Kolkata	KEIP	The Kolcata Environmental Improvement Programme (KEIP), 2002-2007, was designed to take a holistic view of all environmental problems.
Mumbai	Institutional versatility	Advanced Locality Management (ALM) scheme as an example of decentralized management: waste handling and transport contracted out: Waste-to-Energy disposal as well as dumping.
Ranchi	NGO initiated	Nav Bharat Jagriti Kendra (NBJK), an NGO, started up and runs the system as part of the Clean Jharkhand Project: Pocket Development Committees (PDCs) for each 200 households.
Suryapet	Managerial efficiency	Famous as a 'zero litter' city: Clean-up accomplished between 2002-2004 with no extra funding.
Vegalpur Nagarpalika	Managerial efficiency	Litter free: comparable in size to Surypet: peri-urban location to Ahmedabad: Early e-governance system.

nine. An innovation might be a process or procedure, or activities within a certain area of a city. Needless to say, no city is perfect, and there is no single best way in which problems can be solved. Indeed, each Urban Local Body and Metropolitan area will have to improve its performance within its own local context and institutional and political culture.

Urban Governance

Since the early nineties, the problem of urban governance has received much attention worldwide in both scholarly and practical circles. It has been brought into focus by the forces of globalization, the mobility of capital, people and ideas, along with the growing recognition that cities are both the major creators of wealth in most nations and at the same time are the sites of much hardship, poverty and environmental degradation. The need to govern and manage urban areas in an efficient and equitable manner is now seen as a prerequisite for economic growth and poverty alleviation in both the developed and developing world (for a summary see: World Bank 1992). In line with these trends, loosely described as the "Washington Consensus", India embarked upon major economic and governmental reforms in 1991. Trade liberalization, deregulation, unshackling foreign exchange, and public sector reform through privatization and decentralization, were soon to be followed by efforts towards municipal reform and the modernization of city government.

It is generally recognized that formal local government institutions (the elected councils and the civil service) are only one of the actors involved in urban management. Urban local administrations are not only affected by the decisions of senior levels of government (national and state or provincial), as other local actors from the private sector and from civil society have important roles to play as well. For the private sector these might include activities ranging from leadership in economic promotion through providing transit to supplying or sub-contracting municipal services. Civil society includes a host of organizations from professional and service associations to advocacy groups and political associations. They include NGOs, CBOs, cooperatives, sports, cultural and recreational clubs, and religious groups, and may be involved in service provision for social welfare, schooling, housing, recreational facilities, environmental management, advocacy and many other activities (Jouve & Lefèvre 1999).

Public policy in many parts of the world acknowledges this three legged stool of state, market and civil society—hence the growing use of the word 'governance' rather than 'government'. The popularisation of this concept in the nineties is an outcome of three factors. First is a better understanding of how governments actually work in practice; in any city there is always interaction between the government and networks such as Boards of Trade, Chambers of Commerce, service groups and citizens' associations, whether this is formalized or not. Second, the world-wide neo-liberal swing in the 1980s, greatly promoted by the International Financial Institutions (IFIs), led towards increasing involvement of the private and community sectors in what were formerly seen as government preserves. Partnerships, privatization and deregulation mean that many more actors are now involved in service provision, even if it is simply on a contractual basis. Third, the paucity of public funds available to finance urban infrastructure, often coupled with the indebtedness of government, has led to a search for other sources of investment, namely from private and community sectors.

However, there has been very little research on how decisions are actually made within such triads, even in well studied societies such as Canada and the United States (Divay & Wolfe 2002). Formal local government structures have been well researched in the west, as have municipal council decision-making and councillors' voting patterns, but the process of decision-making in terms of para-governmental or non-governmental service delivery has received only cursory examination. Process examples should include: deciding what services or parts of services should be contracted out, purchased or leased, terms of reference, tendering procedures, selection of providers, management of contracts, performance criteria, monitoring service levels, shared responsibilities and liabilities, user satisfaction, and conflict resolution. To a large extent local government contracting all over the world has evolved fortuitously. Recently, questions of accountability and transparency have emerged, not to mention the reconciliation of differing managerial cultures.

The way in which the actors relate to each other is a paradoxical aspect of the governance paradigm. Lowndes and Skelcher (1998), working in the UK, show how each segment of the governance triad operates in a different way, and under a different set of assumptions, as is shown in the Table 1.2. Inter-agency relationships may be competitive, hierarchical or collaborative. The market mode (private sector) is competitive and revolves around contractual relationships and prices; the public sector is hierarchical and has to be the coordinating agency concerned with the public interest; civil society usually operates as a network and consists of actors with complementary interests based on trust, loyalty and reciprocity. If this analysis is correct, it is not surprising that Jouve and Lefèvre (1999) speak of the need for a new brand of "political entrepreneurs" within this conception of governance. Negotiations of partnerships must always bear these different characteristics of the actors in mind.

Table 1.2: Modes of Governance—Market, Hierarchy and Network

	Market (Private Sector)	*Hierarchy (Public Sector)*	*Network (Civil Society)*
Normative basis	Contract Property rights	Employment relationship	Complimentary strengths
Means of communication	Prices	Routines	Relational
Methods of conflict resolution	Haggling resort to courts	Administrative fiat, supervision	Norm of reciprocity Reputation concerns
Degree of flexibility	High	Low	Medium
Amount of commitment among the parties	Low	Medium	High
Tone or climate	Precision and/or suspicion	Formal, bureaucratic	Open-ended, mutual trust and benefits
Actor preferences or choices	Independent	Dependent	Interdependent

Source: Lowndes and Skelcher (1998).

A second paradox is that with more involvement of the private sector and civil society, the need for strong rules and regulations increases, especially relating to contracting out, partnerships, BOT (build, operate and transfer), BOOT (build, own, operate and transfer), BOLT (build, own, lease and transfer) and other degrees of privatization, in order to protect public interest. Thus, while the advocates of privatization call for "less government," at the same time, robust government capable of introducing and administering a fair and just regulatory system is a pre-requisite for successful private and third sector participation.

The focus of the governance debate is on the provision of collective goods and services which are defined as those enjoyed or used by the population at large: infrastructure, economic, social and cultural development facilities, and environmental protection. Also known as public goods, or joint consumption goods, these are needs that are "non-subtractable", because everyone requires them. Roads, water supply, sewage treatment and clean air are examples. However, the need is in more than just equipment and services as such, it is in the ultimate outcomes of their presence: fluid and safe traffic movements for instance, and most especially, high levels of public health.

Following this line of thought, another issue is the attempt to understand how the outcomes relate to, or are influenced by decision-making structures and processes. This is a most challenging task, since there are very few analytical attempts at the evaluation of results except on a programmatic basis. The precise mechanisms of vertical decision-making (that is, above and below the municipal level), are virtually unexplored territory. While powers and duties are laid out in laws and regulations, how decisions are actually made in practice is not really known.

In summary, the contemporary concept of governance involves multiple stakeholders, interdependent resources and actions, and shared purposes and blurred boundaries between the public and private, formal and informal, and state and civil society sectors. This complexity underlines a great need for coordination, negotiating and building consensus (UNCHS 2000: 57), a Herculean task for which there are no clear-cut answers. However, the point here is at least to know the parameters of the questions.

Many international organizations are promoting "Good Governance" as a means of improving the development process and reducing poverty. In fact it is presently one of UN Habitat's two Global Campaigns, the other being Security of Tenure (UN Habitat 2002). This leads us to the question of what are the criteria for good governance and how do we evaluate it? There is wide debate on this subject. Klink (2002) has listed three: efficiency, (economies of scale, territorial spill-overs), equity (redistribution), and voice (flexibility, accessibility, and accountability). In many studies, this type of approach is usually translated into normative terms, listing goals, general principles, criteria, or indicators that may or may not be measurable.

Alternative compilations of measures of good governance tend to respond to the approaches and values of their proponents. For instance, the urban consolidationists stress efficiency, economies of scale and equity; the public choice school emphasizes sensitivity to local area decision-making and public participation; and the environmentalists concentrate on measures of sustainable development, political ecology and performance indices (Bartone, Bernstein, Leitmann and Eigen 1994).

The eminent urban theorist, John Friedmann (1998: 20) extending his reflections upon the nature of the good city, prepared a series of criteria divided into 'good governance', 'good management' and 'good outcomes'. In other words he envisions a city with an open political process, that is well run, and provides a high level of services. (Table 1.3).

Table 1.3: Criteria for the Good City

Criteria of Good City Governance	*Criteria of Good City Management*	*Criteria of Good City Outcomes*
• Inspired political leadership • Public accountability • Inclusiveness • Responsiveness • Non-violent conflict resolution	• Accessibility, transparency, responsiveness • Effectiveness • Efficiency • Honesty	• A productive city • A sustainable city • A liveable city • An actively tolerant city • A caring city

Source: Friedmann (1998: 20)

> "Good *city governance* refers to the political process of allotting resources and "steering" the collective life of the political community. It involves the triad of state, market and civil society joined in various forms of collaborative local action. *Good city management* concerns the administration and use of common resources in bringing about those minimal conditions of urban life that nurture human flourishing. Finally, *good city outcomes* concern those which further the common good of the city, including the strengthening of good governance, thereby completing the circle." (Friedmann 1998: 20).

This conceptualization is perhaps the most helpful to the activist in that it forcefully reminds us, and puts up-front, the process sequence of political choice, administration and management, and the resulting quality of life outcomes. Much of the city-based literature focuses only on the managerial aspects.

UN Habitat, rather than focusing on criteria, has tried to operationalize the idea of a good city. It has prepared an expanded list of goals for the inclusive city, one where all residents would have equitable access to collective goods and services, thus introducing a rights dimension to the debate. These goals, along with their defining components and the means and methods of achieving them are shown in the table 1.4 because this line of argument is the model for India.

Table 1.4: Normative Goals, Means and Methods for the Inclusive City

Normative Goals	*Operational Components*	*Means and Methods*
Decentralization and participatory democracy	Local autonomy	■ Delegation of authority to the competent level closest to the citizens (decentralization, subsidiarity and proximity)
	Leadership for public participation and stakeholder involvement	■ City referenda ■ Public hearings and town hall meetings ■ City consultation and participatory planning ■ Citizens' fora and other mechanisms for negotiation ■ Processes for conflict mediation
	Building democratic culture	■ Enabling legislative framework to protect the rights and entitlements of all groups in society ■ Women's participation
	Enablement	■ Affirmative actions for marginalized groups ■ Procedures for public petitioning
Efficiency	Transparent financial management	■ Participatory planning and budgeting ■ Transparent contracting and procurement systems
	Administration and service delivery	■ Popularization of service standards and complaint procedures ■ Codes of conduct for leaders and officials ■ "Best value" approaches to target setting
	Efficient investment in infrastructure	■ Participatory strategic planning to address the needs of all groups in society ■ Public-private partnerships for service provision

(*Contd.*)

Table 1.4 (*Contd.*)

Normative Goals	*Operational Components*	*Means and Methods*
Equity	Resource allocation	▪ Investment incentives for targeted sectors and geographic areas ▪ Social pacts and fair and predictable regulatory frameworks
	Empowerment	▪ Rules governing freedom of access to local authority information ▪ Civic education ▪ Enabling legislative framework for traditional economic and social institutions and informal sector operators
Security	Environmental management	▪ Environmental planning and management methodologies based on stakeholder involvement
	Disaster preparedness	▪ Partnership disaster prevention strategies
	Crime control and prevention	▪ Conflict mediation mechanisms and taking into account local and ethnic democratic traditions ▪ Safety audits (especially for women) ▪ Partnership crime prevention strategies

Source: UNCHS (Habitat) (2000).

The UNCHS approach was adopted by the Government of India in launching the "Good Urban Governance Campaign" in September 2001. This campaign, headed by the Ministry of Urban Development and Poverty Alleviation, in cooperation with the All India Institute of Local Self-Government, the Commonwealth Local Government Forum, UN Habitat and a host of other players, has developed a National Action Plan designed to realize a vision of the Inclusive city (GOI 2001). The Action Plan framework essentially translates the means and methods column of the UN Habitat principles into the Indian context.

Decentralization

The ideology of decentralization from national or state levels to local government in order to improve services by putting power into the hands of the people who use them is an almost worldwide movement, and India is no exception (Dillinger 1993). Decentralization was originally promoted in the sixties and seventies by international experts as a way of relieving development pressures on primate cities and spreading the benefits of economic growth evenly within countries. It was essentially a national programme for national ends (Rondinelli 1983). Today it is advocated as a way of ensuring the provision of appropriate public services through participatory processes like local decision-making and citizen input. It is deeply embedded in the concept of the Inclusive City. Further, following the Rio declaration on Environment and Development of 1992, and the recommendations of Agenda 21, Chapter 28, decentralization is advocated because local control and empowerment is seen as the key to rehabilitating and protecting urban environments, while promoting development and growth (ICLEI 1996). India subscribes to these sentiments (GOI 2002a & 2002b).

Decentralization of Urban Governance in India

India took the bold initiative in 1992 of passing the 73rd and 74th Constitutional Amendments (CAs), which assign powers of self-government to rural and urban-local governments respectively. In India, local government is the responsibility of the States, although its status is governed by the national constitution. The main division in local government is between rural and urban areas. In rural areas local government may be single-tiered, two-tiered or three-tiered according to the locality. (The more lightly populated and or mountainous states tend to have single tiers). The smallest units are the Village or *Gram Panchayats* of which there are 240,588 across the country. The next level up is the *Panchayat Samiti*, also known as *Taluka* or Block Panchayats, of which there are 5,930, and the third tier is the *Zilla Panchayats* or District Panchayats, which number 459. The last named may contain as many as a thousand villages.

In urban areas, local government is single tiered, but there are three categories of Urban Local Bodies (ULBs), as they are known collectively. These are the *Nagar Panchayats* which are emerging urban local bodies, usually located in the fringe of an agglomeration; Municipal Councils for medium sized cities; and Municipal Corporations for the large cities. There are 3,694 ULBs. All this means that the country has over a quarter of a million local governments, all with directly elected councilors who hold office for five-year terms.

Further, although urban local government is not tiered as it is in rural areas, the constitutional amendments decree the formation of Metropolitan Planning Committees for all urban agglomerations of over one million population, and District Planning Committees for each administrative district. While the composition and precise duties of these committees are left to the states to decide, it has been ordered that they are to be largely made up of elected officials of member municipalities, two thirds for Metropolitan areas and four-fifths for Districts. Clearly, it is intended that policy making will be political. Some of the larger cities already have Metropolitan Authorities, created by special legislation years ago to administer certain area-wide functions. These include New Delhi, Kolkata, Mumbai and Chennai.

It is worth noting that historically the District, encompassing urban and rural territories, was the most important administrative division and still has an important role. The District, a sub-division of the State, is a creation of British rule dating from 1772, and run by the Indian Civil Service. The District Commissioner was responsible for law and order, communications and other civil matters, and the District Collector for assessment and collection of land taxes, and other financial matters. Today the district is still responsible for much of the regional administration in the country; the devolution of decision-making from highly skilled bureaucrats to elected politicians as required under the 73rd and 74th Constitutional Amendments (CAs) has not been simple.

The 74th Constitutional Amendments (CA) of 1992 is the primary legislation prescribing the role and responsibilities of urban local governments, which had previously been limited to maintenance services, water and sanitation, drainage, road sweeping, solid waste, markets and the like. The new approach is based on the principle that urban localities should be self-governing, and not merely administrative appendages of the states, and enjoins state governments to devolve local responsibilities to them. The major functions which are new to ULBs include urban planning, economic and social development, water supply, public health and sanitation, slum improvement and upgrading (Box 1.1). States may also devolve other responsibilities if they so choose. Representation in the ULBs attempts to be democratic: seats are reserved for Scheduled Castes and Scheduled Tribes in proportion to their population, and at least one-third of the seats are legislated to be occupied by women. If a ULB has more than 300,000 residents, community participation must be encouraged by the formation of Ward or Wards Committees. Further, ULBs have constitutionally guaranteed status as state governments can no longer dissolve them on a political whim, and there is a defined procedure and time-frame for suspension and reform. They are thus supposed to be assured long-term stability, and can plan ahead with confidence.

Box 1.1: India: Functions of Urban Local Bodies

1. Urban planning including town planning.
2. Regulation of land-use and construction of buildings.
3. Planning for social and economic development.
4. Roads and bridges.
5. Water supply for domestic, industrial and commercial purposes.
6. Public health, sanitation, sanitation conservancy and solid waste management.
7. Fire services.
8. Urban forestry, protection of the environment and promotion of ecological aspects.
9. Safeguarding the interests of weaker sections of society, including the handicapped and the mentally retarded.
10. Slum improvement and upgradation.
11. Urban poverty alleviation.
12. Provision of urban amenities and facilities such as parks, gardens, playgrounds.
13. Promotion of cultural, educational and aesthetic aspects.
14. Burials and burial grounds, cremations, cremation ghats/grounds and electric crematoria.
15. Cattle pounds, prevention of cruelty to animals.
16. Vital statistics, including registration of births and deaths.
17. Public amenities including street lighting, parking lots, bus stops and public conveniences.
18. Regulation of slaughter houses and tanneries.

Source: 74th Constitutional Amendment, 12th Schedule

At the same time, the State Governments were each required statutorily to constitute a State Finance Commission, in order to devise mechanisms for ensuring adequate local government finance so that ULBs could properly undertake their new responsibilities. The provisions of the 74th CA then had to be enacted in State laws and rendered operational which proved to be a lengthy process. Local government finance in India is through grants from the State governments and by means of local taxation. These may include property taxes, *octroi*, licensing, and other fees and charges.

The urban development departments of the state governments devolve grants to the ULBs following the recommendations of either the state finance commission or the state municipal finance board. In contrast, all expenditures for rural development and tribal development are routed through the District Panchayat. Besides funding the ULBs, in most states except Gujarat and Maharashtra, the state government's urban development department is involved in the development of some city level public infrastructure services, such as water supply, sewerage networks, storm water drains and even road construction, through specific agencies called *parastatals*. Thus, many states have water supply and sewerage boards or a State Road Development Corporation, or a state government supported public transport corporation. The presence of such parastatals means that a ULB may not be required to either construct the physical infrastructure or operate and maintain these facilities. In terms of governance, it also means that such parastatals are not within the ambit of the newly decentralized governance system and there is potential for conflict between ULBs and parastatals. In some states, the state government has not followed this path, yet through its own departments such as the Public Health and Engineering Department, the Urban Development Department or the Public Works Department, it directly undertakes capital expenditure for urban infrastructure. The maintenance and operation of such infrastructure is managed by the state government department or the ULB. Such confusion of jurisdictional responsibilities exists in many of the less developed states of India, whereas in most developed states if the ULB itself is not in charge of these responsibilities, then the state government may have set up a parastatal for a specific function. All these muddled aspects of service delivery need to be thoroughly reviewed and rethought to ensure an effective decentralised governance system as envisaged in the 74th CA. Even the Lok Sabha Standing Committee on Urban Development notes that "In many cases, the capital works are executed by the special agencies though the loan liability is passed on to the municipality" recommends resolution of this problem (GOI 2004: 19). There is virtually no research literature on the nature and *modus operandi* of parastatals.

All policy decisions are supposed to be undertaken by the elected council. The municipal corporation or councillors elect the mayor who is usually the leader of the majority party (and who in the case of smaller ULBs is known as the Municipal Chairman), on a yearly, renewable basis. If there is no Mayor-in-Council (MIC) system, as there has been in Kolkata since 1984, then the mayor must select an Empowered Standing Committee including himself, the deputy mayor and seven (five in smaller places) councillors, to serve as an executive body, since some councils are very large. Mumbai, for instance, has 227 elected councillors. Then there is the Municipal Commissioner, a powerful figure who essentially works as the general manager of the municipality, and is appointed by the state for a fixed term, with an executive role and powers that are fixed. It is the Municipal Commissioner who presents the ULB budget to the Standing Committee. Thus, in spite of the 74th CA, the state governments' control over the functions and priorities of ULBs continues. In such a system, the Mayor is not very powerful and remains only a formal head and not a functional head of the ULB. In contrast, where there is a MIC system, for example in West Bengal, Madhya Pradesh, and Chhattisgarh, the mayor is directly elected by the people and is the supreme head of the ULB with all the executive powers vested in him or her. The Municipal Commissioner is the head of the municipal administration, but directly accountable to the Mayor.

Since 1993, some states have made enormous progress in empowering local governments, while others have been very slow. There is much debate about the institutional capacities of individual municipalities to perform their assigned roles. Both political leadership and managerial skills at the local level are sometimes seen as deficient, and some state governments are reluctant to relinquish control or to make sufficient resources available to make the system work. Further problems are related to inept contracting out, transparency, accountability, disregard for the poor and the informal sector, tokenism, clientelism and absence of or non-democratic ward committees. There is an enormous amount of both scholarly and popular writing on the subject. (For example: Anand 1995, Sukthankar 1995, Jha 1996, Vyasulu 2000, Rao 2000, Morris 2002, Sivaramakrishnan 2004, Chamaraj 2006, Pinto 2005 to name just a very few).

The Central Government's Ministry of Urban Development has of course tried to provide guidance during the transition while also covering transaction costs. It has prepared a Model Municipal Law for use by the states on which it has held workshops, Private Sector

Participation Guidelines as an aid to properly contracting out, a Model National Municipal Accounting Manual to support the modernization of accounting, and has introduced many capacity-building training programmes, often through municipal associations, and often with international donor help. For instance the Financial Institutions Reform and Expansion (FIRE) project, which among other things endeavours to build credit worthiness in municipalities, is supported by USAid, while UNDP helps in capacity building (USAid-India-FIRE 2004, UNDP 2006).

Since progress was slow, an Urban Reform Incentive Fund (URIF) was set up in 2003 to speed the urban renewal which the reforms were supposed to promote (Urban Finance, Dec. 2004). This was replaced in 2005 by the Jawaharlal Nehru National Urban Renewal Mission (JNNURM) to help states and large cities accelerate the process of reform and infrastructure up-grading (GOI 2005). It is applicable to all cities of over one million population, including state capitals and important tourist and historic centres for a total of sixty-three. Cities can apply if seven conditions are fulfilled, of which the first three must be undertaken by the state. These are: repeal of the Land Ceiling Act, rationalization of stamp duty, reform of rent control, introduction of a computerized process of document registration, reform of the property tax (updating tax rolls and improving tax collection), the levy of reasonable user charges, and adoption of a modern accrual based double-entry accounting system (GoI 2005). These measures all lie squarely within the neo-liberal agenda, and some are hotly contested (Baindur 2005, Chamaraj 2006, Mahadevia 2006). In order to qualify for funds, a city must prepare a City Development Plan, which explains exactly what infrastructural projects are to be undertaken, their cost, programming and benefits, and how they will be managed. The mission thus overtly links urban renewal and urban reform.

Self-generated municipal finance is a thorny question. In 1995 the contribution of local governments to the GDP was about 4 per cent (CLGF 2006)—a trifling proportion considering their roles and responsibilities. There is a wide disparity between urban centres in economic terms, and there is concern that the larger centres will tend to attract the most investment, both public and private (Kundu 2001). Efforts are being made to increase local revenues from property taxes by improving the cadastre, updating tax rolls and improving tax collection rates. Since property tax evaluation is based on Annual

Rental Value (ARV) (and not on market value as in North America), the proposal to abolish rent control is part of the effort to make property tax a more robust source of income to municipalities. *Octroi*, a tax on all goods entering a city, has historically been a very lucrative source of local finance, but under the liberated economic measures this is supposed to be abolished. It has disappeared in some but not all cities yet. Direct sources of taxation have to replace indirect. Since 1996, and encouraged under the FIRE programme, Class 1 cities (those over 100,000 population) are permitted to float bonds: Ahmedabad was the first to do this, but the long-term outlook is not clear (Mahadevia 2001, 2005).

Solid Waste Management

Over the years, the national government has strived continuously to bring reforms to local level administration and to the environmental sector. While the two major reforms have been the 12th schedule of 74th CA of 1992 and the Municipal Solid Waste Management and Handling Rules[9] 2000, there have been a series of laws enacted to deal with specialized materials. However, even with these legal mandates, the urban waste management scene remains very poor overall (Asnani 2005).

Box 1.2: Laws Relating to Solid Waste

1962	Atomic Energy Act.
1986	Environment (Protection) Act.
1989	Hazardous waste management and handling rules
1996	Chemical Accidents (Emergency Planning, Preparedness and Response) Rules, 1996.
1998	Bio-medical waste (Management and handling) rules amended in 2000.
1999	Recycled plastic manufactured and usage rules
1999	Solid waste management in Class-1 cities in India—guidelines by Supreme Court of India.
2000	Municipal Solid Waste (Management and Handling) Rules
2000	Ozone Depleting Substances (Regulation) Rules.
2001	Batteries (Management and Handling) Rules.

Solid waste management is one of the most important duties of local government, usually associated not only with garbage, but also with street cleaning, the maintenance of roadside drains and dealing with sanitation residues. On average, according to the Twelfth Finance

Commission 2005-2010, it consumes 60 per cent of municipal budgets (GoI 2004: 141). It too has undergone many changes since the passage of the 74th CA in the early nineties. The continuing filth in cities and the very slow start in the development of effective waste management led the central government to set up a "High Power Committee on Urban Solid Waste Management in India," popularly known as the Bajaj Committee (GoI 1995). Its report made a large number of sensible recommendations; waste segregation at source, primary collection, user fees, employment of appropriate equipment and vehicles, composting and sanitary landfills, and pilot testing of Public-Private Partnerships (PPPs). However, with no real improvement in sight, Public Interest Litigation challenging the failure of the government to provide appropriate services in New Delhi was instigated by Mrs. Almitra H. Patel and others in 1996 (Supreme Court: Writ Petition No. 888 of 1996).

This resulted in the Supreme Court constituting another expert group, the Asim Burman Committee, to examine all the aspects of waste management for Class I cities (those over 100,000 population). Its interim recommendations of 1998 included actions to be taken by ULBs and the support measures to be extended by the Central and State governments. The final report, submitted in 1999, included recommendations advocating simple technologies including composting, easily achievable standards, and a reasonable time frame for compliance (GoI 1988 & 1999). The Supreme Court then issued directives to the various levels of government to take steps to implement them. The Central Government responded with the "Municipal Solid Waste (Management and Handling Rules) 2000" (GoI 2000), which lay down the procedures for the collection, segregation, storage, transportation, processing and disposal of wastes. The rules stipulated that all Class I cities implement them by the end of 2003, a condition not met. At the same time, an expert group under the umbrella of the Central Public Health and Environmental Engineering Organisation (CPHEEO), a technical wing of the Ministry of Urban Development and Poverty Alleviation (MUD&PA), prepared a "Manual on Solid Waste Management" designed for use by any urban municipality (CPHEEO 2000). This most excellent guide is clear and straightforward, and easy to understand. Finally, the Twelfth Finance Commission proposed that for solid waste "at least 50 per cent of the grants provided to each state for ULBs be earmarked" to support the cost of "collection, segregation and transportation only, as the activities to be taken up by the private sector should be commercially viable once the municipality is able to discharge its role effectively" (GoI 2004: 154). Again, as our research shows, these expectations were not met.

The Government of India's Municipal Solid Waste (MSW) Rules 2000 decree the following standards:

1. *Collection of municipal solid wastes*—Organizing doorstep collection of municipal solid waste from houses, hotels, restaurants, office complexes and commercial areas.
2. *Segregation of municipal solid wastes*—Municipal authorities shall organize awareness programs for segregating the waste at source as dry and wet waste and promote recycling or reuse of segregated materials.
3. *Storage of municipal solid waste*—Municipal authorities shall establish and maintain storage facilities such that wastes stored are not exposed to open atmosphere and shall be aesthetically acceptable and user friendly and it should have easy to operate design for handling, transfer and transportation of waste.
4. *Transportation of municipal solid wastes*—Vehicles used for transportation of waste shall be covered and waste should not be visible to public, nor exposed to open environment and shall be so designed that multiple handling of wastes prior to final disposal, is avoided.
5. *Processing of municipal solid wastes*—Municipal authorities shall adopt suitable technology or combination of such technologies to make use of wastes so as to minimize burden on landfill.
6. *Disposal of municipal solid waste*—Land filling shall be restricted to non-biodegradable, inert waste and other wastes that are not suitable either for recycling or for biological processing. Land filling of mixed wastes shall be avoided unless the same is found unsuitable for waste processing. (GoI 2000)

Thus, to meet the mandatory requirements of the SWM rules of 2000, cities have to comply with provisions with regards to eight tasks involved in SWM. These are: storage at source, segregation at source, primary collection, street sweeping, provision of closed containers for intermediate storage, transportation in covered vehicles to the dump site, processing of waste, and provision of scientific landfill sites for dumping. In the first phase, 128 cities were directed to comply with the MSW Rules 2000, supposedly by the year 2003.

Table 1.5: Status of Compliance with MSW Rules 2000 in Cities of India, 2004

Percentage compliance	*Number of Cities*							
	Storage	*Segregation*	*Primary collection*	*Street sweeping*	*Provision of closed container*	*Transportation in covered vehicles*	*Processing of waste*	*Scientific landfill*
100	8	2	8	35	5	27	4	0
75 to 100	12	13	15	33	10	18	2	1
50 to 75	30	29	19	33	23	24	4	0
25 to 50	25	23	31	11	12	8	5	2
10 to 25	25	22	17	3	20	10	7	1
0 to 10	4	5	7	0	1	0	5	1
0	12	25	21	1	43	30	95	116
NA	12	9	10	12	14	11	6	7
Total	128	128	128	128	128	128	128	128

Source: Asnani (2005), Presentation made at workshop in Bhopal on 28th & 29th Sept. 2005

As can be seen from the table, compliance has been abysmal (Table 2.1). By 2004, the most recent year for which we have data, in 25 of the 128 cities (19.5 per cent) there is no source segregation of waste and in 21 of them (16 per cent) there is no primary collection. In only 8 of the 128 cities is there 100 per cent primary waste collection. Street sweeping is the only activity where compliance is reasonably good; 53 per cent of the cities have 75 to 100 per cent compliance, although only 27 per cent do a complete job. In one-third of the cities, there are no closed containers at intermediate collection points, indicating that waste would be spilling over once again on the roads causing nuisance of stench, flies and mosquitoes, and street animals hovering around. Further, 26 per cent are not complying at all with the requirement of transportation in covered trucks. 96 per cent (116) of the cities do not have an appropriate landfill site and 78 per cent have no waste processing facility. A close look at the statistics reveals that SWM activities scheduled at the end of the waste management chain have lower compliance than those at the beginning of the chain.

The table 1.5 also shows that there is no city that has 100 per cent compliance for all the eight waste management tasks and hence, none could be classified in a true sense as a 'Best Practice'.

While some cities have made enormous progress, most have not (Asnani 2006). This point will be taken up in more detail in the following chapter, but it is worth pointing out that the worst bottlenecks overall are source segregation and that of the final disposal of residual wastes. While India has one of the largest rates of recycling in the world, largely due to the enormous number of waste-pickers and their networks of dealers, and has a high degree of composting because a large proportion of the waste is organic, the residues are not disposed off in a sanitary way. Most landfills are simply dumps and subject to fires or scattering by marauding animals and rag-pickers. Under the regulations all urban municipalities must designate sites for sanitary landfills and close unhygienic dumps. A recent reading of the Urban Plans of the towns of the State of Madhya Pradesh showed that not one city had indicated the location of such a facility (Jayant Verma: personal communication 2005). Many options are under debate. The research for this book has thus been undertaken during a period of extraordinary change, where new urban policy initiatives are being implemented and those that are not working require fresh perspectives.

NOTES

1. Description of Bangalore after the deluge. "After the recent chaos in the rains in Bangalore, now it's the turn of the city's major wholesale vegetable market which has been completely ignored by the Bangalore City Corporation. For over a week, the Corporation authorities have simply not bothered to clear a dumping ground overflowing with rotten vegetables, fruits and other hazardous waste. Even vegetable vendors are now complaining of the mess that has been left uncleared." .... "Adding to the rotting vegetable stench and pileup are also medical wastes dumped from nearby hospitals, posing serious health hazards. .... The rains, the festival have all resulted in the garbage pileup. The market is flooded with this garbage. We are allowed to bring only eight trucks into this market to clear the garbage. "For more trucks the authorities are not giving permission. If they gave permission, then we could clear all this," said Shiva Kumar, market supervisor. Despite all the bad publicity that the Corporation got during the recent deluge for not doing their work properly, they seem to have turned a deaf ear to

the citizenry. Basu, Nupur (2005): *New Delhi Television*, November 6. *Source*: (http: //www.ndtv.com/template/template.asp?template =health&id=80969& callid=1)

2. Katakam, Anupama; Lyla Bavadam and Dionne Bunsha (2005) "High Water and Hell", *Frontline*. *Source*: http: //www.hinduonnet.com/fline/ fl2217/stories/20050826006000400.htm.
3. Since July 26, 2005, the metropolis had registered 30 confirmed cases of leptospirosis, 10 dengue, 352 cases of hepatitis and 596 cases of malaria, the state health directorate said, adding that during the same period, the number of cases of fever in Mumbai was 46,075 while that of diarrhea and vomiting was 22,787. All of them were under treatment. *Source*: 'After deluge, its diseases now in Maharashtra', *Rediff News*, August 10, 2005; at http: //in.rediff.com/news/2005/aug/10cholera.htm.
4. As a result pressure on city's municipal hospitals increased while they did not have requisite equipment to deal with the disease. By mid-August, over 120 people died of leptospirosis the city. Agrawal, Dipti (2005): New Delhi Television, August 16. *Source*: (http: //www.ndtv.com/ template/template.asp?template=health&id=77472&callid=1)
5. http: //www.bethechange.org/blog/yoomi/index.php?p=53; http: // www.my-gujarat.com/n/a/arc6-2005.shtml
6. "The dengue scare in West Bengal is worsening, with 925 cases having been reported in just the last one month. These numbers are alarmingly higher from an earlier average of 20-30 dengue cases in the state every year. The first case of a suspected dengue was reported in Kolkata on August 3. Since then, 581 people have tested positive in the city, which has also seen 15 deaths. The dengue scare has also created a rush at the malaria clinics, and the infrastructure is under a sever strain. Predictably, with an epidemic in the town, no one is willing to take a chance." Banerjee, Monideepa (2005), *New Delhi Television*, September 7. *Source*: (http: //www.ndtv.com/template/template.asp?template=health &id=78443&callid=1)
7. Term used to project transformation of Mumbai and Bangalore into globally competitive cities with global level infrastructure.
8. There are many other severe planning failures. But, we do not get into details here.
9. On the basis of these recommendations (Committee Constituted by the Honourable Supreme Court of India 1999) national legislation was adopted with the "Municipal Solid Waste (Management & Handling) Rules 2000" (Ministry of Environment and Forests 2000). Summing up, Kundu and Kundu (2005) argue CA has been very nominal excepting the four states of Kerala, Tamil Nadu, Punjab and West Bengal and also highly uneven across the country.

2

Solid Waste Management in Indian Cities
An Overview

DARSHINI MAHADEVIA, JEANNE WOLFE
and APPEEJI PARASHER

Introduction

This chapter presents a review of solid waste management practices and their problems in a number of Indian cities that are actively trying to come to grips with the problem. It is not comprehensive: to research all the cities of India would be an enormous task, and there are certainly innovative practices which we have missed. It is drawn both from a literature search and from the empirical research findings of the seven detailed case studies presented in later chapters. The research agenda used in the case study work is employed as an organising framework, not only to structure an otherwise unwieldy mass of information, but also to emphasise the complexity of waste management systems, and the way in which they are interlocked with all other aspects of urban development and governance. Each facet has to be considered if an urban area is to achieve the status of a "Clean City" or better still, a "Sustainable City".

There has been a spurt in the literature on "Best Practices" in waste management. Further examination shows that such popular accounts usually focus on one small aspect of the subject, other key details are often absent, and the context, or up-stream and down-stream effects, are frequently neglected. In essence, in much of the literature, any move away from the conventional understanding that only Urban Local Bodies (ULBs) are responsible for Solid Waste Management (SWM), is being noticed and written about, and more often than not labelled a 'Best Practice'. At the same time there have also been suggestions that any multi-stakeholder activity or partnerships are a move towards participatory governance (Sekhar 2001; Godard 1996; Beukering *et al.* 1999), when it is clear that the decision-making roles of government, the market and civil society are by no means equal.

The Research Agenda

The research framework depicted in the Table 2.1 was used as a guide to the field-work and data collection conducted in each city. For some subjects the information sought was simply not available, or the data is not collected by the responsible agency. In other cases, either conditions of confidentiality apply, such as in some contractual arrangements, or agreements are made verbally following common traditional practice and there is no documentation. Further, in each city research emphasis was put on the defining or special characteristics of waste management. Consequently, in the later chapters the separate accounts for each case-study city differ according to the information available and the focus of the initiative. It must be added that almost all agencies, local government officials and employees, private sector contractors and NGO and CBO workers and members were extraordinarily helpful and open to the field researchers; they all seem keen to find solutions to their common problems.

This research agenda is also useful in that it points to the sort of studies that must be undertaken to plan or revise a solid waste management system. Modifying one part of the process without considering the whole system can have unwanted consequences. Integrated Solid Waste Management (ISWM) must be the watchword for planning better systems, which in turn should be thought of as part of an integrated system of urban management.

Table 2.1: Research Agenda for SWM Study

	Research Agenda	*Contents*
1.	Background and project genesis	When, why, how, where and by whom; role of funding organization(s) if any; time taken to implement; identification of potentials and constraints.
2.	Solid waste generation	Its quantity, sources, nature and composition
3.	Actors	Roles and responsibilities of the three levels of government, private sector, NGOs, CBOs, SHGs, informal sector waste pickers, rag pickers, households, individuals and any other participant or institution;
4.	Partnerships, alliances and voluntary associations	Legal status (formal/informal), decision-making system (hierarchic/flat), beneficiary awareness and participation, organizational issues, resource mobilization, extent and nature of partnership(s), sanctions, conflict resolution mechanisms
5.	System of waste management	System of segregation, collection, transportation, re-use, recycling and disposal from different sources such as households, hospitals, vegetable markets, slaughter houses, commercial places, public places, industrial houses
6.	Finance	Recurring costs, capital costs, sources of finance, grants, loans, interest rates, repayment period; how the practice relates to the financial situation of the local government, extent of community contribution and financial sustainability
7.	Up-scaling and replicability	Coverage of projects, up-scaling to city level, possibilities of integration with other projects or programmes in the city, replicability in other places
8.	Environment and Sustainability	Degree of resource conservation efforts, environmental improvement, health impacts. Sustainability of projects in terms of social, community, economic and environmental impacts
9.	Conflicts and issues	Management, jurisdictional, social, and economic conflicts, conflict resolution mechanisms
10.	Conclusion: Innovation in Practice	The way ahead

Background and Project Genesis

The when, why, how, where and by whom new initiatives get started and the relative time required in realizing the efforts is important to know to get a full understanding of any SWM venture. This information, available in varying degrees of detail, is useful to assess the success and replicability of any new endeavour. The literature, as well as the case studies in this book, indicate that the genesis of all the new initiatives for SWM have been devised in response to the local conditions to which the main actors in the initiative responded.

The overall pattern of events and many of the triggers to action have been similar, although subsequent events have not played out in the same way. These are summarized as follows. During the post independence years, urbanisation proceeded rapidly; displacement, migrations of refugees, and the attraction of cities, led to urban over crowding, overloaded infrastructure, and "The Unintended City" (Sen 1976) of slum and pavement dwellers. Local governments did not have the resources to cope, and cities became worn down, squalid, ill serviced and unhealthy, many with half the population living in slums. One result was the founding of numerous NGOs to help fill in the gaps, some of which have become networked organisations. Famous examples include SEWA (Self-Employed Women's Association) in 1972, NSDF (National Slum Dwellers Federation) 1974 and Unnayan 1976, but there were and are many, many more. While some of these were not primarily interested in SWM, their advocacy and social work, especially among rag-pickers, has had a direct impact. The middle classes meanwhile were organising CBOs (Community Based Organisations) to look after their waste collection, sometimes on an RWA (Residents Welfare Association), neighbourhood, street or lane basis.

With the surge of globalization in the late eighties, came world-wide acceptance of the fact that in any national economy urban areas account for far more of national productivity than their share of the population. At the same time, as the Washington Consensus gained ground, it was realized that for cities to become players in the competitive global economy, sound infrastructure and a welcoming environment are absolutely necessary. Against this backdrop, India launched its reform agenda: economic and trade reforms starting in 1991; the 74th. Constitutional Amendment (CA) in 1992 which re-affirmed that solid waste management (along with many other functions) is firmly in the hands of ULBs; the first India Infrastructure Report in 1996 which advocated seeking private sector investment in the municipal service sector; the right of cities to borrow money by launching bonds without government underwriting in 1996; the Supreme Court's issuance of the Solid Waste Management Rules in 2000; the commitment of half funds for the urban infrastructure by the 12th Central Finance Commission in 2005; and finally the launching of JNNURM (Jawaharlal Nehru Urban Renewal Mission) in 2005, which makes funds available to the country's 63 most important cities for infrastructure purposes if they comply with a list of reform conditions—the carrot and the stick.

In the solid waste field, Chennai and the emergence of Exnora[1] illustrates this sequence. Exnora and Civic Exnora activity has attracted the interest of many researchers, the result of which is a substantial amount of information on its beginnings in 1989. Exnora, an international NGO started by non-resident Indians, promoted the idea of forming neighbourhood organisations, namely Civic Exnoras, for managing primary waste collection to clean up the city (Anand 1999, 2000). Municipal efforts to improve services were undertaken in 1996, when the Corporation hired a consulting firm to develop a SWM plan and identify areas for private sector involvement through assistance from the Tamil Nadu Industrial Development Corporation (IPE 2004). The Chennai Municipal Corporation (CMC) began moving towards privatisation of waste transportation, while a major undertaking was made by Exnora in organising rag pickers in door to door collection, certainly a best practices innovation (Anand 2000). Under the impetus of the Solid Waste Management Rules 2000, privatization started in other parts of Chennai by contracting Onyx (Asia Holdings, Singapore) in three zones for collection, and for transportation. Chennai, like many other cities, thus has a mixed regime of waste management involving the public, private and civil society sectors.

Other cities had other triggers as starting points. In Suryapet, the new waste management programme introduced in 2002, was largely propelled by an energetic Municipal Commissioner, Mr. S.A. Khadar, as it was in Vejalpur. In Ranchi the impetus came from an NGO the Nav Bharat Jagriti Kendra (NBJK), under the Clean Jharkhand Project, with foreign aid from the India-Canada Environment Facility, the Ranchi Municipal Corporation having been superceded and ineffective since 1991.

In Mumbai, the Advance Locality Management (ALM) initiative was started in Joshi Lane, Ghatkopar in 1996 by a few citizens who resolved to improve their filthy surroundings. They started to work jointly for segregation of garbage, cleanliness, *vermiculture* and beautification, providing voluntary contributions of Rs. 1 per day. The concept was accepted as policy by the city in July 1997, and expanded to other localities. By the year 2000 the total number of ALM societies had reached 666. In Kolkata the impetus came largely through the

Kolkata Environmental Improvement Programme, the most recent of many attempts to improve the city, which was designed to systematically attack all interrelated problems.

Table 2.2: Dates of Important New Initiatives

	NGO/CBO		*Contract. Out / Privatisation*		*City reforms*
	Earlier	Later	Earlier	Later	Initiatives
Bangalore	Civic 1990 Reds 1979 CEE	Janaagraha 2001	1986, 4 wards	By 2005, 62% private	Swachchha Bangalore 2001
Hyderabad	RWAs, NAs	DWCUA VGDS 1993	Small scale contracts 1995	By 2005, 75% private	Litter free.... 2001
Kolkata	MASS Educ.				Decentralization KEIP 2003
Mumbai		ALMs 1996			Reform 1997
Ranchi		NBJK			CJP 2002
Surapyet				DWCUA	Zero litter 2002
Vejalpur		SEWA-local Rag-pickers coop 1998		Vishnu Khadi Coop	New prog-2004

The genesis of solid waste projects evidently covers a wide variety of events. The essential message is that in order to set up a viable system, a thorough knowledge of what has gone on before, and what has worked or not worked in a particular situation, is essential to designing a new system or upgrading an old one. Policies in every city are strongly affected by local networks of social relations and political aspirations, ingrained customs and local leadership.

Solid Waste Generation

Knowing the amount and composition of solid waste generated is clearly necessary for all management planning. Domestic and commercial garbage, including abattoir and market wastes, along with street sweepings and drainage channel silt, are the responsibility of local governments. Industrial wastes are usually handled by the industries themselves, or they pay the ULB to pick them up, under rules and disposal conditions usually set out by the state Pollution

Control Board, except for small-scale enterprises which tend to leave waste on roadsides. Hospital wastes and other hazardous materials are the subject of special legislation and are supposed to be carefully handled in purpose-built facilities.

Information on the total amount of waste generated by cities is widely 'guesstimated', based on estimates of average per capita waste generation given in the MSW rules of 2000. These estimates are by city size, and the larger the city the higher the per capita waste: between populations of 100,000 to 5 million, waste varies from 200 to 500 grams. The composition of waste as estimated by National Environmental Engineering Research Institute (NEERI 1996), (on which the rules were based), showed that compostable matters range from 30 to 57 per cent, inert materials from 40 to 54 per cent and recyclable wastes between 5 and 10 per cent of the total solid waste in Indian cities. In some jurisdictions estimates have been made by surveying a sample of households and measuring their output. This was the technique used in Bangalore and Ranchi. More recently, some cities have acquired truck-weighing machines at their disposal sites and amounts have become better known, as in Hyderabad and Kolkata. However, as shown in the most recent figures from the Central Pollution Control Board, (Table 2.3), reported amounts per capita vary significantly between cities, and we are at a loss to explain why.

Table 2.3: Waste Generated and Composition for Specific Cities, 2005

Cities	*Total Tonnes per day*	*Generation grams per cap. per day*	*Compostable per cent*	*Recyclable per cent*
Mumbai	5320	450	62.44	16.66
Delhi	5922	450	54.42	15.52
Kolkata	2653	580	50.56	11.48
Chennai	3036	620	41.34	16.34
Bangalore	1669	390	51.84	22.43
Ahmedabad	1302	370	40.81	11.65
Hyderabad	2187	570	54.20	21.60
Kanpur	1100	430	47.52	11.93
Lucknow	475	220	47.41	15.52
Surat	1000	410	56.87	11.21
Pune	1175	460	62.44	16.66
Jaipur	904	390	45.50	12.10

(Contd.)

Table 2.3 (*Contd.*)

Cities	*Total Tonnes per day*	*Generation grams per cap. per day*	*Compostable per cent*	*Recyclable per cent*
Bhopal	574	400	52.44	22.33
Nagpur	504	250	47.41	15.53
Varanasi	425	390	45.18	17.23
Ludhiana	735	530	49.80	19.32
Vadodara	357	270	47.43	14.50
Madurai	275	300	55.32	17.25
Coimbatore	530	570	50.06	15.52
Indore	557	380	48.97	12.57
Kochi	400	670	57.34	19.36
Patna	511	370	51.96	12.57
Vishakhapatnam	584	590	45.96	24.20

Source: Central Pollution Control Board (http: //www.cpcb.nic.in/pcpdiv_plan4.htm), accessed on March 7, 2007.

Many of the figures do not match those found in the case-city studies (Table 2.4) where some informants indicated that their figures were estimates. Perhaps it is because some totals may be calculated only from the number of truck-loads moved, or perhaps some cities omit private sector activity, or recycled materials taken away at source, or include industrial wastes. In any event, it is clear that there is not a standard way of reporting. Monitoring and recording need to be improved if forward planning is to be ameliorated, which requires the production of standardised comparative data series.

Table 2.4: Waste Generation, Case Study Cities 2005

	Total Tones per day	*Generation grams per cap. per day*	*Compostable %*	*Recyclable %*	*Inert %*
Bangalore	2214	527	43	38	19
Hyderabad	2200	578	52	21	27
Kolkata	4000	450	42	49	8
Mumbai	7800	630	40	15	37
Ranchi	325	247	51	7	40
Surapyet	43	400	52	23	30
Vejalpur	60-70	400	70	13	16

The composition of solid waste is also an issue. Vegetable matter is usually about half the total. The moisture content of solid waste in India is 50 to 60 per cent, much higher than that of North America or Europe, and its calorific value is low. This means that incineration or using it for the thermal generation of electricity requires drying and or pelletization, an expensive proposition that is exacerbated in the monsoon season. This is why the solid waste handbook favours biological and vermi-composting for the disposal of biodegradables, notably kitchen and garden wastes. The recyclables consist of paper, metal, plastics, rags, slaughterhouse residues, and again the reported proportions vary, probably because much of it, especially paper and cardboard, is pre-sold to itinerant buyers by householders, or collected and sold by waste-pickers before it enters the system. For these reasons, for ULBs to imagine that they are going to make a profit on these materials may be illusory, despite their relatively high selling prices.

Table 2.5: Recyclable Materials as a Percentage of Total SW, Case Study

	Paper	*Glass*	*Plastics*	*Metal*	*Rubber*	*Total*
Bangalore	16.5	2.9	6.7	1.5	9.9	38%
Hyderabad	8.04	2.71	8.36	1.4	1.59	21%
Kolkata	8.0	na	na	na	na	42%
Mumbai	7.2	1.28	4.8	0.85	—	14%
Ranchi	8.0	1.0	8.0	1.0	1.0	19%
Surapyet	—	—	—	—	—	na
Vejalpur	—	—	—	—	—	8%

Table 2.6: Price of Recyclable Materials, Hyderabad 2005

	Rate: Rs/Tonnyl
Paper	3,750
Plastics	11,333
Metal	11,333
Glass	750

Source: Muncipal Corporation of Hyderabad

For planning purposes, more attention has to be given to the future increase in waste generation, a consequence of both rising standards of living and urban population growth. Authorities have estimated per capita increase in waste to be anywhere from 1.5 to 5 per cent per year (Shekdar 1999 & GoI 2004). Urban population growth rates vary from city to city, but from 1991 to 2001 the average for India was 2.7 per year. In addition, urban growth also means a greater territorial extent of urban areas, meaning a greater area needing coverage and likely resulting in increased transportation costs.

The different sources of waste have not been given much attention, apart from household waste, although this sort of monitoring is improving. For instance, in Bangalore, Beukering *et al*, (1999) reported this lack, though this has changed and records are kept, while in the larger cities special arrangements are made for the collection of industrial, hospital and other dangerous wastes as required under the SWM Rules. Galab *et al* (2004) report on each of the bulk generators of wastes in Hyderabad, the hospitals, educational institutions, offices, hotels and markets, each of which has different mixes of garbage, and each of which is handled in different ways according to the rules of the city.

In sum, the level of complexity in waste management depends on the density of settlements, income levels, land-use, tenure rights and physical access: the last two are especially important in slum settlements where non-tax payers often receive no services. Due to the variance in the amount of residential waste generated, (high income areas generate more per household than low income, and low-income areas have very few recyclables) the actions needed to manage them vary.

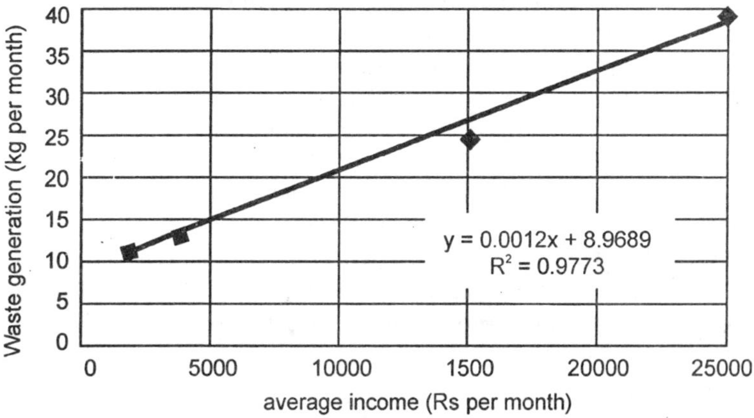

Fig. 2.1: Correlation Between Per Capita Income and Waste Generation

Source: Beukering *et al* (1999).

Each city is not only segmented by income, but also by caste and religion, each with different consumption habits. Each land use category, whether housing, commerce, health care and educational facilities, industry, recreation, markets, slaughterhouses, or construction sites, all produce different mixes of waste, so ideally for each of these separate information and informed discussion of the options is warranted, with the idea of maximising recycling and reuse, and minimising the residues going to dumps.

Source identification can lead to good ideas about efficient disposal: in Suryapet hotel and restaurant food waste is collected daily by pig farmers and directly fed to their animals, as it is from a few establishments in Hyderabad. Some institutions and residential colleges have started their own compost or recycling operations, which significantly reduces the amount of waste to be handled. Examples include Gaddiannarum fruit market in Hyderabad, which has its own composting unit (Galab *et al* 2004: 47), Jawaharal Nehru University (JNU) and the Indian Institute of Technology (IIT) Delhi which have both engaged NGOs to devise their own collection and compost systems, achieving 80 per cent landfill diversion (Gupta 2004), and Bhabha Atomic Research Centre which processes all biodegradable waste through its NISARG RUNA plant producing gas through biomethanation and high quality organic manure (BARC 2007).

Some of the larger cities, such as Bangalore and Hyderabad, are instituting Geographical Information Systems (GIS) as an aid to improving municipal management. These will eventually contain spatially-referenced data on land-use, building types, infrastructure networks, environmental conditions, a fiscal cadastre, and in fact all the data required to monitor and plan city-service operation, maintenance and management, including optimisation of transportation routes. For instance the solid waste plan of the Kolkata Environmental Improvement Project (KEIP), proposes to build a GIS-based solid waste management information system, which would undoubtedly be part of an overall city GIS (KMC 2005). With such a system operational, it would become possible to plan a more effective tuning of waste collection and processing to the diverse sources of generation.

Actors

Solid Waste Management in India falls under the domain of ULBs. However there are many sets of actors other than the ULBs in waste management activities. Baud (2000) has classified them into four main groups: (1) the public sector (national authorities, state authorities, local authorities and local public departments) constituting a central set of players; (2) the private sector (large and small registered enterprises carrying out collection, transport, disposal and recycling); (3) the small scale, non-recognised private sector (waste pickers, itinerant buyers, traders in waste materials and non-registered small-scale enterprises); and (4) the local community and its representatives (NGOs, RWAs and other CBOs). To these should be added international organisations, both aid agencies and financial institutions who have emerged as important actors in MSW, as financiers of innovative approaches and also shapers of waste management policies. The way funds are disbursed influences practice. The role of each of the actors is discussed separately.

Table 2.7: Actors and their Roles

Public Sector National State	Policy-making, decision-making, regulations, oversight and monitoring, overall logistical support, investment in public infrastructure, landfill site evaluation, environmental health
ULB	PRIMARY RESPONSIBILITY FOR SERVICE DELIVERY
Ward Committees	Oversight and monitoring, some service delivery
Private Sector Contractors	Investment, labour management, equipment, recycling, energy recovery Service delivery: collection, sweeping, transportation, some re-cycling
Power cos. etc	Re-use of waste: pelletization, thermal, biogas, and composting plants
NGOs	Co-ordinating and networking at grass roots level for community mobilisation, community capacity building and education, setting up innovative models, training workers
CBOs Neighbourhood SHGs	Local level activities, financial participation, managing collection, ensuring waste segregation at the household source. Labour, in collection and sweeping
Informal Sector	Waste collection and recovery
Waste pickers	Regular routes, regular customers
Rag pickers	Forage around collection points, transfer stations and dumps
Households and establishments	Separation at source
International agencies	Advisors, donors, lenders

National Authorities

'The Ministry of Urban Development is responsible for national urban policy and for the administration of the 74th Constitutional Amendment (CA). One of its agencies, the Central Public Health Environment and Engineering Organisation (CPHEEO) is the technical arm for infrastructure standards and the author of the "Manual for Solid Waste Management 2000" that was distributed so widely to all ULBs. Other departments with important roles in waste management include the Ministry of Poverty Alleviation and Housing which is responsible for providing opportunities for the poor through a variety of programmes, including income-generation and slum improvement. The Ministry of the Environment and Forests through the Central Pollution Control Board (CPCB) monitors environmental conditions, devises standards, and keeps records of the progress in solid waste management practices across the country. The Ministry of Non-Conventional Energy Sources invests itself in solid waste issues through its National Programme on Energy Recovery from Urban and Industrial wastes, and offers financial incentives for the development of such schemes. The effect of the Supreme Court rulings, and the role of the National Finance Commission in proposing the earmarking of state grants to ULBs for SWM has already been discussed. Finally, the Central Government is instrumental in facilitating bilateral and multilateral agreements with the international development agencies and banks.

State Governments

Although the Central Government has enacted important legislation for decentralisation in the 74th CA, implementation is the responsibility of the states. The functional responsibilities and fiscal powers of ULBs are laid out in the revised state municipal acts. While the functional aspects are usually quite clear, the fiscal powers of ULBs are fairly limited. Cities depend heavily on transfers from the state government for all their development expenditures, some being general purpose grants and some being specific programme grants. States are required to set up State Finance Commissions to guide the allocation of funds to lower levels of government, but this has not always occurred. It should also be mentioned that some states have not yet adopted the constitution of Ward Committees as provided for by the

74th CA. Lastly, the jurisdiction (the physical boundary) of the ULBs is decided by the state government: ULBs are forced to accept whatever area is bestowed upon them. Most states have an Urban Development department responsible for these matters and sometimes a state Infrastructure Development Fund which makes loans to ULBs for projects requiring capital investment.

Another important role of each state is the enactment and implementation of pollution control laws, through the State Pollution Control Board (PCB). The Board oversees and monitors the management and handling of municipal, industrial, hazardous, and bio-medical wastes. It also governs the location and setting up of landfill sites in the cities. Each year it is supposed to report to the Central Pollution Control Board on the status of solid waste management in the ULBs of the state. In addition most states have an Environmental Department responsible for environmental monitoring.

Urban Local Bodies (ULBs)

The ULBs are the key players in solid waste management as laid out in the 74th CA where the ULBs are projected to be solely in charge of all urban services, which will be funded through their own resource mobilisation efforts. The transition has been slow and sometimes difficult, largely due to differing perceptions of the capabilities of the elected officials, particularly in technical matters, and partially due to the reluctance of the state politicians and bureaucracy to give up the old ways. The states reserve the right to recall a mayor, or to remove councillors: Kumar (2004) has documented several cases of this sort of harassment. In states containing mega-cities, rivalry rather than complimentarity between the state and the city is sometimes a hindering factor. Municipal government has traditionally been held in scorn: local administrations have often been superseded (suspended), sometimes for many years. Chennai's government was reinstated in 1996 after 25 years in limbo, while Ranchi, the state capital of Jharkhand, has been superseded since 1991.

Waste management is covered under the 6th function of the 12th schedule of the 74th CA as "public health, sanitation conservancy and solid waste management," a mandatory function of the ULBs (Ministry of Urban Development 2004). Except in metro cities, SWM is the responsibility of a public health officer assisted by engineers (Kumar and Gaikwad 2004). The engineering wing of the ULBs supports the health department in the removal of construction debris and silt in drains, transportation and sometimes also taking care of the disposal facility (Supreme Court 1999).

Table 2.8: Actors in Residential Solid Waste Processing

Tasks	*Actors*	*Process*
1. Storage at Source	Households	— collection usually 6 days/wk. — may take waste to primary collection site
2. Segregation	Households Waste collectors (formal & informal)	— separation of biodegradeables and dry waste — recyclables sold to itinerant buyers — resold to dealers — resold to enterprises
3. Primary collection	CBOs, RWAs, Coops, SHGs, ULBs (rag pickers)	— tricycles or carts with separate bins — waste taken to intermediate containers — or to local aerobic or vermi-composters (scavenge around bins and collection points)
4. Street sweeping & storm drain cleaning	ULBs — sub-contractors — private sector	— work gangs of 14-20 people and foreman — work gangs, some mechanization, but limited
5. Provision of closed intermediate bins	ULBs	— containers plundered by stray animals and rag pickers — supposed to be emptied daily
6. Transportation to treatment site	ULBs — sub-contractors — private sector	— trucks — lifters and lorry drivers
7. Processing	ULBs —sub contractors — private sector	— composting — compost sold to farmers — palletizing for energy use, biogas (rare) — incineration (rare)
8. Disposal	ULBs — rag pickers	— *dump site*: many informal — almost no sanitary land-fills — work (& live) at dump sites

The work force of ULBs has often been criticised for being too large, lazy, ineffectual, and improperly managed. Even the Twelfth Finance Commission notes that "most of the urban local bodies are over-staffed and have to pay for a large but idle workforce" (GoI 2004: 141). This has been one trigger for the advocacy of private sector service delivery, supposedly a more efficient employer. Hyderabad for instance has not hired any new solid waste labourers since 1984 or Bangalore since 1985. However, most large cities have well-organised and powerful labour unions who vociferously protest any form of privatisation which they regard as an incursion on their territory.

It has been commonly held that the ULBs are not able to deliver services efficiently, not only because of ill-equipped human resources, but because of lack of finance. However, Gidman *et al* (1995) argue that ULBs overlook the demand side approach, which would move towards a style of service delivery geared to, and involving clients, which should lead to greater efficiencies. Instead ULBs have traditionally employed a supply side model and hence only emphasize public service provision without any thought about cost effectiveness.

Despite all these problems, real or perceived, it is the ULBs that have to manage waste. Not only must they manage the process, labour force, collection points and equipment as they have done in the past, they must also manage all the procedures of contracting out, whether to NGOs and CBOs or to the private sector, either as joint ventures or privatization of some activities. Some cities are managing this very successfully. For instance Hyderabad, after a period of trial and error (Broekema 2004), now operates in a very transparent way, tendering publicly, giving details and announcing contracts through its website.

Ward(s) Committees

The 74th CA[2] emphasises not just delivering more powers to third-tier government, but also proposes the creation of ward committees in cities of over 300,000 people to foster more democratic participation (Supreme Court 1999). Ward Committees are usually chaired by a council member, and made up of people prominent in local housing associations, NGOs, CBOs and the like. However, the 74th CA does not specify their functions and powers, leaving this decision to the state legislative assemblies (Jayaraman n.d.; Pinto 2005). There seems to be some confusion between previously existing citizen-based organisations, such as electoral wards, Resident Welfare Associations (RWAs), and neighbourhood committees, and these more recent wards. For instance, Mumbai has 24 wards and 16 wards committees (some have been grouped together), while the number of electoral wards is 227. It accommodates already active CBOs and NGOs as members of wards committees, a move driven by the civil society organisations in 2000 (Navtej 2005). The new ward committees have been given powers to redress complaints regarding public services and are financed up to Rs. 5 lakh (Navtej 2005). In Delhi, as per section 50 of the DMC[3], each zone has a ward committee and representatives of each form a standing committee to look after matters of common interest (Kundu and Kundu 2005). Only Kerala, West Bengal and Tamil Nadu, have one committee for one ward. In Kolkata, wards are grouped into boroughs, which were in existence long before the 74th CA: they have extensive service delivery responsibilities, especially for SWM for which they have a significant budget.

Ward committees have the potential to become a very important institution at the local level in improving SWM services, but progress in setting them up has been very uneven (Sekher 2005; Asnani 2001). Residents can bring, theoretically at least, accountability to local administrations and have a voice in various local development works (Kundu and Kundu 2005; Planning Commission 2001; Mendes *et al* 2005). Still, there has not been effective decentralization in many states: there is great variation between cities in the levels of functional devolution. (Aziz *et al* 2002; Gupta, n.d).

The National Advisory Council, which deals with civil society issues, has proposed several improvements to make ward committees more responsive to public needs (NAC 2005). These include every ward having a committee (not groups of wards as often happens), elected representatives, significant devolution of funds and responsibilities, control of employees and the like. In fact this is a restatement of the "subsidiarity principle," adopted by the European Union (among other places), which states that services work best when undertaken at the smallest scale at which they are functional (Baindur 2005).

Private Sector

The private sector here refers to contractors and companies formally engaged in waste collection, recycling, reuse and disposal. While private sector participation in urban service provision is often hotly contested, municipal experts argue that it brings in investment which alleviates local fiscal stress, imparts efficiency and effectiveness, and if a service is parcelled out between public and private operators, injects an element of competition into service provision (Baud *et al* 2001; NIUA 2003). Cointreau (1984) pragmatically views it as an opportunity for private investment mobilisation and not a panacea for all ills.

Private sector activity can be roughly divided into three groups, although there is some overlap: contractors engaged in waste collection, transportation and dumping; companies engaged in large scale disposal, namely composting operations, waste-to-energy, biogas plants and landfills; and finally, companies specializing in the disposal of biomedical, electronic and industrial residues among other hazardous wastes.

Private contractors responsible for the collection, sweeping, transportation and dumping of waste are usually small-scale, local entrepreneurs employing day labourers. Cities such as Hyderabad and Bangalore have contracted out 60 per cent of their municipalities to private contractors for sweeping, door-to-door collection of waste from households and commercial establishments in addition to the transportation of waste. Nasik Municipal Corporation (NMC) is reputed to have privatised all its services (Yashada 2004). ('Nasik' is also known as 'Nashik'). Chennai and Mumbai have engaged private contractors for transportation of waste to dumping sites, which has led to cost reductions. For instance, in Chennai, the cost of handling a tonne of waste, which was Rs. 1000, has been reduced by 50 per cent.

The private sector involved in large-scale waste composting, energy production and scientific landfills includes some giant players like Excel, Sriram, and Salco. They work in alliance with the public authority through long-term mutually beneficial agreements. Generally the ULBs have leased them sites and assured a regular supply of waste, while in return the costly disposal problem is much reduced (Ravikaran 2005).

The third group is involved in the transportation and disposal of highly dangerous specialized wastes demanding separate treatment as required by the legislation relating to hazardous and biomedical wastes. For the most part, this is just in its infancy. Biomedical wastes are of special concern because they still turn up in abundance at most municipal waste dumps. In the case of major industries, contracts are often negotiated between producers and the disposal company and the waste may not enter the municipal stream. For the ever-growing stream of electronic wastes, many imported, there is not as yet any formal procedure although this is under debate (MPCB 2007).

Private sector involvement in public service provision requires strong regulation to safeguard public interests. Ironically, while neo-liberalism calls for less government, at the same time it requires stronger regulatory frameworks to ensure a level playing field. Like other Private Sector Participation (PSP) projects there are regulations and guidelines laid out by the central and state governments, that the private sector has to abide by. There has been a fair amount of trial and error in devising equitable contractual arrangements, as has been demonstrated in Hyderabad (Broekema 2004), and clearly municipal expertise is increasing in this field as knowledge is shared. Transparency is increasing, and this is especially so with an increasing number of documents posted on the internet. Examples of such cities following this route include Bangalore and Mumbai.

Non-Governmental Organisation (NGOs)

NGOs play three different sequential roles in the development process: relief and welfare, local self reliance, and promoting a sustainable development system (Korten 1987). In the context of solid waste, the NGO role falls into the first two. Many NGOs work with the socially and economically vulnerable groups, especially slum dwellers, such as street children, rag-pickers, unemployed youth, single parents, the elderly and handicapped. (Huysman 1994; Hunt 1996). Activities taken up by NGOs are, organising and training rag-pickers for door-to-door collection of waste and the use of recyclable waste for income generation, both leading to the organisation of self-help groups of the poor for organised waste management tasks. If recyclable waste is collected and recycled by groups such as rag-pickers organisations, then the burden of waste transport and disposal are reduced (Ghosh 2000).

Examples of NGO involvement in helping the poor include: organising women rag-pickers and training them for vermi-composting by Stree Mukti Sangathana (SMS) Mumbai; training of rag-pickers for door-to-door waste collection by the Self-Employed Women's Association (SEWA) in Ahmedabad; organizing and training street children and unemployed youth in waste collection and sale by Clean Environs, Waste Wise (Mythri Foundation) and the Centre for Environmental Education (CEE) in Bangalore[4].

In regular residential areas NGOs have also played key roles in SWM in community mobilization, organising collection and transport, and promoting environmental cleanliness. Such activities include public education, awareness and anti-littering campaigns, teaching waste segregation at the household level, and motivating households to contribute fees for waste management. There are many cases of promotion of neighbourhood self-reliance by NGOs. For example, in Pammal Town (a peri urban area) in Chennai, the lack of adequate service provision gave rise to voluntarism—wherein civil society organizations were formed for the provision of self-help basic services to local residents (Dahiya 2003). Other examples are the Exnoras in Chennai (Anand 2000), ALMs in Mumbai, and the Bhagidari system in Delhi (Kundu and Kundu 2005). In Ranchi the present solid waste system was started by an NGO, the Nav Bharat Jagriti Kendra (NBJK), that was funded by the Indo-Canada Environment Facility, a bilateral agency.

CBOs, RWAs and Neighbourhood Associations

Community Based Organisations (CBOs) Resident's Welfare Associations (RWAs) and other neighbourhood or street groups exist at the lowest level of the waste management institutional system, their role being important in local collection and monitoring efforts at the ground level, and in bringing people together to work collectively towards cleaner, litter-free neighbourhoods. They are essentially of two types. First are the residents associations usually associated with the wellbeing of their own community or housing colony, such as the RWAs, who group together to improve their own environment.

Second are the groups of slum dweller and rag-pickers who are mobilized, not only to improve their living conditions, but also to create decent work for themselves under one or another of the government employment programmes. Most important of these is the Swarna Jayanti Shahari Rojgar Yojana (SJSRY), which under its Urban-Self Employment Programme (USEP) operates the Development of Women and Children in Urban Areas (DWCUA) initiative. These programmes are financed by the central government and the states on a 75: 25 ratio.

The SJSRY is predicated on the foundation of community empowerment to provide support mechanisms, and cooperation with local government, which, it might be recalled, is now responsible for local economic and social development planning. DWCUA will subsidize groups of ten or more women to set up a work unit by offering Rs.2000 per head for training, Rs.600 each for a tool kit, and Rs.1000 per member, to a maximum of Rs.25000, to set up a Thrift and Credit society. Such groups are becoming more important in obtaining SWM contracts for collection and sweeping from the ULBs, undertaking waste collection in more affluent areas, composting, and in ameliorating their own settlements through the formation of Neighbourhood Groups (NHGs) and Community Development Societies (CDS).

Active involvement of communities puts a positive upward pressure on the system. Agrawal (2001) notes that there is an increasing emphasis on community participation through group formation in all forms of development interventions. Goetz and Gaventa (2001) view this as "developing new socio-cultural identities in the process and new political stature as citizens." An Urban Waste Expertise Programme (UWEP) study defines community management as "a situation in which a community takes the responsibility for, obtains authority over and carries out control on operation, management and maintenance of a service benefiting its members; which does not mean that a community is responsible for every aspect of a service" (Anschutz 1996).

CBOs, RWAs and NHGs basically work in coordination with NGOs, ULBs and their workers. They usually do not go much beyond the neighbourhood level in their activities (Hordijk 2000 and Lee 1998, in Baud and Post 2003). Osmani (n.d.) notes that the voluntary groups and community organisations can deal effectively with local problems, but only to a limited extent because of their absence of decision-making powers and lack of legal status. Anschutz (1996) sees community based SWM efforts faced by a variety of constraints including cost recovery, operational and management problems, social problems and the low participation of households themselves. Toxic Link (2005) has highlighted the low upscalibility of community efforts due to inadequate encouragement from municipal authorities. Many municipal workers think community activism will usurp their rightful jobs.

The most successful 'clean neighbourhood' groups are characteristic of middle and high income localities (Kundu 2005). In Chennai, numerous RWAs are found in such areas, while in low-income areas there are none: they tend to have Self-Help Groups (SHGs), funded under various government programmes (Baud and

Dhanlakshami 2005). Only where NGOs are engaged in slum development as in the Slum Networking Programme (SNP) of Ahmedabad, do resident associations exist. In metropolitan cities, some success in attracting funds and involving local NGOs and community groups has occurred. For, instance the 'Bhagidari System[5]' (partnership), in New Delhi directly links the ULBs with the RWAs and civil society organisations (Kundu and Kundu 2005).

Waste Collectors

Waste collectors are itinerant collectors and buyers, who largely service households, institutions, commercial establishments and the like. They pick up all recyclable waste that they can re-sell. Some buy materials from householders and enterprises for small sums. They usually have regular routes, and regular purchasers for their collected materials. Some specialize in one commodity, like paper and cardboard, or plastics. This reduces the amount of waste that has to be picked up by the local government by considerable amounts, although no one has yet quantified this, but on the other hand it deprives the public authority of marketable wastes.

Rag-Pickers

Rag-pickers are essentially scavengers. They are an important part of the waste management chain, not from an ecological point of view, but logistically, since they reduce the amount of garbage to be handled by 5 to 10 per cent of the total produced. Rag-pickers sell their goods to intermediate buyers, who then sell them to middlemen who pass them on to recycling industries. All pickers work informally and are usually not part of the municipal system. It is a dirty and dangerous occupation, and a cause for ill health from toxic materials, vermin, insect and bacteriological infections, cuts and abrasions, and the inhalation of smoke and fumes from oft-burning dumps. It often involves washing and cleaning the waste for it to be marketable, for instance, plastic bags. Further, since the cleanest and most valuable recyclable waste has already been sold off to waste collectors, the pickings are slim.

The rag and waste pickers are usually socio-cultural minority groups and poorest of the urban poor. Women and children dominate this activity, eking out are bare living. Such work further isolates them from the main-stream, which reinforces their low social status. Picking is done from bins left out to be emptied, collection sites, transfer points and dumps, and even on the top of moving transport trucks. Pickers live in the most marginalized urban locations, often on the edge of a dump, where they can store and recycle collected wastes. Their conditions are usually much worse than those of other slums, and often the target of demolitions, as they have been in Mumbai and Delhi. With development and the increase in employment opportunities in cities, it is expected (and it is necessary) that such occupations be eliminated. The social capital of the urban poor engaged in waste recycling has been interpreted as an asset for urban development institutions (Beall 2001). But the romanticisation of urban poverty must be avoided: the NGO activities previously described, along with other interventions under the DWUCA (Development of Women and Children in Urban Areas) programme were undertaken to provide dignity, health and stability to these citizens.

Householders, Enterprises and Institutions

All the literature on solid waste, not to mention common sense, emphasizes that the fundamental key to effective management is separation of wastes at source. If wet, biodegradable waste can be kept separate, then it can be composted to produce clean, uncontaminated, valuable fertilizer. If not, not only is compost full of broken glass, bits of metal, plastics and the like rendering it dangerous to handle or requiring expensive sieving, but the dry waste is contaminated with rotting vegetable matter and is rendered difficult and dangerous to sort. Paper becomes worthless, and everything else must be washed, using up valuable water. Separation at source by householders, businesses, industry and institutions seems like a straightforward task to accomplish, but it is fraught with difficulties. Many people do not understand why it should be done, find it distasteful, do not have the space to store two sets of containers, think they are already paying for this service, or are simply disinterested. The need for public education is urgent. It is in this area that environmental advocacy NGOs play a most important role.

International Agencies: Financial Institutions and Donors

The role of international agencies in financing urban improvement projects has been especially important in the mega-cities, not only as sources of funds, but also as shapers of policy. The Kolkata

Environment Improvement Plan (KEIP) is an example of this, as is the Ranchi project already noted. Almost all of the largest cities have loans or assistance from the International Financial Institutions (IFIs) and aid agencies such as the World Bank, the Asian Development Bank (ADB), the EU, USAID, DFID, JICA, CIDA, the Ford Foundation and the like for infrastructure projects, which usually have a solid waste management component.

Institutional Arrangements: Partnerships, Alliances and Voluntary Associations

With the participation of a multitude of stakeholders in solid waste management, each making their own day-to-day decisions, understanding institutional and informal interrelationships is the key to an enlightened planning process. Most of the present initiatives in SWM are found to have multiple actor partnerships and alliances between the municipal corporation, and SHGs, RWAs, CBOs, NGOs, private enterprises, and the not-to-be-forgotten informal sector (Table 2.3). These alliances have evolved and are evolving as a result of the interests of each group, which must be understood, as must the set of conditions (agreements) among them. It has been found that service alliances exist as a continuum of relationships between formal and informal, from written contracts or agreements, through memoranda of understanding (MoUs), to verbal promises. Since the interests of all parties is involved, it also becomes important to know the conflict resolution mechanism in case of any dispute or inadequate fulfilment of service, if this exists.

Sekher (2001) and Beukering *et al* (1999) have categorised the various relationships that exist in the waste management sector as: (1) between two or more agents in the formal sector, (2) between formal and informal agents, and (3) among informal agents. Baud (2000) views urban partnerships from two different perspectives; 'top down' which means contracting out or privatization, requiring an enabling framework, and 'bottom up' which promotes the interests of the urban poor engaged largely in waste picking and recycling for a livelihood. She further categorises these partnerships into three types: (1) between local government and private sector companies, (2) between communities and the private sector and (3) between CBOs, NGOs and local government.

Table 2.9: Examples of Partnerships and Alliances

City/Jurisdiction	Partnerships Between		Nature of Partnership	Nature of Work
	Actor 1	Actor 2		
Ahmedabad	AMC	SEWA (NGO)	NA	Has trained 4000 rag-pickers to collect garbage from housing societies
		CEE (NGO)	NA	Awareness Programmes
		Private Partner	Formal	Recycling of Garbage—Excel Industries
		Contractor	Formal	Collection of waste from fish and meat markets (Shri Khadi Gram Udyog Parishad)
Bangalore	BMP	BATF	Formal	Provides training for rag-pickers
		Contractor	Formal	Transportation
		KCDC	Formal	Obtains MSW for vermi-composting, free of cost as an agreement
Chennai	CMC	Civic Exnoras	NA	Employs ragpickers in waste collection (neighbourhood level)
		Contractor	Formal	Waste Collection
	Exnora (NGO)	Civic Exnora (CBO)	Formal	Assists citizens in the formation of Civic Exnoras
	Civic Exnoras	Ragpickers	Informal	
Hyderabad	MCH	Contractor	Formal	Engaged in street sweeping and waste transport
		RWAs/NHC	MoU	Employ workers for door-to-door collection of waste. MCH provides tricycles free of cost and collects garbage from community bins

(*Contd.*)

Table 2.9 (*Contd.*)

City/Jurisdiction	*Partnerships Between*		*Nature of Partnership*	*Nature of Work*
	Actor 1	*Actor 2*		
Kolkata	KMC	Contractors	Formal	Sweeping, Collection, Transportation and Disposal
		Medicare Incin Pvt.	Formal	Biomedical Waste—collection and disposal
		KK Plastic Waste	Formal	Production of polyblend for use with bitumen for road surfacing
		PUBLIC (NGO)	NA	Publicity campaigns
Mumbai	MCGM	ALM	Formal	Waste Collection
		SMS	NA	Vermi-composting
		Contractor	Formal	Transportation and disposal
	SMS	Ragpickers	Informal	
Nashik	NMC	Contractor	Formal	Collection and Transportation
Ranchi	Funding Agency	NBJK (NGO)	Formal	Clean Jharkhand Project started with funding from ICEF Canada. Involved in collection and disposal of waste.
	RMC	NBJK (NGO)	NA	RMC provides vehicles for waste transport
Suryapet		DWCUA Groups	Formal	Sweeping, Collection, Transportation
Vegalpur Nagarpalika	VN	SEWA (NGO)	Formal (annual)	Collection of segregated waste and disposal of storage containers
		Khadi (co-op)	Formal	Cleaning around community dustbins
Vijaywada	VMC	DWCUA	Formal	Contract with DWCUA group under SJSRY scheme

For the purposes of this analysis, a finer grain has been adopted: alliances will be discussed under the following categories, which reflect the complexity of the relationships: (1) ULBs and private companies, (2) private contractors and sanitary workers, (3) ULBs and NGOs (4) NGOs and CBOs (5) CBOs and households (6) NGOs, SHGs (self-help groups) and local government.

Local Government and Private Companies

The entry of the private sector into the field of solid waste was largely prompted by the publication of the first India Infrastructure Report in 1996 as a way to upgrade and manage infrastructure efficiently, although some of the larger cities were already contracting out a variety of services (GOI 1996). A recent survey of the thirty-five cities with over a million inhabitants, reports that 23 out of the 25 respondents have already partly privatized some solid waste management services (FICCI 2007). It is partial in the sense that either not all the SWM activities have been privatized, or privatization has been implemented in only certain wards or zones. The most common activities contracted out are door-to-door collection, transportation from collection bins, street sweeping, and transportation to the dump, and this is mostly by small-scale local contractors.

In Hyderabad and Bangalore, privatization was introduced partly because of the costs of hiring more workers (according to the National Institute of Public Finance and Policy, on average, 60 per cent of a ULBs expenses are for labour). Instead, they have just continued to extend the number of privatized areas as the cities have grown. In Hyderabad, in the early days of private contracting, entrepreneurs either bid on contracts for cleaning a given area, including lifting and transporting the waste, or they bid on a basis of cost per tonne of garbage removed, and either way, the contract was given to the lowest bidder. Neither way was particularly effective; output norms were not known, accountability was lax, and monitoring was in its infancy. Further, as Broekema (2004) has relayed there were newspaper accounts of price-fixing between city officials and contractors in Hyderabad in 1998.

New ways of contract tendering were devised. Today, in both the Municipal Corporation of Hyderabad (MCH) and Bangalore Mahanagar Palike (BMP), the privatized areas are grouped into "packages" of more or less equal units, which are then put out to tender under strict rules. Bidders need to furnish an Earnest Money Deposit (EMD) which is intended to assure their financial viability (the BMP asks Rs. 300,000 per package), and contracts are awarded on both technical feasibility criteria and the lowest bid. Most importantly, requests for proposals (RFPs) or tender documents are widely publicised in the newspapers and on the web, as are the names of the successful bidders. The contract documents are standardised and include penalties for lack of due diligence, protection of the workers (often day-labourers) in terms of protective clothing, medical aid in case of injury, and minimum wage (the BMP sets this at Rs. 1800 per month or Rs. 60 per day), and a system of monitoring the quality of the work done.

The publicising of contract documents is proving very useful, both for those who have to write them, and for citizens who want to monitor them. Some of the more progressive states, such as Tamil Nadu and West Bengal, and the larger cities have started publishing tender documents on their websites, and some states are making model contract documents available to their ULBs for guidance (see for instance Maharashtra 2007). In fact, the passage of the Right to Information Act in 2005 by the central government, and the adoption by state and local governments of similar legislation, has had an extraordinarily salutary effect in improving transparency in governance.

Contracts between ULBs and the private sector for collection, cleaning and transportation are usually for a short duration, seven months, because labour laws in many states insist that persons continually employed for more than 210 consecutive days are entitled to the fringe benefits of a regular employee, such as insurance, medical care and holidays. In the large cities, there are many more bidders than there are jobs, and sometimes resort is made to a lottery to distribute them to those who qualify. A contractor may or may not be lucky. This means that there is no incentive for a contractor to invest in new trucks and equipment, since repayment is chancy if the contract is not renewed. The consequence of all this is that most are using old, sometimes rented, often highly polluting, vehicles, while large-scale enterprises are not interested on bidding for short term jobs. On the other hand, this system does keep the field open to small-scale start-up entrepreneurs.

Sanctions for poor performance of private sector contractors are now built into most contracts. These can be exercised as deductions from the initial deposit (EMD) or reduced payment for work poorly done. Monitoring the quality of work by the ULB is done in some cities by insisting that the contractor get approval signatures from residents or ward committee members each day for work well done.

Public-private partnerships for the large-scale disposal of waste have been started or contemplated in some places: these include; bio and vermi-composting, reuse of plastics as a strengthening additive in road surfacing, pelletization and the thermal generation of electricity, biomethanation (the production of biogas), incineration, and the engineering and management of scientific landfills.

Table 2.10: Some Examples of Private Sector Firms Active in Solid Waste Disposal

Composting	*Plastics to Polyblend*	*Pelletization & thermal electricity*	*Methanation*	*Landfills*
Excel Industries, Kolkata Ahmedabad	KK Plastic Waste, Bangalore Kolkata	M/s SELCO Int., Hyderabad	IL&FS, Delhi*	BOOT projects, Bangalore* Delhi*
Vernar Organics, Mysore		M/s Sriram Energy Systems, Vijayawada	M/s Bio-energy Pvt. Lucknow**	
Terrafirma, Bangalore		Timapur Waste Man. Co Delhi*	BARC, Nisarguna	
Sunrise, Bangalore			Arashi Hi-tec Biopower Pvt. Coimbatore	

Notes: *proposed **not presently working
Source: Ministry of Non-conventional Energy Sources

In general, the companies get the waste free from the municipality, and the plant site at a token rent. BOT and BOOT ventures will return to the ownership of the City when the contract expires, usually in 25 or 30 years. In return of course, the ULB is relieved of the dumping costs of the wastes exploited. The high-tech operations which produce Refuse-Derived Fuel (RDF) are superficially seducing, but raise a number of environmental and economic questions. These will be discussed later in the section on issues and conflicts.

Partnerships between government agencies also exist in composting and recycling of waste. In Bangalore, the Karnataka Compost

Development Corporation (KCDC), a state company, obtains solid waste free of cost under an agreement with the BMP. The latter has supported the former also by allocating space in local parks for composting and vermi-composting[6]. The compost is sold locally, and to farmers through the Karnataka Agro Industries Corporation.

Agreements between Contractors and their employees

In the collecting, cleaning and transportation business, the number of workers a contractor must employ for a unit of work, and stipulations about the provision of protective clothing, health insurance, provident fund contributions, and the rate of pay for labourers is written into most municipal contracts. In reality, these are often ignored. In Bangalore for instance, labourers are supposed to be paid Rs.1800 per month, but in some areas are only making Rs.1000 or less. Galab *et al* (2004: 51) report that in Hyderabad labourers are supposed to receive a monthly minimum wage of Rs. 1300, but that the average is Rs. 1100. The other conditions regarding safety and insurance are completely neglected. Clearly, with such a shortage of regular work, rent seeking is occurring.

Alliances between ULBs and NGOs

NGOs have typically been movers and shakers in getting improved solid waste management started in communities, neighbourhoods, housing colonies and slum settlements, usually though setting up local resident groups, who later take charge of their own affairs. Such activities have often been supported by donor funds and residents' own contributions, as in Ranchi today, and the ULB was only responsible for carting waste away from collection points. There was often no financial relationship between the two. Since the publication of the SWM Rules in 2000, when ULBs started to become much more pro-active, the NGOs role of physically organising waste collection has declined, except in slum areas, but has increased in the key area of public awareness and education. Many NGOs are now contracted to undertake this work for municipalities.

There are many agreements, partnerships and alliances mentioned casually in the literature which are not explicit. One example is from Bangalore[7], where NGOs such as Clean Environs, Waste Wise (Mythri foundation) and CEE (Centre for Environmental Education) are assisting street children and other unemployed people to collect and sell wastes. But it is not clear whether these initiatives have been institutionalized, or indeed if there is any provision in the municipal laws to formalise such systems (Ramachandra & Saira 2003). NGOs are keen to play a greater role in urban management, and be properly recognised for their work. Recently, in the aftermath of the 2005 flooding, a coalition of civil society organisations in Mumbai formed an 'NGO Council,' which has now been recognized by the city, to advise and assist in its most urgent problems. The first problem tackled was solid waste, and an MoU was signed between the NGO Council and the Brihanmumbai Mahanagar Palika (BMP) in late December 2005 formalizing this relationship, said to be the first ever of its type (Karmayog 2006). This has led to the formulation of a solid waste management policy for the whole city and the adoption of strong municipal rules (MCGM 2006).

Agreements between ULBs and CBOs, RWAs and the Like

Formal contractual alliances between civil society and municipalities used to be rare, but this is changing rapidly. Partnerships between local government and CBOs are usually for door-to-door collection and primary transportation to community bins, and are often mediated by an NGO, who in the case of self-help groups, may have done the worker training. Examples include; partnerships between Vijaywada Municipal Corporation (VMC) and self-help DWCUA groups, between the Municipal Corporation of Hyderabad and the VGDS (Voluntary Garbage Disposal Scheme) groups, and between the Municipal Corporation of Greater Mumbai and ALMs (Advanced Locality Management) groups, which also include neighbourhood vermi-composting.

In Mumbai, the ALMs were required to be first registered with the corporation and then take up their activities. The corporation in turn appointed an officer on special duty for a particular ALM to assist cooperation and collaborative work between the community and the authority (Redkar 2005). A recent survey of ALMs groups in Mumbai found a declining interest in ALM activities, some respondents stated that only one or two people were actual members and that some formerly active members had got too old or incapacitated to participate (Karmayog 2006a). It is evident that every effort has to be made to keep citizen engagement positive. In fact the ALMs groups are to be replaced by LACGs (Local Area Citizens

Groups), which will serve larger areas, cover the whole territory of the city, have expanded responsibilities, and be formally a part of the city machinery, although doubtless some ALMs will mutate to become LACGs (Karmayog 2006b). While Mumbai has opted for larger neighbourhood groups, since the LACGs are to serve 5,000 to 10,000 households, Ranchi has opted for Pocket Development Committees (PDCs) of at least 100 but not more than 300 households under the Clean Jharkhand Project (CJP) (NBJK 2006).

In almost all these examples, the CBO workers are paid by householders, either by a fee or through an RWA, and may supplement their incomes from recycling dry waste themselves. In most places, the ULB provides the required tricycle or cart, some start-up funds, sometimes a site for composting, and in some cases provides households with two waste bins, one for biodegradable organics and one for inorganic recyclables, in order to facilitate separation at source.

Partnerships between NGOs and CBOs

Partnerships between NGOs and CBOs grew as a result of the withdrawal of government from social programmes under the terms of structural adjustment (Baud 2000). NGOs have been providing technical assistance and support to local communities to build alliances with local government (Mitlin 1999; Baud 2000; Anschutz 1996). For example, in Chennai, Exnora assists Civic Exnoras in SWM at the neighbourhood level. The NGO Exnora is in direct link with the local government, not the Civic Exnoras (Anand 2000). Snell (1998), in his study on NGOs and CBOs in Hyderabad has stressed on the quality of the co-operation of the local government with the NGOs, the CBOs and the citizens themselves for efficient and sustainable waste management, pointing out that such partnerships are still relatively new.

Relationship between CBOs and Households

The relationship between CBOs and households is surprisingly variable according to both the literature and our research findings. In areas where the CBO was resident-founded, and where the householders understand the fundamental necessity of waste separation, relationships are amicable and smooth. In other areas, perhaps those without a history of resident cooperation, there is resentment when householders do not keep their garbage separated, or when they refuse to pay their cess because they believe they have already paid for the service in their taxes, or object to neighbourhood composting. Since the workers usually have to collect the garbage from 100 or more households per day, six days a week, and those who are not city-workers depend on the fee for their income, these are not happy situations.

Three-way Partnerships between NGOs and DWCUA or Other SHGs, and ULBs

NGOs mobilise slum dwellers and rag-pickers and train them to become work groups as already described in the section on actors in SWM. The important thing is that most NGOs are usually multidimensional in their approach. Training includes health, nutrition, child care, literacy, legal rights, information about useful public programmes and how to apply for them, and thrift and savings, in addition to the skills required for work and income generation-activities, in an attempt to lift them out of hardship. NGOs negotiating contracts for economically deprived groups with ULBs or private sector employers cannot be seen as just another commercial bidder, they must be given special consideration because paid work for their members is one way out of extreme deprivation. This has been the case in Ranchi for example, where the NBJK's Clean Jharkhand Project (CJP) works with the Ranchi Municipal Corporation in the deployment of the Safai Mitras for door-to-door collection of waste, while also working as a spokesperson for the community PDCs on municipal issues (NBJK 2006).

Informal Pickers, Informal Traders and Recycling Enterprises

The great hordes of informal waste pickers and traders have very complicated relationships and chains of exchange. The large cities all have certain districts, usually located in clumps along main roads, where much trading in used goods is concentrated (Jain 2004). Materials that are sold and re-sold find a home if they have a use, and if not find their way into recycling industries of paper, plastics, glass and metal, with useless residues ending up in dumps. Galab *et al* (2004) have traced these waste flows in Hyderabad in great detail. From waste pickers, materials go to itinerant buyers, to retail traders, to wholesale traders and thence to the recycling enterprises. Each intermediary makes a certain profit from their labours. Those observers who advocate that all of SWM should be totally privatized may not be aware that this sector is: there are no government actors in this chain.

Table 2.11: Returns to Actors in the Waste Reycling Chain in Hyderabad

Street pickers	Rs. 250 per week
Dump pickers	Rs. 389 per week
Itinerant buyers	Rs. 1,222 per week
Retailers	Rs. 817 per week
Wholesalers	Rs. 30,000 per week
Recyling industries	Average 10% profit/ year

Source: Galab, Reddy and Baud 2004

System of Waste Management

Waste Collection

Before the MSW handling rules 2000, most householders were responsible for taking their own garbage to the local collection point, usually an overflowing community bin. Since then, door-to-door collection has become a common feature of waste management practice as prescribed by the rules. Householders are supposed to segregate their waste, which is picked up by workers using either tricycles (*ghanta gadis*[8]) or push-carts. The collection is organised by the ULB itself, by a private[9] contractor, or by an NGO, CBO or SHG.

When the ULB is involved, it engages its own staff whereas private contractors either have their own staff (mainly casual labour) or employ rag-pickers in door-to-door work. NGOs generally do not have any dedicated staff of their own; usually they organize rag-pickers for door step collection. Exnora (Chennai), SEWA (Ahmedabad), Nav Bharat Jagriti Kendra (NBJK) (Ranchi) are a few examples of NGO organised waste collection. CBOs may employ DWCUA groups where their women members are organised as SHGs under the urban employment programme SJSRY. Collectors are usually identified by an official tag or badge, and often a uniform, so that householder know they are *bona fide* workers.

Door-step garbage collection in residential areas is usually done in the morning hours, often from 6.30 to 10.30 am, with the collector's arrival being announced by a bell or whistle. On an average one worker covers 100-150 households in a day. The waste collection has been so good in some cities such as Nasik and Suryapet that there are no community waste collection bins in the city, the ones that exist are only at vegetable markets, slaughter houses and religious places (Yashada 2004b; Sengupta 2004).

Table 2.12: Primary Waste Collection

City	*Actors*	*Collection Efficiency*	*Remarks*
Ahmedabad	SEWA (NGO)	80%	The NGO has trained & engaged 4,000 rag-pickers.
Bangalore	BATF Contractor	Na	BATF provides training to rag-pickers and organizes door-to-door collection. 183 of 297 wards privatized
Chennai	Civic Exnora (NGO)	Na	Civic Exnora employs rag-pickers in waste collection.
Hyderabad	Private Contractor	90%	75% of work is privatized in terms of sweeping of roads. RWAs/ NHCs (in slums) employ worker for door-to-door collection of waste. MCH provides tricycles free of cost and collects garbage from the community bins.
Kolkata	KMC	Na	Door-to-door for 90% of city
Mumbai	ALM MCGM	90%	ALM employs rag-pickers in waste collection, 9% door-to door. MCGM community bins for the rest
Nasik	Private Contractor	95%	
Ranchi	NBJK (NGO) (*safai mitras*)	Na	Total waste collection is done by NBJK in 21 wards out of 37 wards. ULB does the rest
Suryapet	ULB	100%	
Vegalpur	SEWA	95%	*Arogya bhaginis*
Vijaywada	CBO (SJSRY)	Na	

Waste Segregation

Normally household and commercial waste can be segregated into two broad categories, dry waste, much of which can be recycled, and wet waste for composting. Segregation at source has three advantages; it makes both the wet and the dry wastes more readily re-usable, it thus reduces the amount reaching the landfill site, and hence reduces transportation and dumping costs. However, it is not widely practiced in the manner suggested in the SWM Rules 2000. Some ULBs have provided households with two bins, such as Vejalpur and Suryapet, and the tricycles or carts have separate bins to receive the materials. In some ULBs the collectors sort the wastes on the cart, carefully putting aside all they can sell, and dividing the rest between compostable and residual. Road sweepings are usually put into community bins or other collection points, and thus mixed with whatever else is there. This is not good for either compostables because it is full of grit, or recyclables because it is dirty and often contains cow dung and faecal residues. If it could be kept separate, it might make useful cover-layers in sanitary landfills.

Table 2.13: Percentage Waste Segregation

City	*% segregated*	*By whom*
Bangalore	some	Households served by CBOs and NGOs
Chennai	some	Households served by Civic Exnoras and other CBOs
Hyderabad	Small amount	Households served by VDGS or NGOs
Kolkata	none	Waste is sieved at treatment plant: pickers at community vats
Mumbai	9%	Households served by ALMs and other CBOs
Nasik	none	Sorting at compost plant
Ranchi	none	Sorting by employees of PDCs
Suryapet	90%	Householders
Vejalpur	90%	Supposed to be hh: often done by *arogya bhaginis*, who sell the recyclables

Street Sweeping and Storm Drain Cleaning

Almost all street sweeping and cleaning is done manually by gangs of labourers or DWCUA groups either employed by a contractor or by the ULB. The Supreme Court (1999) guidelines give the norms[10] for street sweeping. Sweepings and silt from the drains are usually deposited in community bins. Sweeping in residential areas takes place at the same time as door-step pick-up if this exists. Most sweeping is undertaken in the morning, and for commercial streets, before the shops open. Night time sweeping for main roads is becoming more common, as in Mumbai, Hyderabad and Bangalore. Mechanical sweeping is rare except in the largest cities: equipment is expensive.

Waste Transportation

Waste collected from households and street sweepings are usually dumped at communal collection points, which may be dustbins, concrete enclosures, garbage houses, placer bins or containers. If there is no collection, residents dump it there themselves. Most collection points are very messy, with garbage strewn about by cows, dogs, birds and scavengers. From these primary collection points the waste is loaded, often manually, into trucks. It is then either transported directly to a disposal site, or to a transfer station for later movement to a landfill site or processing plant in more efficient large-capacity trucks. Hyderabad was the first city to have transfer stations, but these are now being introduced in Mumbai, Bangalore and Kolkata.

Waste trucks are supposed to be covered, but this is rarely the case. Litter free towns, such as Suryapet and Nasik have dispensed with community bins, except at markets and the like, and the waste is loaded into trucks directly from the collectors carts and *ghanta gadis*. Large cities such as Mumbai, Ahmedabad, Bangalore and Hyderabad are using demountable containers (community bins) and hydraulic lifter trucks for transportation of garbage.

Table 2.14: Secondary Waste Transportation

City	*Actors*	*Remarks*
Ahmedabad	AMC	na
Bangalore	BMP & Contractor	Transportation of wastes to disposal site is done in two ways through the same trucks by engaging 82 trucks of the corporation and through contractors by engaging 129 vehicles for layouts and markets and 72 vehicles on contract for transportation of waste.
Chennai	CMC	na
Hyderabad	MCH & Private	75% of work is contracted out. Waste is segregated at transfer station.
Kolkata	KMC Contractors	54% of work is contracted out
Mumbai	BMC	na
Nasik	Contractor	Garbage is transported between 3 pm to 5 pm directly to compost plant.
Ranchi	RMC	RMC provides its trucks and drivers, which are supervised by NBJK
Suryapet		Not required
Vegalpur	ULB	Community bins overloadeed
Vijayawada	CBO	Waste transportation by DWCUA group under SJSRY scheme

Waste Disposal

In the majority of the case studies, the bulk of waste disposal is primitive. It is dumped in low-lying areas outside the city. Most disposal sites, being devoid of a leachate collection system or gas monitoring and collection equipment, can hardly be called sanitary landfills.

Compaction and levelling of waste and a final covering by earth are rarely observed practices. Altogether present activities have been identified as the relocation of waste from one point to another (Toxic Link 2005)[11]. Nasik was the first ULB to build a scientific landfill site in conformity with the SWM Rules 2000 (Yashada 2004). However many more are being planned and developed.

Table 2.15: Waste Disposal Sites in Case Study Cities

City	*Methods*	*Location*
Bangalore	Dumping	Many sites. Sanitary landfill planned at Mavillipuram
Hyderabad	Dumping	Mansurabad (Autonagar), Jamalikunta (Golconda) Damgamguda, sites
Kolkata	Dumping	In marshes NE of the city. Sanitary landfill planned at Dhapa
Mumbai	Dumping	Deonar, Mulund, Malad, Gorai. Site for landfill under study
Nasik	Landfill	n.a.
Ranchi	Dumping	Mukkadam. New site found at Zheeri
Suryapet	Dumping	At the treatment unit, Ambedkar Nagar Rd. Landfill planned at Kasarabad Rd.
Vegalpur	Dumping	Open plots and water bodies

Alternative methods of disposing of waste are developed or being developed in line with the SWM Rules 2000, but even where they operate successfully, do not yet fulfil their potential for significantly reducing the amounts that have to be land-filled.

Composting is the most cost effective way of disposing of biodegradable wastes, which usually make up about half of the total volume generated. However, the success of composting rests heavily on the kind of waste available. Un-segregated waste either has to be sorted, or the compost sifted, in either case making for a less saleable product. Generally, waste produced in India has a high organic content suitable for vermi-composting[12]. The process can be undertaken at the neighbourhood level, or at an industrial scale.

Neighbourhood initiatives usually operate through CBOs or NGOs, or by the ULB contracting the work to a company or a NGO (Table 2.16). For example, in Bangalore, two NGOs namely CEE Kalyana Nagar Residence Association and Residents Issue for a Safe Environment (RISE) have initiated community-based bin composting serving 980 and 1200 households respectively (Zurbrügg *et al* 2003).

Table 2.16: Waste Disposal: Composting and Waste to Energy

Treatment Procedures	*Existing*	*Under study, planned, being introduced, or expanding an existing treatment process*
Bio-composting	Mumbai, Delhi, Nasik, Coimbatore, Bangalore, Kolkata, Ahmedabad, Jamshedpur, Rajkot, Pune	Delhi, Kolkata, Asansol, Madurai, Jabalpur, Chennai, Coimbatore, Pune, Meerut, Nagpur, Hyderabad, Patna, Jaipur, Indore, Surat, Mumbai
Vermi-composting	Mumbai, Indore, Pune, Nagpur, Bangalore	Jabalpur, Jamshedpur, Coimbatore, Pune, Mumbai
Methanation	Mumbai	Delhi, Mumbai, Nasik, Jabalpur, Pune
Pelletisation	Rajkot, Hyderabad	Delhi, Varanasi, Nasik, Jaipur, Ahmedabad, Chennai, Coimbatore, Pune, Hyderabad, Bangalore, Surat, Cochin
Gasification		Delhi
Incineration	Varanasi (BMW), Indore, Hyderabad	Kolkata, Pune

Source: FICCI 2007

Decentralized composting is low-tech, and inexpensive to set up once a site is identified. Composting is either aerobic, using windrows which are turned by hand, or through vermi-culture which is usually carried out in concrete pits. Many operations use bio-composting to start the process, often fuelled by simple cow dung and water slurries, and then finish the process off with vermi-culture. The best operations are paved, roofed, fenced, surrounded by trees, and the composting beds covered with netting to prevent animals, birds and vermin from entering. Neighbourhood operations are often run by DWCUA groups who have been trained in the work by NGOs, and who are able to make a living by selling the compost. If correctly run, there is no odour or nuisance.

The location of a neighbourhood composting facility faces hindrances at two levels. One is because of the NIMBY (not in my backyard) syndrome and second is the scarcity of land. Currently, urban planning does not include such needs in its city plans (Toxic Link 2005). Neighbourhoods request land from the ULBs because the legal responsibility for SWM rests with them. Municipalities have provided land in some cases. In some residential societies, as in Basera Colony, Mumbai, RWAs have provided land. For the activities of

NGOs like Naya Savera and Vatavaran, educational institutions such as IIT and JNU have provided space (Gupta 2004).

Table 2.17: Waste Composting

City	*Com-posting*	*Actors*	*Remarks*
Ahmedabad	na	na	Excel Industries Ltd. has set up a plant in the city for manure generation.
Bangalore	Y	o KCDC o NGOs o Private Company[13]	State enterprise Rag-pickers, with the help of NGOs run neighbourhood level vermi-composting Terra Firma and Sunrise
Chennai	Y	Civic Exnora	Neighbourhood level composting
Hyderabad	Y	MCH and RWAs	MCH provides one time cost for setting up of vermi-composting centre and the recurring cost for maintenance will be generated by the agencies through collection of service charges and selling vermi-compost.
Kolkata	Y	KMC Excel Ind.	*In situ* composting at Dhapa (garbage farms) Processes 70MT/day, sold as 'Bhu-Dhan'
Mumbai	Y	SMS-Composting	Numerous locations. Also some ALMs do composting
Ranchi	Y	CJP	Vermi-composting plant at Chiuttu, produces 192 MT/year
Suryapet	Y	ULB	Vermi-composting at Ambedkar Nagar Road
Vejalpur	Y	SHGs	Vermi-composting being started with SMS help
Nasik	Y	On BOT basis	NMC's newly constructed compost plant has a capacity of 300 metric tonnes and was built at a cost of Rs. 5 crore. Nearly 5,000 trees have been planted around the compost plant.

The oldest large-scale composting plant in the country is a parastatal, the Karnataka Compost Development Corporation (KCDC) in Bangalore, the only plant still functioning in India from a series of ten built in the early 1970s. Today it is a subsidiary of Karnataka Argo Industries Co-operative and its Board of Directors includes representatives from the Karnataka State Cooperative Marketing Federation and the BMP, which greatly aids in marketing its products. Under agreement the BMC supplies waste cost-free to KCDC. It processes up to 350 MT per day (88,000 MT/yr), which goes through a first process of aerobic decomposition in windrows, then is vermi-composted. Despite many early technical and financial difficulties, and problems of quality control and marketing, today the business makes a profit and sells compost in three south Indian states. Two more plants are planned for the Bangalore urban area[14]. Two other private companies also operate in Bangalore, Sunrise and Terra Firma Biotechnics. The latter produces 13 tonnes of manure per day from 96 tonnes of waste (Drescher 2002). The establishment cost of the plant was Rs. 6 million and the producis sold to a fertilizer company.

The largest private company in composting is Excel Industries Ltd., of Mumbai, pioneers in agro and industrial chemicals. Excel is also active in Kolkata, Ahmedabad, Puri and Indore and there are other companies that employ its expertise. Excel's technology is one involving waste sanitization at points of collection, and controlled aerobic composting and mechanical screening of digested compost at the treatment plant. The biological innoculum used for sanitization is claimed to be non-toxic, eco-friendly and non-hazardous, kills odours and pathogens and repels birds and vermin, making waste safe to handle. It has been suggested that some products, *bioculum, sanitreat and herbocel*[15] developed by Excel could be used in composting waste at community and household levels (Gandhi 2005)[16]. For large installations it is estimated that a plant capable of handling 500 tonnes of waste per day requires an initial investment of Rs. 60 million.

Of the case study cities, only Hyderabad has invested in very high-tech solutions, namely the pelletisation and thermal electricity generating plants operated by M/s SELCO International Ltd., and M/s RDF Power Projects Ltd., the electricity being sold to Andhra Pradesh Transco. The economics of these are not clear since they were set up as two of the sixteen demonstration projects partially subsidized by the Ministry of Non-Conventional Energy Sources (MNES 2004a). It is worth noting that the Ministry also subsidizes municipalities for Waste-to-Energy (WTE) projects. A financial incentive of Rs. 15.00 lakh per MWe is payable to a Municipal Corporation or Urban Local Body, for providing garbage free of cost at a project site, and land on a long-term basis (30 years +) at a nominal lease rent. However, this incentive will be reduced to 50% in the case of projects for generation of power from fuel or fuel from waste (MNES 2004a). Similar conditions apply to the project for the generation of 5.7 MW power from municipal solid waste in Vijayawada by M/s Shriram Energy Systems Pvt. Ltd., Hyderabad (MNES 2004a). A BOT (Build-Operate-Transfer) project is underway in the same city, where the corporation has leased land to Excel and Sri Ram Industries (Ravikaran 2005).

Bio-methanation is a well established non-polluting procedure, and as demonstrated by BARC, can be operated without any outside energy consumption, if powered by solar panels (BARC 2007). None of the case study cities have adopted this technology as yet, although there are many examples of small-scale bio-methanation plants in India. So-called advanced technological options used in some developed countries, such as incinerators, pyrolysis plants, Refuse Derived Fuel (RDF) and thermal gasification are now being adopted in developing countries as well (Berthier 2002). As was shown in table 2.16 some of these are under debate for some cities in India.

Financing

The major concern at the time of the adoption of the SWM rules in 2000 was the cost of providing the services. Estimates suggested that these would be higher than the ULBs could afford. In fact the Twelfth Finance Commission (2005-2010) reported that on average, urban local bodies spend about 60 to 70 per cent of their budget on this important activity (GoI 2004: 141). Our studies showed that the costs are not so excessive, they varied between four and sixteen per cent of the total municipal budget, although the figures shown in the following table must be interpreted with extreme caution: in some cases they may be overestimated because they include other health expenditures such as the maintenance of public toilets, and in some cases may be underestimated if they omit major capital outlays on treatment facilities, or administrative salaries.

In general, costs in large cities are greater than those of smaller: this is understandable since transportation distances are longer thus making costs higher. In per capita, in large cities, the costs are between Rs.200 and Rs.280 per year, in smaller towns it is under Rs.100, but the ULBs have not yet made appropriate arrangements for the landfill of residues, a point which will be returned to later. Kumar (2005) estimates the expenditure of a city with 1 million population is around Rs.10 crores, which seems low. Suryapet is most interesting in it became a dustbin-free, 'Zero Garbage' town in the space of two years between 2000 and 2003 with no financial aid from the centre or the state and no increase in taxes.

Table 2.18: SWM Budgets and Wages

	% of total ULB budget 2002-2003	*Annual per-capita cost*	*Municipal workers av. pay per mo.*	*Contractual labourers pay per mo.*	*CBO, SHG & DWCUA earnings per mo.*
Bangalore	8.4	Rs. 217	Rs. 5000	Rs. 1800	Rs. 1000
Hyderabad	10.9	Rs. 199	Rs. 4000	Rs. 1300	Rs. 1200
Kolkata	16.6	Rs. 279	na	na	Rs. 1000
Mumbai	7.5	Rs. 265	na	na	Rs. 750 min.
Ranchi	na	na	na	na	na
Surapyet	5.52	Rs. 67.1	Rs. 5000	na	Rs. 1375
Vejalpur	4.0	Rs. 20.3	Rs. 4000	Rs. 1800	Rs. 1250

The largest outlay in SWM operations is for labour, especially in door-to-door collection, sweeping and primary transportation. For instance in Mumbai, the corporation spends Rs. 400 crores annually on SWM of which Rs. 250 crores (62%) is spent on wages for the 25,000 employees (Jain 2004). Regular municipal labourers earn up to Rs.6000 per month, and rightly benefit from health insurance, sick leave, pensions, protective clothing and holidays.

A second alternative is contracting out. Contractual labour is supposed to be paid at a minimum rate of Rs.1800 per month and receive fringe benefits, according to most labour laws, but these rules are usually blatantly flouted by contractors, as explained earlier. Similarly, day labourers are supposed to be paid Rs. 60 per day. Our studies show that cost of labour is why primary collection, road sweeping and transportation are cheaper when contracted out to the private sector. For instance, in Chennai, overall waste handling costs per tonne by the public sector are Rs.1050, and Rs. 648 per tonne by private contractors (Devi and Satyanarayan 2001). For secondary transportation this may not be the case. Small local contractors usually have old or hired general-purpose trucks which have to be filled manually, while the larger municipal corporations may have hydraulic trucks with lifting gear and sometimes compressors, so can get the job done more efficiently: this is the case in Mumbai.

The third labour option for primary collection and sweeping is for ULBs to contract self-help groups on a neighbourhood basis for this work. This is the least expensive option since some of the workers' income is derived from various grants under SJSRY or the Ministry of Social Justice and Empowerment programmes for the first few years, and from user fees as whatever community system that is adopted becomes operational. Not only is it economic, but equally important, it is socially responsible in offering disadvantaged groups the opportunity for decent work. Further, the workers are trained by the NGO in charge, often have access to child-care and health clinics, and are able to supplement their meagre but steady incomes by trading the dry recyclables collected from households, bringing their income up to about Rs. 2000 per month.

Raithi (2005) has used an econometric optimisation model to compare the three systems using data from the Municipal Corporation of Greater Mumbai. She showed that the cost per tonne of waste management with community participation and local composting, (the ALMs system) is Rs.1518, with public-private partnerships including Excel Industry is Rs.1797, and when handled solely by the municipality is Rs.1908. This is a very important finding.

The big capital cost that ULBs are facing, or will face, is that of residual refuse disposal in sanitary landfills by environmentally friendly means, according to the SWM rules. Most presently dump wastes on low-lying ground in an unorganised manner with disastrous environmental consequences. The alternatives have already been outlined: bio and vermi-composting for bio-degradables, methanation, refuse-to-fuel through drying or pelletisation and use in electricity generation, incineration through pyrolysis, thermal gasification, and the construction of sanitary landfills. Of these, composting, methanation (biogas production) and landfill solutions have been found to be the most cost effective (IPE 2004). It should be noted that no matter what technology is chosen, landfills are necessary for the ultimate disposal of left-over residues. The real challenge is to minimize these.

Composting can be introduced at the neighbourhood level: the big advantage of this is that transportation costs are minimal, which reduces costs of trucks, man-hours and fuel, wear and tear on the roads, and reduces traffic-induced degradation of the environment. Neighbourhood composting usually covers its own costs through sale of the product if it is done by a SHG. Composting can also be large scale, such as the Excel Industries plant at Mumbai or the KCDC plant at Bangalore. Bio-decompostion reduces the need for landfill sites, and provides valuable fertilizer. But ideally, it requires that wet waste is pre-sorted to ensure an uncontaminated product, and to eliminate the need for costly sieving.

Most other high-tech solutions are not cost effective yet. Waste-to-energy projects have been beset by difficulties, and would not operate without subsidies from the Ministry of Non-Conventional Energy Sources (MNES)[17]. It has two programmes focussed on solid waste (the largest part of its operations are in promoting renewable energy resources: solar, wind, hydro and geothermal). These are: (i) the National Programme on Energy Recovery from Urban and Industrial waste, and (ii) a UNDP/ GEF assisted project on the development of bio-methanation. It is also, with other government authorities, advising ULBs how they can qualify for carbon credits under the Clean Development Mechanism as set up under the Kyoto Protocol, if they can measurably reduce their GHG emissions. The World Bank through the Carbon Finance Business Unit is currently working with Karnataka, Tamil Nadu and Andhra Pradesh on this possibility for a new income stream. Interestingly, preliminary estimates show that composting and biomethanation hold by far the most promise (Hanrahan *et al* 2006: 38).

Scientific landfill sites are expensive to build and operate. Estimates suggest that the cost for dumping one tonne of waste is about Rs.200 but can be as much as Rs.1155 per tonne (CMAM 2004-2005: 17). Such calculations, amortized over 25 years, include: cost of land, site preparation, fencing, utilities and roads, costs of engineering the site with liners, membranes, leachate-collecting equipment, pumps, ground-water monitoring devices, operating equipment such as bulldozers, spreaders and the like, in addition to operating costs. Futher, there is often a problem of finding a suitable site because of land availability and suitability, and because of the regulation that no site should be within a 20 km. radius of an airport because of bird-strike hazard. The rule of thumb is that 15 acres per 100,000 population is required for a landfill with a life expectancy of 25 years. The establishment of new sites has been hotly contested by local residents in a great many places, especially if the site chosen is not within the municipal boundaries of the ULB concerned. In some cases this has even led to closure of a facility.

Solid waste management expenses are mostly covered by the municipal budget. Proposed revenues include: income from recyclables, collection fees or a conservancy tax, bulk collection charges, fines for littering, waste-to-energy schemes and effective property tax collection. The recovery of solid waste management costs are usually grossly over-estimated in planning studies because it is often assumed that the public authority is somehow going to recover the value of all the recyclables, although experience shows that households and enterprises either give or sell anything marketable to itinerant dealers, or the regular waste collectors may pick them up themselves as an income supplement. The non-biodegradable waste left over is usually of little value.

Table 2.19: Some Cost Recovery Variables

	Door-to-door collection charges 2005	*Bulk collection charges (commercial complexes, markets, hotels, restaurants, marriage halls..)*	*Fines for littering*	*Property tax collection rates*
Bangalore	will start '07	User fees		63%
Hyderabad	VDGS Rs.10/mo.	User fees: Rs.1200 to Rs.1800/mo.	Rs.50	
Kolkata	None, but prop.	User fees	Yes	90%
Mumbai	Rs.10-30/mo.	User fees	Yes	
Ranchi	SHG Rs.20/mo.			
Surapyet	None		Rs.100	85%
Vejalpur	Rs. 11 / mo.	Construction debris Rs. 300 per truck		66%

Door-to-door collection charges are made in many communities that are served by NGO-CBO and self-help groups, varying from Rs.10 to Rs.30 per month, and are collected either by the workers or by their supervisors, or sometimes by the local RWA. These usually go directly to the groups doing the work and do not enter the municipal budget. Even though the amounts are trivial, householders sometimes refuse to pay, because they think that SWM is a municipal service that they have already paid for in their taxes.

The charging of fees for bulk waste collection is becoming more popular, and trucks making special trips is usually a more efficient way of collecting it. Charges are usually made monthly for regular users such as shopping centres and hotels, or by weight for building and construction debris. Fines for littering and roadside dumping have been introduced in many jurisdictions. These have the double effect of both improving careless behaviour, and being a source of income.

A final point about revenue streams is more general, namely the overall management of ULB budgets. The main local revenue source is the property tax, along with the *octroi* in states where it has not been abolished as proposed, and the bulk of the rest is made up by state allocations and various grant programmes. Many ULBs have outdated or incomplete property tax rolls and a weak record of tax collection. While many larger municipalities are making an effort to improve this situation through the Financial Institutions Reform and Expansion (FIRE) programmes, there is a long way to go yet (USAid-India-Fire 2004). The introduction of GIS (Geographical Information Systems) should improve the tax rolls since all taxable properties will be clearly identified. Tax collection rates could be improved through incentive schemes. Kolkata for instance has totally modernised its tax administration with remarkable results. Between fiscal years 1992-93 and 2005-06 it increased its tax revenue by 345 per cent, and its non-tax revenue (rents, fees, charges, fines etc.) by 854 per cent. Innovations in financial management include, application of computers in accounting and revenue mobilization, issuance of building permits and municipal documentation such as providing birth and death certificates, the introduction of self-assessment tax on land and buildings, and improvements in billing and collection of property taxes. It also set up a Municipal Accounts Committee on the lines of the Public Accounts Committees in Parliament and the state legislature.

Some cities and towns are going to have to make major capital infrastructure investments to solve their problems, including not only for recycling and landfill sites as already discussed, but also equipment in the form of trucks, lifters, waste containers, transfer stations and the like. Such municipal investments may be supported by central and state grants, loans or bond issues. As mentioned earlier, the 12th Central Finance Commission[18] (CFC) has earmarked Rs 5,000 crores for municipalities with a population above 100,000 (except for megacities) for the period 2005-2010, of which 50 per cent is to be devoted to SWM. Municipalities are required to prepare a comprehensive SWM scheme to include composting and waste to energy programmes, to be undertaken in the private sector (NIUA 2005). Loans can be made through HUDCO, the Life Insurance Corporation of India, or from State Urban Development Funds as in

Tamil Nadu and Karnataka. Since 1995 infrastructure projects may be supported through municipal bond issues. Some cities such as Ahmedabad, Bangalore, Hyderabad, Nasik, Vishakhapatnam and Chennai have already done this for various purposes (Mahadevia 2003). More recently, the Government of India[19] has permitted Foreign Direct Investment (FDI) inflows into city hardware and have also issued guidelines for issuing municipal bonds for raising finance. There are also private sector intermediaries such as Infrastructure Leasing and Financial Services Ltd. (IL&FS) who enters into concession agreements with state and local governments, and the Infrastructure Development Finance Company (IDFC) who brokers public infrastructure projects and capital markets (Mathur and Ray 2003).

Launching bonds or long term borrowing demands stable and predictable sources of income for the ULBs choosing this path, which in turn requires that the taxation system be updated and modernized (Vaidya and Vaidya 2004). Financial credit-worthiness in turn requires a credit-rating system, a field which is largely dominated by CRISIL (Credit Rating Services of India Limited), a company founded in 1987: since the mid-nineties the majority share holder is US giant, Standard and Poor's. CRISIL's role is somewhat ambiguous. On the one hand, its original role was credit ratings, risk management, industrial restructuring, economic impact assessment and the like, but today it also does management ventures, including for instance, Pune's City Development Plan (Barse 2006). This runs the danger of putting it in the position of evaluating the investment outcomes of something it has proposed, an uncomfortable arrangement.

Up-scaling and Replication

While the Solid Waste Management Rules 2000 give a clear outline of what should be done, the ways of carrying out each operation are many, as the growing number of studies show. The City Managers' Association of Maharashtra's recent Compendium is an excellent example (CMAM 2004-5). The up-scaling or replication of methods that work in one area must be pursued with caution, according to context Essentially, there are three ways to think about this: up-scaling as an increase in the number of households served, extension into different kinds of interventions, and replication in other urban centres. Examples of each will be discussed.

Decentralized community-based solid waste management practices, such as those promoted by Exnora, Stree Mukti Sangathan (SMS), the Centre for Development Communication (CDC) and many other NGOs and CBOs have received world-wide acclaim and are copied on other continents, and yet they only account for a small portion of waste management operations in India, perhaps 3 or 4 per cent all told. The advantages of decentralised community-based initiatives are well known: environmental sustainability because they promote segregation, reuse and recycling, with bio-degradable and recyclable wastes to be treated separately, and only inert residues needing landfill treatment, thus minimising the space needs for these expensive installations; ecological sustainability because they usually include local composting, thus reducing transportation costs, road-wear and traffic pollution; financial sustainability, because most residents show a willingness to pay for community controlled activities where they are involved in their management; social sustainability through creating a sense of belongingness in the areas where people live and work; civic pride in a clean and green city which also boosts the local economy, especially in tourism; and finally, social justice in providing regular, safe, dignified work to innumerable waste-pickers, who otherwise scavenge in the fringes of society (Agrawal 2005). Clearly, decentralization is a model worth replicating for all these reasons.

An important in-depth investigation into up-scaling people's participation in urban solid waste management has recently been made by Toxics Link (2005) through the examination of the activities of sixteen NGOs and CBOs working in the field and visits to thirty-one cities. While all sixteen organisations have up-scaled, the major constraints to expanding the community-based model are: indifference on the part of municipalities; lack of consensus among city politicians about the validity of community leadership; waste-pickers who are immigrants being seen as a threat to national security; private sector service providers seeing community efforts as a threat to their potential business; the municipality setting up a "free" waste collection service with no reference to what has been functioning, thus destabilizing systems in place and de-motivating people from contributing; the ill-defined *locus standi* of Residents Welfare Associations, which have no legal status; the rotation of office bearers, both community and municipal, who need briefing and sensitizing to the operations in hand to assure the continuity of efforts; the unsatisfactory or below par performance of municipal staff at the interface between community and ULB work (Toxics Link 2005).

These are problems that could be overcome with stronger directives about how door-to-door collection and community waste management are to be handled. The Maharashtra City Managers are very clear about this. In planning a system, they strongly recommend both integrating rag-pickers into waste management by giving house-to-house contracts to NGOs or SHGs (an activity they interestingly classify as a public-private partnership), and pursuing the option of decentralized smaller treatment plants for composting or bio-methanation in order to reduce transport costs (by as much as 60%), and have a system which can be built incrementally, is inexpensive, more manageable, less risky, and can generate steady income for the needy (CMAM 2004-5).

However, some progress in decentralization is being made. One example is that of Mumbai, where on the urging of the NGO Council, a city-wide system of Local Area Citizens Groups (LACGs) is being created. LACGs will each take care of 5000 to 10,000 people, for a total of around 2000 groups, and will have clearly defined territories, covering the whole of the city. There is a clear set of rules institutionalizing them as an integral part of the MCGM structure and functioning. Existing ALMs can transform into LACGs, or if small, may merge together to become one. Each of the 227 wards will thus be made up of about ten LACGs (Karmayog 2006 b). While there has been some resentment felt about replacing the ALMs that were successful, there is also the strong potential for very real decentralized citizen-controlled action. The project is still in its infancy, and it remains to be seen how well it will work out. One of the drivers is to have in place a complete set of CBOs in place capable of responding to crises such as the floods of 2005.

Only two ULBs in the case-study work had almost complete compliance with the SWM Rules 2000. These are Suryapet with a population of 90,000, and Vejalpur Nagarpalika in the urban fringe of Ahmedabad, with 113,000. In each case this was achieved in the space of two years, between 2002 and 2004, without any additional funding or aid, although neither solved the landfill problem. Clearly small size is a factor in getting things done on the ground, but these examples also underline another important variable: good leadership and management practices.

At another level, replication of effort is to do with management practices overall. Kolkata is interesting in this regard because of its innovations in modernizing urban management, focussing on efficiency and transparency. The continuing environmental and social crises of the city have been the subject of a whole series of planning and remedial projects since the fifties. More recently, in 2003 the Kolkata Municipal Council, along with the GoWB and foreign aid from a variety of sources, has embarked on the Kolkata Environmental Improvement Programme (KEIP), which has a large component devoted to modernizing municipal government, in addition to infrastructure building and a large slice of funding for SWM. Reform of city management, especially in terms of administering the property tax and thus increasing self-generated revenue has been remarkable.

Thakur (2006) has recently investigated the financial situation of 17 cities using data from their recently completed City Development Plans prepared as part of the JNNURM (Jawaharlal Nehru National Urban Renewal Mission) in an effort to find out why some municipalities have better financial performance than others. He shows that there is little consistency in financial reporting, accounting methods, or in the proportion of revenues derived from local sources as compared to those from state and central grants, nor is there any explanation of why these go up or down. Further, it is very difficult to know the precise expenditure on any one service. He shows how the proportion of expenditures on establishment costs, operation and management costs, and debt service differ wildly between municipalities, but finds no explanation. Some municipalities have larger functional domains, for instance some are responsible for water, while in other cases water is provided by the state or a parastal, but this does not explain the differences. Our experience with the case study ULBs was similar. Clearly, forward planning depends on the ULBs having a good grasp of their financial situation and prospects.

Environment and Sustainability

In planning for solid waste management, every practice in the chain of activities must be examined to see if it is environmentally sound, and if it contributes to sustainability before it is adopted. The impacts of new ventures in SWM need to be studied in greater detail than they are now, from the point of view of local environmental conditions, health impacts on the population, waste handling, recycling, uses of recycled wastes, impact of disposal methods and landfill sites, employment generated, cost reduction for the ULB, business generated, and the like. This is a very important area of future research. While solid waste facilities are not subject to environmental impact assessment, the location and operation of landfills and other disposal facilities are subject to review and approval by the state pollution control board.

In terms of environment the most direct impact of mismanagement of solid waste is on public health. In the zero garbage Nasik city, there has not been an outbreak of cholera, gastroenteritis or malaria epidemics since the SWM project has been implemented. In fact, cases of malaria have come down to practically zero. Based on the analysis of nearly 9,000 water samples over the year, it has been found that the percentage of potable water has gone up to 98 per cent. The sanitary condition of the city has, by and large, improved substantially. Nasik generates about 270 m tones of garbage per day. As per Nasik Municipal Corporation's (NMC's) estimate, farmers directly use 20-25 m tonnes as manure. Stray animals consume 8 to 10 m tonnes of garbage. Of the remaining, about 220 to 230 m tonnes are collected by the corporation. The collection efficiency thus works out to 95 per cent (Toxic Link 2005).

There has been growing consciousness of the concept of sustainable development both nationally and internationally. Researchers have used this term in various contexts and aspects such as sustainable livelihood, sustainable environmental development, sustainable economy and social sustainability (Mahadevia 2000). Urban sustainability is the challenge of integrating social, economic and environmental perspectives into all municipal activities, including SWM.

The first important initiative towards sustainable SWM practice is the 74th Constitutional Amendment (CA) itself, in that it reinforces the institutions of local self government, promotes direct dialogue with community groups through ward committees, forges partnerships between stakeholders, raises awareness of people about SWM practices, and hopefully leads to more effective management. This is in line with Agenda 21, the Rio Declaration on Environment and Development, adopted by more than 178 Governments including India at the United Nations Conference on Environment and Development (UNCED) held in Rio de Janerio, 1992. Local Agenda 21 emphasizes that sustainable practices must begin at the local level.

Environmentally, the greatest promise for sustainability in SWM is to minimize waste at source, followed by separation at source in order to maximize recycling, reuse and composting of biodegradable materials as laid out in the SWM Rules 2000, for all the reasons previously explained. It should be promoted as resource recovery rather than as getting rid of unwanted materials: in fact, a sea-change in thinking on the part of managers is required. Ecologically sound waste management has to reduce environmental hazards and destruction caused by randomly strewn garbage, indiscriminate dumping, and leaking and smoking landfills. The new National Environment Policy has identified industrial and municipal waste as the major cause of soil pollution (GOI 2006). It estimates that environmental factors are responsible for almost 20 per cent of the burden of disease in India, and recognizes that reducing air pollution, protecting water sources, protecting soil from contamination, and improved sanitation are urgent issues.

Financial sustainability of waste management services does not seem like an impossible dream. Our case studies and the extensive literature on the subject suggest that for primary collection and transport, a modest user fee of Rs.30 per month in all but the poorest areas is an acceptable sum for a well provided service. This is in line with the "polluter pays" principle advocated by the National Environment Policy. If this is coupled with the employment of CBOs and SHGs for this task, the latter supplement their incomes by recycling dry wastes, this part of the operation becomes self-sustaining. The municipality then can look after the transportation and disposal of the residues from its normal self-generated budget, assuming that it can introduce more effective ways of administration.

Social sustainability can be improved by finding ways in which to integrate waste pickers into the community. The estimated numbers are enormous: figures for a few cities are, Mumbai 100,000, Delhi 75,000, Kolkata 40,000, Ahmedabad 30,000, Chennai 30,000, Bangalore 25,000, and Jaipur 10,000. The Alliance of Waste Pickers in India is concerned that municipalities in their haste to comply with the SWM Rules 2000, give out contracts with no thought about the future of their kind. They point out how in Nasik, 1000 waste pickers have been displaced and rendered destitute by the privatization of door-to-door collection and demand that the SWM Rules 2000 be amended to encourage ULBs to integrate them into the system (SWACHH 2005). Since local social and economic development are now part of the mandate of ULBs, following the list of duties enumerated in the 74th CA, this seems reasonable. Agrawal (2005) shows how the 'Swatcha Doots' were incorporated into the system in Nagpur through the work of the NGO, CDC (Centre for Development Communication). He calls this a successful "Public Private Peoples Partnership," a good term to be introduced to the lexicon.

Finally, urban sustainability requires real political and civil society commitment to the amelioration of the quality of life. Awareness programmes are essential, and these could well be run by established NGOs hand in hand with the ULBs. These should include not only residents, but all municipal employees and contractors from managers down to the lowliest truck operators, garbage handlers and sorters so that they realize the critical importance of their work. At the same time, benchmarking and sustainability indicators on the three parameters of social, economic and environmental variables, suitable to the Indian context, should be developed, so that municipalities can measure their progress towards sustainability. The present system of reporting progress to the state pollution control boards only monitors the mechanics of conformity with the SWM Rules. It is a good start, but does not reflect the outcomes of these activities. Good governance is an amalgam of the activities of the positive efforts of government, civil society and the market.

Conflicts and Issues

Many of the conflicts and issues being debated at the present time stem from two sources. First are those that devolve from how the Solid Waste Management Rules 2000 are being or are to be implemented, and second are those that are generated by the mixed messages that come from government policies. They are both widely discussed in the research literature.

The problems of implementation focus on how things should be done. An example is in Chennai, where about 14 per cent of the households refused pay user charges to civic Exnora (Anand 2000). Further discussion was whether Civic-Exnora should charge the residents anyway, when it is the legal responsibility of the corporation to collect garbage and keep the streets clean, and for which the municipal corporation is already collecting taxes (Appaswamy 2005). (Added to this was the question of how much funding an NGO with overseas connections needs). A third issue was if Onyx, the private contractor engaged for collection and transportation of waste, substitutes for corporation personnel, then the amount paid to Onyx is an additional expense and whether that is worthwhile (Appaswamy 2005). These sorts of questions can only be sorted out through a proper airing of the issues, municipal council, ward committee and resident association discussions, and a transparent disclosure of how the system works and why it was chosen.

Issues related to institutionalization and formalization of services have not been addressed adequately. For instance, in Mumbai the technical as well as organisational support to ALMs was provisional and not institutionalised into regular municipal functions. MCGM expected ALMs to segregate waste at source and create in-house system for bio-degradable waste, while ALMs expected MCGM to provide separate carriers for dry and solid waste (Redkar 2005). Formal written contracts between the ULBs and the third sector, as are presently being introduced in Mumbai through the LACGs initiative (Karmayog 2006b), and have been in Nagpur (Agrawal 2005) are essential.

One of the major issues in SWM is the disposal of the inert waste. Following the SWM Rules 2000, the disposal should be done in a scientific manner in sanitary landfill sites. The rules also suggest that their location should 20 kms. away from the urban fringe. But, the issue of location raises several dynamics. When some cities have tried to identify a landfill site and acquired land for the purpose, as the NMC did, they face protests from the local population who do not want such a facility in their vicinity. In such a densely populated country, where would land with no habitation in the vicinity be available? The best solution for this is to identify a landfill site in the city master plan itself, perhaps in an industrial park, and then plan for the later redevelopment of the area as a green space. In Hyderabad the new landfill sites under consideration are at a considerable distance. Increased transportation costs are likely to lead to more indiscriminate dumping.

Another hotly contested issue is the burning of wastes, whether for waste-to-energy projects, pyrolysis (thermal gasification) or any other forms of incineration. The production of dioxins, furons and other Persistant Organic Pollutants (POPs) is well known. The anti-incineration movement, as for example represented by GAIA (Global Alliance for Incinerator Alternatives), is world-wide. This is where the ambiguity of urban policies arises. On one hand the SWM Rules

2000 advocate low-tech non-polluting solutions for dealing with solid waste as does the National Environment Policy (GoI 2006). On the other hand, MNES gives grants to both companies and municipalities to employ high-tech methods, including incineration, to destroy wastes. A 'National Master Plan for Development of Waste-to-Energy in India' is being prepared by the National Bio-energy Board of the Ministry of Non-Conventional Energy Sources (MNES 2004). This is largely focused on industrial sectors. A more promising approach is that of the Waste Management Circles promoted by the National Productivity Council of the Ministry of Commerce and Industry, which encourages the use of the wastes of one industry to be used as raw materials in another (WMC 2007). Further, at the same time as many composting enterprises have trouble in finding markets for manure, and in rural areas soil composition is deteriorating, the Ministry of Agriculture is subsidizing the chemical fertilizer industry. It should also be remembered that India is the signatory to many international agreements which condemn burn technologies.

Table 2.20: International Agreements on Waste Management

Accord	*Date*	*Subject*
Basel Convention	1992	Trans-boundary movement of hazardous wastes
Kyoto Protocol	Opened 97, in force 2005	Green House Gas emissions
Stockholm Convention	2001	Persistent Organic Pollutants
UNEP Global Assessment	2002	Mercury
Dhaka Declaration on WM adopted by SAARC	2004	Incineration
Kathmandu Declaration	2004	Environmental dumping

Nevertheless burn technologies continue to be promoted despite the poor performance of those that have been built. A plant at Lucknow which was also economically unviable, was closed for polluting in May 2005 by the Supreme Court, who also sought a stay on any further subsidies for proposed and future WTE plants (Krishna 2006). Meanwhile in Delhi, a contract was signed in March 2005 with Infrastructure Leasing and Financing Services Ltd. to build an electricity-producing incinerator at Timarpur, the location of a previously-failed plant. And this was despite a MOEF (1997) "White Paper on Pollution in Delhi with an Action Plan" which then expressed the view that "the incineration plant at Delhi and the briquette plant at Bombay support the fact that the thermal treatment of municipal solid waste is not feasible."

Another major issue is the closure and or remediation of old dumps, of which there are doubtless tens of thousands around the country. No inventory as yet exists. Some states are beginning to attack this problem. For instance, Karnataka has closed the Tumkur dump, using ADB funds (KUIDFC 2004). Dumps are great producers of methane worldwide, and it is perhaps through the scientific closure of dumps, with appropriate retrofitting to entrap Greenhouse Gasses that India could most profit from carbon credits under the Clean Development Mechanism if acceptable bench line evaluations can be made to determine emission reductions (Goswami 2006).

Conclusion: Innovation in Practice and the Way Forward

The targets of the SWM Rules 2000, set for compliance by 2004, were not met in any city in India. However, many promising initiatives are bearing fruit, and the pace of change, although still very slow, seems to be accelerating in the more developed states. A 'Task Force on Integrated Plant Nutrition Management using city compost' was constituted under the aegis of the MOEF in 2003 to sort out the anomalies in managing composts and other problems. It has proposed further grants both to help set up 1000 composting plants across the country and to modernise those existing, and has made substantive recommendations on marketing and technical aspects (CPHEEO webpage). The Supreme Court, dismayed by the slowness of change, set up a monitoring system in 2004 to help unplug bottlenecks. 'The Supreme Court Monitoring Committee on Hazardous Wastes' conducts site visits, has sub-committees in several states, and is astonishingly pro-active (SCMC 2005). Further, at the same time, the Supreme Court directed the 63 major cities to prepare SWM action plans and submit them to the CPCB. Simultaneously, the CPCB is to prepare a fresh timeframe for implementation of similar action plans all over the country (CPHEEPO webpage).

Some of the states are taking much more active roles. Karnataka can be used as an example. It had set up the Karnataka Urban Infrastructure Development and Finance Corporation in 1993 as a funding mechanism for urban infrastructure. KUIDFC has more recently researched and prepared a report on the status of urban infrastructure in the state and set up a special solid waste management cell (KUIDFC 2004). It has conducted SWM studies for 18 ULBs, gives them technical advice on the design of landfills, and provides training for ULB personnel. A state policy on integrated waste management has been produced (Karnataka 2007). Some other states such as Tamil Nadu, Andhra Pradesh, Maharashtra, West Bengal, and Gujarat have followed or are following similar trajectories, often with international help, and urban management is becoming a very real part of their operations.

At the city level, the mega-cities, Mumbai, Chennai, Bangalore, Kolkata and Hyderabad are all making progress. For example, Kolkata has published a Master Plan on Solid Waste Management (KMC 2005), plan a new sanitary solid waste disposal site at Dhapa and has secured loans from JICA to undertake the work. For the smaller towns, in many cases the situation remains abysmal as the survey of ULBs in Karnataka shows (KUIDFC 2004).

From our examination of the SWM field a few generalisations can be made. First of all, reliable information is available only for a few cities. The new reporting requirements to state and the central pollution control boards should improve this, but it only covers the technical aspects. It does not cover the financial or environmental domains, which are so important in forward planning. In other words, there is very little hard data research available on SWM in Indian cities; what is available is only on some aspects of the system. More complete financial and environmental reporting is required at the municipal level if SWM planning is to be improved.

Second, privatization of parts of solid waste management operations are increasing, with very evident saving in costs. Two observations are made here. One is the question of the payment of private sector contractual labour, which has been shown to be much lower than that of municipal labourers doing the same task. The State Government of Tamil Nadu has exempted the Chennai Municipal Corporation from the purview of the Contract Labour (Regulation and Abolition) Act 1970 vide its order No. 40 MS No. 99 dated 8th July 1999, allowing for the hiring of contract labour for sweeping and scavenging at below legislated norms. This is not a sustainable practice. The other question relates to the integration of informal sector waste pickers into the regular work force, which is an urgent necessity, as has been shown in many studies, and which can be readily achieved by employing organized groups through NGOs and contractual processes, as has been shown in many centres. Dovetailing SWM with employment makes both social and economic sense. Kolkata, for instance, has set up a continuous 'benefit monitoring evaluation' program for the KEIP project.

Third is the question of user charges and special SWM taxes. It is evident that many people can afford them, and sliding scales such as used in Hyderabad can avoid unaffordable burdens on the poor (Mohanty 2002). The argument has been made that if people pay for a service, they value it more, monitor its outcomes more closely and subscribe to its intent, if they get good service in return. These are good reasons to adopt user charges for door-to-door service. Similarly, user charges for bulk generators and fines for littering and encroachment both raise money and deter unsocial behaviour.

Fourth is the question of landfill sites, the opening of new sites and closing others. Since each municipality is responsible for its own waste, large cities should find sites within their own territories. (It is virtually impossible to identify sites more than 20 km. from housing settlements as the rules suggest. The rules should be modified). These could well be within existing industrial parks, where they might be able to be part of a WMC scheme, and where if they are strongly buffered and well maintained, they would cause little additional harm. If strong recycling and reuse rules are implemented their space requirements will be minimized. For smaller ULBs the idea of combined landfills for many small municipalities, as is being proposed in Tamil Nadu, is cost effective and should be encouraged. It should become a mandatory condition for the approval of all City Development Plans (under JNNURM), municipal Master Plans, and the plans for Special Economic Zones (SEZ) that landfill sites are identified and notified as soon as possible to avoid a repetition of the mess that exists. For existing landfills needing or nearing closure, an inspectorate should assess their potential for remediation and closure and its possible financing under the Clean Development Mechanism carbon credits system. Bio-remediation and biosorption techniques serve as robust potential tools in environmental clean-up (Ghosh *et al.* 2005).

Finally, is the question of management. Effective solid waste management needs someone in charge, perhaps a Commissioner with additional powers, and a dedicated work unit in a separate cell. In all but the largest municipalities it is not a distinct service, and the Medical Officer of Health is responsible for the operation and the Chief Engineer for the equipment and transportation. With all their other numerous duties, this is not an efficient way of running such a key service. The importance of vigorous and dedicated leadership cannot be over-stressed, as has been seen in the case of Suryapet and Vejalpur. In Bangalore, the problems of getting integrated solid waste management off the ground, despite a whole series of studies and activities including help by international agencies such as Ausaid (2000 to 2002) and the Dutch government's UWEP (1999 to 2003) has been attributed to the rapid turnover of managerial officers, and the lack of collective memory of what had or had not worked before. Each new appointment had to learn the situation anew and much valuable time was lost (Rosario and Sheinberg 2004).

In conclusion, sustainability must be measured through ecological, socio-economic and public health criteria (Baud and Post 2003). Ecological sustainability demands minimizing the waste generated; maximizing reuse and recycling; and disposing of remaining waste in a controlled fashion in order not to exceed the capacity of local sinks. The socio-economic criteria are financial viability and affordability for local authorities, consumers, entrepreneurs and self-help groups; employment providing a living wage and reasonable job security to SWM workers; legitimacy and transparency by all actors; and effective monitoring and enforcement of standards. Public health criteria include the reduction of epidemics and the minimization of occupational health hazards to workers in SWM and in particular waste pickers.

NOTES

1. Exnora is derived from Exellent, Novel and Radical.
2. The 11th and 12th Central Finance Commissions (NIUA 2005) have mandated the release of grants to only those ULBs that have implemented the facets of the 74th CA (Kundu and Kundu 2005). The JNNURM, announced in December 2005 too has imposed conditionality of implementation of the 74th CA to be eligible for central government contribution under the programme.
3. In Delhi, the powers given to ward committees are: to exercise all such powers and functions of the Corporation that are to be performed within the concerned zone other than those involving 2 or more zones; of the various others one sanctioning of plans (upto Rs 1 crore), (Kundu and Kundu, 2005). In case of Mumbai the administrative sanctioning by ward committee is upto Rs 5 lakh (Navtej 2005).

4. Exploring Possibilities of Achieving Sustainability in Solid Waste Management, at http: //144.16.93.203/energy/paper/Solidnew/issues.html
5. The Indra Prasth Extension RWA, for example, has developed a network for segregation of solid waste into bio-degradable and non-degradable category, on a large scale, before being collected into separate bins. A number of other RWAs in south Delhi too are hiring people to segregate garbage ande thereby facilitating in organising a better sanitation system in the city (Kundu and Kundu, 2005)
6. Furedy (1992a) and Shah and Sambaraju (1997) in "Initiatives for Source Separation and Urban Organic Waste Reuse" from C. Furedy (n.d.) *Sustainable Consumption and Municipal Solid Waste Reduction in Developing Countries of Asia*, York University, Ottawa, Canada. Source: http: //www.gdrc.org/uem/waste/initiatives.html
7. http: //144.16.93.203/energy/paper/Solidnew/issues.html
8. *Ghantagadi* is a Hindi language term, used in some other Indian languages as well, and means a vehicle (cart, tricycle or tractor used in waste collection) with a bell.
9. Not all areas of the city are privatized for collection. This will be clear when individual city cases are presented in this book.
10. High Density Area—250 to 350 running meter of road length
    Medium Density Area—400 to 600 running meter of road length
    Low Density Area—650 to 750 running meter of road length
    (*Source*: Supreme Court 1999)
11. www.toxicslink.org/pub-view.php?pubnum=55
12. Indian garbage is generally unsuitable for waste to energy projects. Composting is comparatively a better option (Appaswamy 2005).
13. Terra Pharma Biotechnics.
14. "Exploring Possibilities of Achieving Sustainability in Solid Waste Management" article on web at— http: //wgbis.ces.iisc.ernet.in/energy/paper/Solidnew/issues.html
15. Bioculum—Biological Inoculum for composting; Sanitreat—for sanitisation of household waste with odour control agent; Herbocel—a liquid to keep away flies and insects.
16. Presentation made at the Bhopal workshop on 28th and 29th September 2005, by Gandhi H.V. Excel Industries Ltd.
17. http: //mnes.nic.in/business%20oppertunity/pgtew.htm
18. In Economic Survey 2004-05, *Source*: www.indiabudget.nic.in.
19. Ministry of Urban Development India (2001): *Recommendations for National Action Plan for Good Urban Governance* from the website.

3

Centralized and Decentralized Solid Waste Management Practices in Bangalore

NATALIE KEHLE, DARSHINI MAHADEVIA and APPEEJI PARASHER[1]

City	*Bangalore*
Urban Local Body	Bangalore Mahanagar Palike (BMP)
Population — BMP, 2001	4.29 million
Population — Bangalore Urban Agglomeration (BUA), 2001	5.69 million
Population Growth Rate, 1991-2001—BUA	3.25% per annum
Area — BMP	226.2 sq. km
Area — BDA*	1,279 sq km
Other ULBs important in BDA	Krishnarajapuram, Mahadevapura, Dasarahalli, Bytarayanapura, Rajarajeswannagar, Yelahanka and Bommanahalli CMCs and Kengeri TMC
Local Authorities in BDA	BMP and eight surrounding municipalities
Local Authorities in BUA	BMP, 7 CMCs, one TMC (of 2 in Figure 3.1) and ten Panchayats
Water Supply and Sewerage in BMP	Bangalore Water Supply and Sewerage Board
Storm water drains	Bangalore Water Supply and Sewerage Board
Total wards in BMP	100 with each 40,000 to 50,000 population

Note: * BDA is the planning authority over the Bangalore Metropolitan Area whereas the ULBs are the institutions of urban governance, thus taking up the task of SWM.

Source: BDA (2000)

Introduction

Note: In Jan. 2007 the BMP territory was expanded to encompass 8 more municipalities and 110 urban villages. Its size increased from 226sq. miles to 800sq. miles. The data in this analysis refers to its extent in 2005.

The City of Bangalore has been able to create a unique solid waste management system that is both centralized and decentralized at the same time. It is characterized by the privatization of much of the city for the collection and transportation of waste within a centralized system, while also taking advantage of the city's active civil society in the delivery of solid waste services in a decentralized system. The municipal government has developed multiple partnerships for the delivery of solid waste services, mostly for the door-to-door collection and transportation of waste with private contractors and for composting activities. The most notable citywide development is the publicly owned compost plant, the most successful one of its kind in the country. The city corporation operates the largest vermi-composting unit in India. At the same time, Bangalore's uniquely active civil society has given rise to numerous decentralized solid waste schemes throughout the city. Community-based and non-governmental organizations (NGOs) run these decentralized schemes, which the municipal government has in some cases integrated into the centralized system.

Background

The industrialization of Bangalore has been rapid since 1951, vitally improving employment opportunities and attracting migrants from all over the country. Migrants became mostly attracted to employment prospects in construction, business, and industrial activities, and to the developed educational and medical facilities. After economic reforms, Bangalore has become famous world over as the 'Silicon Valley' of India as it is a major centre for the location of Information Technology (IT) firms, both domestically and internationally based. It continues to be the capital of Karnataka and hence has gained a great amount of attention from the state government in city development efforts. The IT boom has attracted educated youth from all over the country and in doing so has converted Bangalore from being a typical middle class city to a 'split city', where IT firms and their employees are geographically concentrated in certain districts while the rest of the city maintains its middle class and industrial identities. The city has been growing very rapidly in population and it is expected that the BUA will pass the 10 million mark by 2021 (Jalal 2005).

Bangalore's economy is reputed to be dependent on the IT sector. The number of IT professionals in Bangalore is estimated at 285,000 with over 1,584 IT firms established, 512 of which are multinationals [*The Hindu* 2005(a)]. However, estimates based on National Sample Survey (NSS) data from 1999-2000, (which constitute a small sample of the city), suggest that of the total males employed just 4.34 per cent were employed in computer and other high end services (Mahadevia 2005). According to the BDA (2000) estimates, the IT sector accounts for just 15 per cent of the total employment while 60 to 70 per cent of the city's economy is still informal sector based. Bangalore continues to be a global competitor as an investment location for IT and biotechnology companies with roughly four new companies registering every week (Department of Information Technology and Biotechnology 2005).

Bangalore is also recognized as a centre for public sector industries, as by the late 1970s Hindustan Aircraft Factory, Indian Telephone Industries, Hindustan Machine Tools and Bharat Heavy Electricals Limited (the Big Four public sector units), had all established in the eastern and northern outskirts of the city (Nair 2005: 81).

Municipal Organization

The Bangalore Mahanagar Palika (BMP) (Bangalore Municipal Corporation) has a city council that consists of 100 elected members, or councillors, each representing a ward. Both the Mayor and Deputy Mayor are elected from among councillors for a term of one year. The BMP has 14,000 employees and is headed by the Commissioner. The Commissioner is the Chief Executive of the BMP, appointed and deputed by the State Government of Karnataka, and is responsible for performing duties and functions.

As laid down in the Karnataka Municipal Corporation Act, 1976, and restated in the 74th CA, keeping the city clean is the responsibility of the civic administration—that is, the BMP. The specific clauses are:

The Karnataka Municipal Corporation Act, 1976 **(CPCB 2005)**

- *Section 58*: The watering and cleaning of all public streets and public places in the city and removal of all the sweepings.
- *Section 255*: Provision for removal of filth.
- *Section 256*: Public notice ordering removal of the deposit of rubbish and filth.
- *Section 258*: Provision for daily cleaning of streets and removal of rubbish and filth.
- *Section 261*: Maintenance of establishment for removal of rubbish and filth.

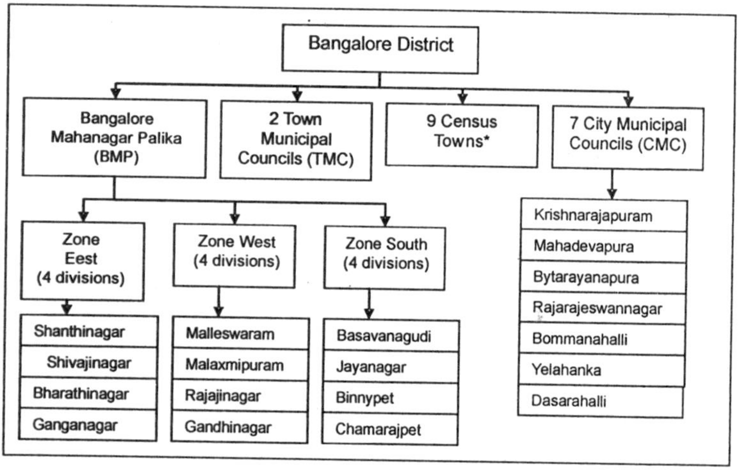

Fig. 3.1: Bangalore District Composition

* Non-Statutory Towns
Source: Janaagraha 2005 and Census of India 2001

The three departments involved in SWM[2] activities are the Health Department, the Engineering Department and the newly formed Sanitary Engineering Department. The Health Department of the BMP handles the collection and transportation of waste, the Engineering Department collects silt from drains, and the Sanitary Engineering unit is responsible for waste disposal. This last unit engages in planning and designing future engineered sanitary landfill sites and transfer stations for the city.

In the case of the Health Department, the city is divided into 297 health wards (independent of the electoral wards) grouped into three zones with 10 ranges per zone for functional convenience. There are approximately 2-3 health wards per range and of the total, 183 health wards are under a private contract system while the remaining 114 health wards are cleaned by *Pourakarmikas*, or municipal workers.

Each health ward has one senior health inspector, and 2 to 3 junior health inspectors. A deputy health officer heads each zone (Zone East, West and South) and there is a medical officer of health for each range, see Fig. 3.2.

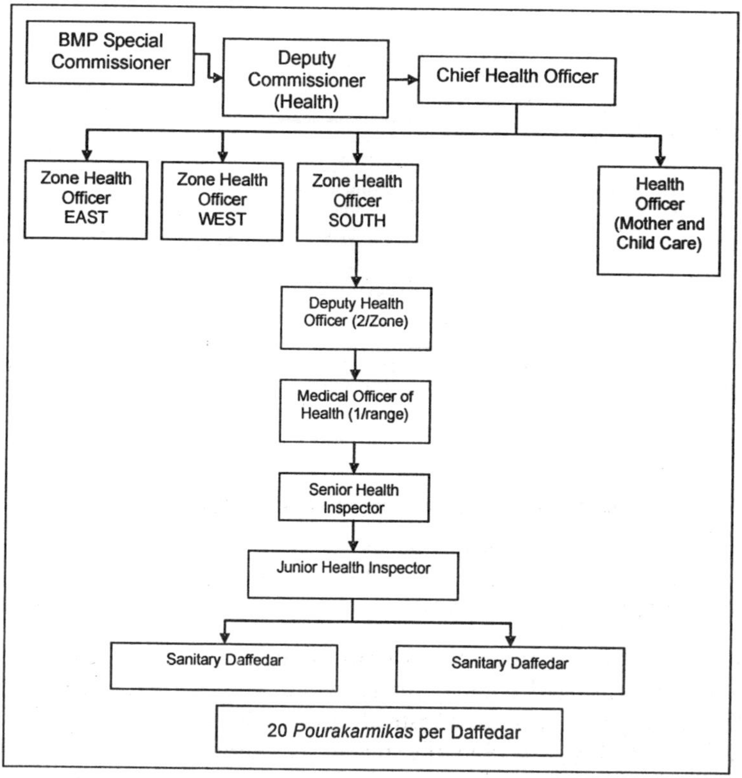

Fig. 3.2: BMP Health Department Organization Relating to SWM

Source: The Official Website of Bangalore City Corporation 2005

As in the case of the Health Department, the BMP Engineering Department is divided into three zones. These zones are further subdivided into 12 divisions (four divisions per zone). Each zone is headed by a chief engineer (for Zonal East, West and South) and all are led by the BMP engineer in chief. Each division has an executive engineer, assistant executive engineers and assistant engineers (Fig. 3.3). Also see Fig. 3.1 for the division names.

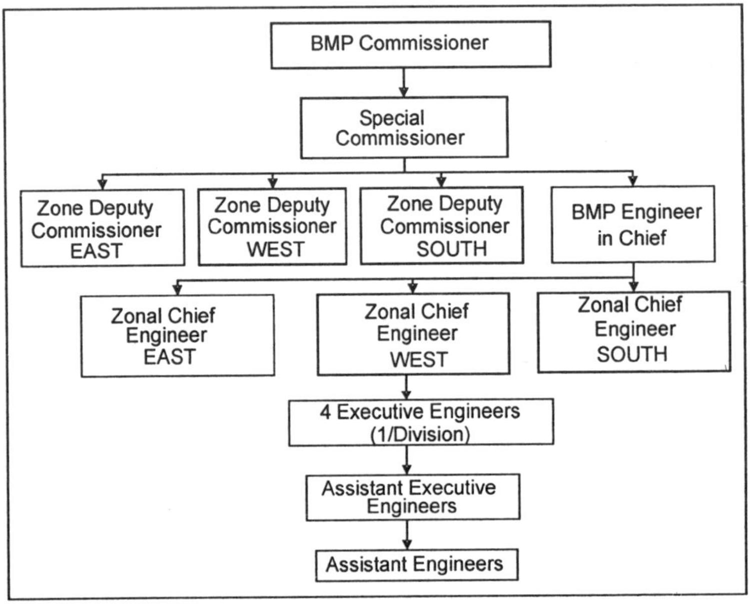

Fig. 3.3: BMP Engineering Department Organization

Source: Janaagraha Official Website 2005.

Status of Ward Committees

In the 1990s, realignment and ward expansion occurred to create more uniform ward populations providing an outcome of roughly 40-50,000 people per ward instead of the former wide variation of 20,000 to 100,000. In 1996, the state government officially expanded the BMP, annexing a number of suburbs that were already urbanized, resulting in the expansion of the population to over 4 million (2001 Census of India). This made the BMP the fifth largest city in India. Under 74th Constitutional Amendment (CA) in Karnataka it has been suggested that in the future, definitions of ward committees would include within each about 300,000 people. If such a proposal were adopted, ward committees in Bangalore would encompass about six wards each. This covers too large a population to form a proper participatory system. According to Heitzman (2004: 24), a major element of true democratic governance of local bodies is accessibility for citizens. Currently, few ward committees have been formed in Bangalore, and of those that have been instituted many are not being used for the purpose for which they were formed. The roles and responsibilities are not clear to either public officials, citizens of Bangalore, or indeed to the committee members themselves.

Solid Waste Generation[3]

It is estimated that the country produces approximately 30 million tonnes of urban solid waste annually, with a per capita between 0.1 kg to 0.6 kg per day. According to The Energy Resources Institute (TERI), a typical Indian metropolitan city increases its per capita waste generation by 1.3 per cent per year (Beukering *et al* 1999). Bangalore, like other urban centers in India, has a high composition of organic matter in its urban waste, consisting of between 30 and 70 per cent.

According to Beukering *et al* (1999), Bangalore generates a total of about 3,600 metric tonnes (MT) of solid waste per day, similar to the BMP's estimate of approximately 3,500 tonnes per day (City Managers Association Karnataka 2005).

Waste from industries is among the highest contributors (39 %) to the total waste generated. But most of this is internally recovered for recycling and reuse with only a small per centage ending up with city waste. The other major producers of SW are hotels (29 %), households (18 %) and markets (10 %) (Table 3.1).

Table 3.1: Source and Quantity of Solid Waste in Bangalore

Source	*Estimated quantity of solid waste (MTs/day)*		*Per-centage*
Households	650		18
Commercial Establishments	1436		39
• Markets		• 369	
• Hotels		• 1066	
Institutions	128		4
• Hospitals		• 20	
• Offices		• 15	
• Educational Institutions		• 92	
Industries	1399		39
Total	3613		100

Source: Beukering *et al* (1999)

Though household waste is not the main contributor, it is important to note that domestic waste levels and composition change with economic development. In addition, current rapid population growth is a cause for concern. In this regard, domestic waste should not be overlooked. A survey by Beukering *et al* (1999) revealed that the level of waste generation increased by 1 kilogram for each increase in income of Rs. 1,000 per month (see Fig. 3.4 Beukering *et al* 1999). This is essential to keep in mind when planning for future projects as India continues to evolve economically.

Bangalore's urban solid waste composition is highly organic. It is estimated that 43 per cent of total urban waste is biodegradable and about 36 per cent is of recyclable material such as paper, glass, plastics, metals etc. (Table 3.2).

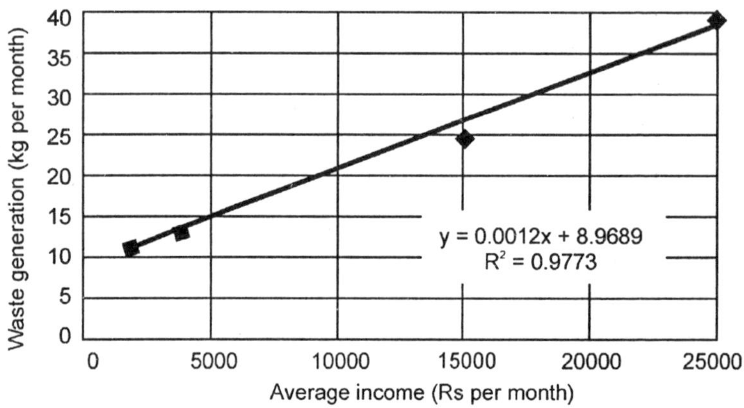

Fig. 3.4: Correlation Between Per Capita Income and Waste Generation

Source: Beukering *et al* (1999).

Bangalore produces approximately 25 MT of bio-medical waste per day, or 1.0 per cent of the total waste generated in the city. Infectious waste constitutes between 22 and 60 per cent by weight of that total, or 5.5-15.0 MT per day. Private hospitals generate about 1.0 kg per bed, government hospitals 0.8 kg per bed while clinics and other health centers generate 0.5 kg per bed (Table 3.3) (CEE 2005). Bio-medical waste is to be managed as per the Bio-Medical (Management and Handling) Rules 1998[5]. It is obligatory to segregate this waste from the rest of municipal waste and manage, treat and dispose it as per the said rules (GoI 2000).

Table 3.2: Bangalore's Urban Waste Composition

Material	*Estimated quantity (tonnes/ day)*	*Percentage of total waste (%)*
Glass	103.8	2.9
Plastic	220.8	6.7
Paper/cardboard	594.2	16.5
Metal	54.7	1.5
Rubber	357.9	9.9
Organic waste	1,206.1	33.4
Other biodegradable	332.2	9.2
Hazardous hospital	97.5	2.0
Miscellaneous[4]	645.9	17.9
Total	3,613.1	100.0

Source: Beukering *et al* (1999).

This legislation had the potential to both provide a healthier environment for the thousands of formal and informal employees who have direct daily contact with Bangalore's urban waste and the general public, while also serving as a mechanism to divert some of the city's waste from the centralized landfill. Though there are examples of some major health institutions, like St-John's Hospital, abiding by the new legislation and separating its bio-medical waste from city

Table 3.3: Quantity of Bio-Medical Waste Generated in Bangalore

Type of Institution	*No. of Institut -ions*	*No. of Beds*	*Quantity of waste generated (per day in kg)*
Major Gov. Hospitals (500+beds)	14	10,500	8,400 (0.8 kg/bed)
Major Corporate Hospitals (100-499 beds)	14	6,053	6,053 (1.0 kg/bed)
Government Primary Health Centres (30-50 beds)	29	390	195 (0.5 kg/bed)
Corporation Maternity Homes (30 beds)	31	708	354 (0.5 kg/bed)
Nursing Homes (10-99 beds)	350	8,750	7,000 (0.8 kg/bed)
Clinics-Govt. Primary Health Units & Dispensaries	150	/	75 (0.5 kg/clinic day)
Clinics General Practitioners	5,000	/	2,500 (0.5 kg/clinic day)
Diagnostic labs & blood banks	100	/	80 (0.8 kg/lab day)
Total	/	/	24,657 kg/day

Source: Kuttaiah (2003).

waste, there remain too many cases of smaller centres that do not follow the rules, thereby exposing thousands to health risks. Hundreds of nursing homes spring up in residential homes, avoiding government radar. A more stringent monitoring system is needed (Kuttaiah 2003).

Finally, according to the State of Environment Report, Bangalore contributed to 14,227 MT, or 18 per cent, of the total slaughterhouse waste generated in the state in 2001. This type of waste falls under the purview of MSW (Handling and Management) Rules 2000. Presently, there are no facilities for treating this waste; however, the government is considering the modernization of the slaughterhouse at Tannery Road or the possibility of relocating it (Government of Karnataka 2003).

Actors and Partnerships in SWM

There are multiple actors in Bangalore's SWM system, a development some researchers have called multistakeholdership (Sekhar 2001). As a result, several partnerships have been institutionalized within the municipal management. NGOs', CBOs' and the private sector's involvement in urban governance, specifically in the delivery of municipal services, has taken a prominent role in Bangalore, which helps manage massive demographic and economic demands (Heitzman 2004: 123).

Private Sector

The private sector is involved in waste collection, transportation and recovery in the city. Each of these tasks are elucidated in the following sections.

Collection and Transportation of Waste

Privatization in the collection and transportation of waste is not a new concept in Bangalore. In 1986, four wards were privatized and the number has been growing ever since (discussion with BMC officials). The BMP has not recruited any new workers in the departments involved in SWM since 1985, only expanding the number of privatized areas. Today, 183 of the 297 health wards (62 per cent of the city) are under a privatized SW collection system.

Table 3.4: Status of Wards Maintained by BMP and Through Outsourcing

Year	*Total No. of Wards*	*Maintained by*	
		BMP	*Outsourcing*
1986	87	84	03
1989	87	67	20
1993	87	59	28
1996	273	124	149
2000	296	114	182
2003	297	114	183

Source: Ananthaswamy (2005).

The privatized areas have been grouped into 61 daytime packages and one night package. The contract period is valid for two years, but has recently been extended for another 1-2 years. Any contractor is eligible for a maximum of four packages, and no more than 50 per cent of the available contracts. All contracts are awarded on a tender bid basis. Tenders need to furnish an Earnest Money Deposit (EMD) amounting to Rs. 300,000 per package. Tenders are selected by the BMP Commissioner. The EMD is returned after the conclusion of the Bid process (Office of the Deputy Commissioner 2003). The solvency is Rs. 1 million for each package. Contracts are awarded based on technical feasibility and the lowest financial criteria. There is no separate profit margin designed under the contract. The contract is awarded based on the technical feasibility and the lowest financial bid. Also, it is important to note that the technical bids, though approved by engineers, are neither designed or managed by professionals.

The work entrusted to the private operators is as follows:

(a) Sweeping of roads, footpaths and pavements
(b) Cleaning of open road side drains and shoulder drains
(c) Door-to-door collection of waste
(d) Transportation of garbage and debris
(e) Silt removal
(f) Conservancy
(g) Cleaning of vegetation and weeds on the roads and footpaths
(h) Cleaning of public toilet and urinals
(i) Cleaning and transportation of waste at night

The 61daytime packages cover:

— Door-to-door collection of waste from 650,000 households
— Collection of waste from 62,800 shops
— Sweeping 4,363 km of road
— Cleaning 254 public toilets and urinals
— Cleaning 56 burial grounds

The night package covers:

— Cleaning 23.9 km of major roads
— Cleaning public toilets and urinals on major roads

The contract documents of the BMP are standard, and lay out various terms and conditions to protect both households and workers. Some of the relevant clauses are:

- *Clause 5*—Irrespective of a reference to the law courts, as the case may be, the contractor shall proceed to execute and complete the works with all due diligence pending settlement of the said dispute or difference. This clause is particularly important in case of any dispute arising between the contractor and the BMP; in the absence of which, waste collection and transportation would be significantly affected.
- *Clause 6, 7 and 8*— These clauses are designed to protect the interest of the workers engaged with private operators. As per the clauses, contractors are responsible to provide medical aid and compensation to workmen in case of any bodily injury incurred during the work. The contractor is also liable to provide all the safety equipment such as gloves, shoes, gum-boots, masks etc. However, such things are hardly visible except the uniform, which is provided to the workers.
- *Clause 9*—This clause ensures the minimum wages for workers employed with the contractor. The BMP has fixed a minimum wage of Rs. 1,800 per month (Rs. 60 per day). These payments are made in presence of the officers of BMP or through an account payee cheque.
- *Clause 13 and 20*—These clauses ensure the monitoring of the work carried out by the contractor and necessary fines in case of inadequacy in work. As per the clause, the contractor

> shall obtain signature of a minimum of two citizens or representatives of RWAs, in the absence of which payment shall not be made to the contractor. Apart from the regular monitoring, special *adhoc* visits by any of the senior officials (DHO, MoH or health inspectors) is done by the BMP. However, as of now (as per the discussions) no such cases have been reported.

Bio-medical waste management collection has been privatized. Special firms, approved and registered with the Karnataka Pollution Control Board (KPCB), collect, sometimes treat, and finally transport the waste to their hydroclaving and autoclaving units. The KPCB works in coordination with hospitals, its associations and the private operators. The firms collect the charges directly from the hospitals and nursing homes. Karnataka has 10 institutions providing common bio-medical treatment facilities, usually termed "common treatment facilities". In Bangalore there are two main private firms; Medicare Incine Private Limited and Mridi Eco Industry who engage in the handling of bio-medical waste (Kuttaiah 2003). Businesses involved in the management of bio-medical waste are facing challenges in finding an economically viable business plan that will provide the services that satisfy the BMP.

Waste Recovery

Currently, two mechanisms for waste recovery within the Bangalore SWM system are composting and recycling. There are three centralized composting plants in Bangalore: the Karnataka Composting Development Corporation (KCDC), Terrafirma and Sunrise. The KCDC is a publicly owned unit that will be studied further in the centralized SWM system. Terrafirma and Sunrise, however, are both privately run composting firms that work within the Bangalore SWM system.

There is only one known partnership between the BMP and a recycling unit, which deals with plastics (paper, metals, rags and some plastics are handled by numerous dealers in the informal sector). The BMP, with the World Bank, has decided to use a polyblend, from recycled plastic, to upgrade over 40 per cent of roads in the city. Over 140 km along 41 roads were to be upgraded at a cost of Rs. 1,400 million under the Karnataka Municipal Reforms Project in 2005. So far, 300 city roads have already been asphalted with plastic waste. A partnership is developing between the BMP and the city-based KK Plastic Waste Management Private Limited for supplying the company's 'KK Polyblend'. According to a BMP technical advisor, the durability of the roads already laid out with shredded plastic waste is about three times better than conventional asphalting. This will result in a decrease in road repairs and will help manage the landfill problem. Each km of road with an average width requires over two tonnes of polyblend, thereby reducing plastic going to the landfill drastically. Both the Central Road Research Institute and Bangalore University Center for Transportation Engineering have tested the strength of these roads and confirm the quality [*The Hindu* 2005(b)].

Civil Society

Bangalore's civil society plays a role in linking public administration and corporate management with citizens on decisions regarding equity in resource allocation, environment, gender, poverty, etc. Many NGOs and CBOs active in the 1990s formed several environmental health advocacy and development projects throughout the city.

As in the case of urban solid waste management, some organizations have developed SWM systems at the local level, each taking the responsibility for service delivery and mobilizing human and financial resources (Heitzman 2004: 123). A few of these civil organizations will be looked into further in a later section.

Informal Sector

Like other cities in India, in Bangalore the majority of waste recovery is carried out by the rag-pickers and itinerant waste collectors who sell the waste to the recycling units which are, for the most part, privately operated. It is still common practice for people at the household level, in all economic classes, to store and sell recyclables to itinerant buyers for small sums. Therefore, what the household deems valuable is taken out of the waste, and can be considered the first sieving through of the recyclables. *Pourakarmikas*[6] and waste collectors are the second stage of sieving; they pick out the most valuable of recyclables and waste. They are allowed to keep what they want to sell and the rest goes to the pushcarts. Once at the transfer point, rag-pickers then go through the un-segregated waste, usually with nothing more for protection than a metal poke for rummaging. At this point, all materials are mixed together, including wet kitchen waste, hazardous or bio-medical household waste, metal, plastics, broken glass, paper etc. This makes it hard to recover saleable material since they get dirty or contaminated, and thereby lose any value. These rag-pickers are mostly women and children and are not recognized within the waste management system.

The recyclable waste eventually makes its way to recycling plants where a lot of usable products are manufactured. Waste processing is also performed mainly through informal waste recycling centres spread across the city. For example, hundreds of privately run recycling centres are found along the Mysore Highway. A study done by Beukering *et al* (1999), estimated that out of the 2,214 tonnes[7] of waste generated per day in Bangalore, around 20 per cent is reused and 23 per cent is sent for recycling. That is, about 40-45 per cent of the waste generated in Bangalore does not go to the landfill. The BMP has a weak relationship with the informal waste recovery system, which is outlined in Table 3.5 and explained in the next section.

Special Local Initiatives—Swabhimana, BATF and Janaagraha

In recent years, the corporate sector, the municipality and civil society have been working together, rather than conflicting, in addressing poor municipal services and deteriorating urban facilities. The development of some of these organizations is outlined here.

The term 'stakeholder', having origins in corporate management, has been increasingly applied to urban management in Bangalore. Today, with the help of the 74th CA in the early 1990s, citizens of Bangalore have been gradually taking ownership of the city. A forum created in the early 1990s called the Citizen's Voluntary Initiative for the City (CIVIC) was a response to both the 74th CA, which mandated the devolution of funds to ULBs, and to the enormous interest from international agencies in promoting the ideology of 'good governance' in cities across Asia (Nair 2005: 116).

In the mid-1990s, a state sponsored initiative named 'Swabhimana' was formed to develop new non-governmental institutions to enhance citizen participation in ward level politics in co-ordination with the Public Affairs Centre (PAC) of Bangalore (Nair 2005: 318). Launched in 1995 *Swabhimana* is a coalition of NGOs and has helped launch a 'citizen-local government initiative for a cleaner, greener and safer Bangalore'. The initiative was also an attempt to invite corporate groups to participate in the maintenance of the city, by sponsoring parks, roundabouts, light fixtures, and fences to restore a deteriorating city (Nair 2005: 114). '*Swabhimana*' with the Centre for Environment Education (CEE) were large contributors in the planning and development of the later citywide SWM programme, '*Swachchha* Bangalore', meaning 'Clean Bangalore', by implementing pilot projects throughout the city.

In 1999, the corporate sector intervened. The Bangalore Agenda Task Force (BATF), a 15 member nominated body of professionals from the IT and biotech industries, management and financial experts, aimed to make Bangalore the 'best city in India by 2004'[8]. It was formally created by the state government and works with the BMP, the BDA, and the local agencies responsible for water, sewage, telecommunications, electricity, transportation and the police, under the belief that the success of corporate governance should be applied to city management. BATF works mainly with stakeholders to upgrade the city's infrastructure, enhance municipal services and build institutional capacity. Its mandate is to 'upgrade the infrastructure and systems in Bangalore through Private-Public Partnership to enhance the quality of life in the city'. In particular, the BATF recommends specific measures to promote IT, introduce modern management tools, adopt international standards, and to involve the corporate sector while empowering citizens and providing urban advancement. The BATF's inaugural sampling survey revealed five problems that required immediate attention: in order they were: the condition of the roads, garbage, mosquitoes, pollution, and public toilets[9] (Nair 2005: 335).

Is the BATF successful? Ghosh (2005) points out that the role of private sector participation in urban governance should be evaluated within the broader context of the politics of urban planning. She notes that the BATF policies do not address the economically poor. In fact, many other researchers have highlighted that by taking up issues in urban governance BATF is insidiously making its way into the local, state and national level politics. It chose to work with the municipal corporation only when those infrastructure services directly benefitted their businesses. For example, BATF got public toilets constructed in Bangalore with funds from the corporates; but by giving such small funds, through the BATF, it is alleged that the corporates intended to divert other public funds for infrastructure that would benefit them. In spite of the fact that the BATF survey highlighted environmental degradation as the overwhelming problem, the organisation prioritised other infrastructure projects such as flyovers, power, telecom and public transportation. Moreover, they interfered with the BDA Master Plan so as to influence decisions about the IT corridor, rather than focussing on SWM (Ghosh 2005).

The creation of Janaagraha was a result of this new 'ownership' attitude. Janaagraha, a local NGO that started in 2001, urges citizens to participate in governance through ownership. Through a Ward Vision Campaign, Janaagraha involves citizens in the neighbourhood planning process, which will be looked into further in the community participation section of this study. Groups such as Janaagraha, '*Swabhimana*', and other NGOs that have popped up throughout the city since the 1990s, have been striving to improve on civic amenities through participatory governance.

Prior to the BMP's implementation of the present citywide SWM programme, '*Swachchha* Bangalore', members of '*Swabhimana*' including NGOs like the Rag-Pickers Education and Development Scheme (REDS), and the Centre for Environment Education (CEE), and officials of the BMP, were asked to formulate a waste management plan for the city. This plan included the development of participatory waste management systems to be implemented citywide (Beal 1997 and Heitzman 2004: 130). They chose 10 model wards in which to start. Later, the BMP phased in similar solid waste programs in all the wards under the '*Swachchha* Bangalore' programme.

Before the implementation of '*Swachchha* Bangalore', there were approximately 45 community-based waste management schemes in Bangalore (Rosario 1994). These systems involved door-to-door collection of waste in segregated form, composting of biodegradable wastes at a community site, and having the remaining waste collected by the BMP workers at community bins. These communities had organized and trained local rag-pickers for the tasks, while being supervised by CBOs or NGOs. The initial costs for these small-scale schemes generally came from donor agencies, or generous community members, while the O & M costs were recovered from fixed monthly payments from the participating community. Some of these schemes still exist, while others were replaced by the citywide system.

After the publication of the MSW Rules 2000 the BMP and the BATF, through a joint initiative, established a Public-Private Partnership (PPP) by launching the sustainable cleanliness programme called '*Swachchha* Bangalore'[10]. Bangalore started to adopt this unified city system in 2000 for its waste collection. The system is the same for the areas under the BMP as for the privatized areas, and the same for the slum[11] and non-slum areas of the city.

Beukering *et al.* (1999) have captured various partnerships within SWM system in Bangalore (Table 3.5). This illustrates the complexity of the SWM system in Bangalore, highlighting the various activities, motivational factors and relationships behind actors involved in the collection, transportation, processing and disposal of city waste. Several conclusions can be drawn here, the most important of which is the BMP's weak relationship with the informal sector in waste recovery and processing, which is a cause for concern.

Partnerships with the BMP

Table 3.5: Existing Partnerships in SWM

Players	*Partnership*	*Activities*	*Relationship*
BMP-KCDC	Gov-Gov	Composting	Formalized
BMP-NGO, CBO	Gov-Public	Small-scale composting	Networking
BMP-Contractor, Recycling Units	Gov-Private	Waste collection/ transport/disposal & waste processing	Contractual & commercial
NGO, CBO-Citizen	Public-Public	Environmental health awareness	Mutual Cooperation
Rag-pickers, Itinerant Waste Buyers (IWB)—Dealers, Wholesaler	Informal-Informal	Waste recovery	Commercial
Dealer, Wholesaler—large and small Recycling Units	Informal-Formal or Informal-Informal	Waste Processing	Commercial
Rag-picker, IWB, Dealer, Wholesaler-NGO, CBO	Informal-Public	Capacity building for waste recovery and social development	Developmental (through service)
HH, Institution, Commercial firms, Industry-IWB, Dealer, Wholesaler	Consumer-Informal	Recyclable collection	Commercial
Rag-picker, IWB, Dealer, Wholesaler-BMP	Informal-Gov	Externalized involvement in waste recovery, complimenting municipal activities	Weak Relationship

Source: Beukering *et al* (1999).

Solid Waste Management System

Bangalore has a complex SWM system since it follows both a centralized and decentralized path within the formal and informal sectors. As has been shown, the centralized system is generally operated and managed by city and regional agencies within the formal sector. The decentralized systems are run by the active civil society—mainly NGOs and CBOs whose work feeds into the centralized system. This section analyzes the day-to-day operation of the two SWM paths in more detail.

Centralized Solid Waste Management

The municipal government, mainly the BMP's Engineering and Health Department, manages the centralized system but also supports the decentralized system. The following are some of the agencies that play a role in waste management in Bangalore—Bangalore Mahanagar Palika (BMP), Bangalore Development Authority (BDA), City Municipal Corporations (CMC), Town Municipal Council (TMC) and the Karnataka Composting Development Corporation (KCDC), NGOs such as Centre for Environment Education (CEE), CBOs mainly Resident Welfare Associations (RWA), and private compost firms like Terrafirma and Sunrise (FEDINA 2005).

Primary Waste Collection (HH to Transfer Point)

There are over 10,000 waste workers, mostly female, in charge of door-to-door collection of waste and sweeping roads. Of these, approximately 6,000 are contractual workers, while the remaining 4000 are BMP workers[12]. Each waste collector arrives at the household between 6: 30 am and 1: 30 pm everyday with a pushcart and bell. Un-segregated waste is placed into plastic holding containers and pushed on to the next house. The pushcarts hold four containers and there are a few auto-rickaws that are used that carry 12-20 containers. There are approximately 8,000 pushcarts used within this system. Though segregation at source is supposed to be encouraged, it is not practiced. The BMP, at the launch of the programme, purchased 500,000 yellow nylon bags for dry waste which is to be collected periodically, while the wet waste is collected daily. These bags are not used to a significant amount [Janaagraha 2003(a)].

While on duty, it is mandatory for workers to sweep or pick up any debris or garbage found on the road. Sweeping is carried out daily with an average road length cover of about 0.5 km per worker. Once collected, the waste is brought to common points where it is transferred to a truck.

Secondary Collection and Disposal (Transfer Point to Dumping Ground)

As mentioned earlier, Bangalore has become a bin-free city. Currently, waste is being picked up at collection points throughout the city where pushcarts or smaller tipper trucks are emptied directly into 9 tonne capacity tipper trucks, for final disposal. There are about 350 BMP and privately owned trucks used within the system. Waste goes directly from collectors to trucks for disposal. This reduces the number of times the waste is manually handled, reducing the exposure to health risks and the period for which waste is left on the roadsides. As per the Supreme Court Rules, it has now been made mandatory by the corporation to transfer garbage in covered vehicles. In Bangalore, these trucks are usually either covered with a plastic sheet or have a covering body of fabricated iron mesh. All the waste collected in the trucks goes directly to the dumping ground. Figures on daily waste dumping at the disposal sites are not available.

The BMP is currently working on a GIS-based database on waste generation in different wards. This will improve SWM strategies by targeting wards under significant stress, by planning for more efficient transfer routes and also by giving a better idea of the total waste generated in the city.

Bangalore has no official landfill or dumpsite, besides the KCDC composting site (discussed below). While just 34 to 36 per cent of the total waste is disposed of through centralized composting, the rest of the waste is illegally dumped along public highways like Magadi Road, Mysore Road, Old Madras Road and Hunnur-Bagalur Road and in open fields. This unregulated dumping is deleterious to both public health and the environment. Local public opposition has hindered landfill operations in all the allocated public sites. The only place where the BMP can legally send garbage is to the KCDC, Terrafirma and Sunrise facilities. Approximately 350 MTs of waste is taken daily to KCDC (Table 3.6).

There are currently no transfer stations (TSs) in the city, but plans are being made. Three main TSs will be set up in Bangalore so that larger capacity trucks can take more MT of waste per trip to the dumping ground, thereby saving on transportation costs.

Table 3.6: Waste Disposal Practices in Bangalore

Waste Disposal Destination	*Estimated quantity (MT/day)*	*Truckloads of waste (per day)*	*Percentage of total waste generated*
KCDC	350	100	15
Terrafirma	150-200	20	6-8
Sunrise (only 6 months/yr)	300	50	13
Illegal Dumping • Dumping Yard[13] • Ad-hoc sites on city's outskirts	1080	180	31
Not Collected	850	/	35
Total	2,400-2,450	350	100

Source: FEDINA (2005) and Janaagraha [2003(a)].

The BMP is also planning for an integrated sanitary landfill site, 20 km from the city, in Mavillipuram. This landfill site will be operated on a Build-Operate-Transfer (BOT) basis. The cost of this is estimated to be Rs. 195 per MT of inert waste. All private contractors are supposed to then dispose the waste without any additional cost as per the contractual agreement. The engineered landfill site has been designed for a capacity of 200 MTs per day. The life of this site is estimated to be 25 years, and about 16o million MTs of garbage will be deposited in that time. The first phase of the landfill site has an area of 100 acres; the second and third phases have 169 acres and 165 acres respectively. It will also include a leachate treatment plant and a compost yard. The BMP is coming up with a waste-to-energy project for the production of 8 MW of electricity, consuming about 1,000 MT of waste per day. The budget estimates of 2005-2006 show a capital expenditure of Rs. 6,300 million and Rs. 7,900 million if miscellaneous expenditures are included. This landfill site was started by the end of 2005.

There is a lack of data on the current state of affairs of the '*Swachccha* Bangalore' system, as the BMP is not open to sharing their information. Given this lack of transparency, it is difficult for stakeholders, such as citizens, NGOs and CBOs, to evaluate and monitor the successes and weaknesses of the new system, and for future planning. As it is centrally run, the system also greatly depends on the motivation of the people currently in power. When first developed, the Commissioner in power was very keen on the idea and pushed for more allocation of funds to the initiative. The current Commissioner lacks the same interest and the programme has slowed down.

Composting—Karnataka Compost Development Corporation (KCDC)

Bangalore stands out from most major cities in India because of the use of a large-scale composting plant, including the largest vermi-composting site in the country. The KCDC plant located in the south of the city, in Haralakunt, is the oldest city-wide composting site in the country still in use. However, the KCDC can hardly be called an outright success because it has had problems arising from financial, political, cultural and environmental constraints throughout the years of its functioning and it will continue to struggle in the future. Since its inception, the KCDC has processed more than 600,000 MTs of urban waste and produced and supplied nearly 200,000 MTs of organic manure (SANDEC 2002).

Due to the mounting urban solid waste problems during the seventies, the GoI subsidized 24 large-scale composting plants in key cities throughout the country. By 1975 only 10 were actually built and within a few years only the KCDC in Bangalore was operating. All the plants at that time, including the KCDC, faced technical problems and financial losses causing major failures. The main problem was that the WHO-based technology employed had no mechanism to deal with un-segregated waste; mixed compostable and inert waste was being brought into the compost facility. The WHO-based system was crushing the compost mixture, while instead there needed to be a way to segregate the inert waste from the compost. As a result, small pieces of glass and metal were found in the final product, making the manure unmarketable.

Several factors were attributed to the KCDC's continuation' including changes to the technology, marketing strategy and financial support from various sources (Lardinois and Marchand 2000). The technological changes included: not crushing the un-sorted waste, composting the mixture, sieving out the inert waste, and more recently, vermi-composting the organic material, thus making a marketable product.

Table 3.7: KCDC Annual Summaries for 2002-03 and 2003-04

Particulars	*2002-03*	*2003-04*
Production—Compost Produced (MT/yr)	11,298	17,517
Profits before tax & before debt repayment (Rs)	605,000	685,000
Waste Processed (MT/yr)	27,691	43,577
Waste Received (MT/yr)	76,341	88,256
Value of Production (Rs)	2,220 million	1,950 million

Source: KCDC Annual Report 2003-04.

Finally, the corporate organization of the KCDC helped establish financial stability. The plant was first set up by the Karnataka Government and eventually became managed by an autonomous body (KCDC Annual Report 2003-04). Today, the KCDC is a subsidiary of the Karnataka Agro Industries Corporation Limited (KAIC) and its board of directors includes people from the BMP, Karnataka State Co-operative Marketing Federation (KSCMF) Ltd. and KAIC Ltd. (KCDC 2004). Having such members, significantly the KSCMF Ltd., as shareholders is one of the key reasons for the KCDC's success. The BMP has a vested interest in its success for continued sound SWM, while the KSCMF Ltd. as shareholders are dedicated to marketing a profitable product. Today, the KCDC is one of the Public Sector Enterprises yielding a profit (Lardinois and Marchand 2000). During its inception in 1975, the initial capital cost for the setup and land acquisition was Rs. 500 million. Of the initial costs, Rs. 300 million were provided by the shareholders while the rest came from the government. As mentioned above, the KCDC has overcome many financial and technical problems. The most pressing constraint has been the financial feasibility of a composting unit producing a marketable organic manure. The Karnataka Department of Agriculture has played a major role with the Karnataka State Co-operative Marketing Federation (KSCMF) Ltd. in selling and promoting the organic manure to farmers within and outside the state, but it remains an ongoing challenge. Currently, the KCDC sells compost in three south Indian states and predominantly sells vermi-compost to florists (KCDC Annual Report 2003-04).

The un-segregated Bangalore waste is supplied free to KCDC by the BMP. Roughly 350 MT per day arrive by both BMP trucks and contractual trucks to discharge the waste. Impressive waste heaps, or windrows, of approximately 13 feet wide, 10 feet high and various lengths are shaped by front-end loaders. Expensive concrete platforms reduce the amount of leachate that penetrates the earth. This makes it easy to remove leachate runoff by sprinkling course compost to absorb the liquid, which is then laid back on the heaps. This mixed waste is cultured by spraying a simple cow dung slurry mix (2 baskets of cow dung to 200 litres of water) and letting it sit for the first stage of degradation (SANDEC 2002).

In the first stage, which lasts 10-15 days, the material goes through a high temperature period where the temperature reaches 70 to 75C°, killing all pathogenic microbes in the process. The middle temperature stage (40-45 C°) lasts for another 30-35 days, bringing the total time needed for the mixed material to go through complete degradation to approximately 60 days (SANDEC 2002). A mechanical rotary with a trunk turns and aerates the heaps on a regular basis to regulate the temperature and hastens decomposition process.

Aerobic windrows go through four rotary screens of 50 mm, 25 mm, 8 mm and 4 mm that segregate the composted material from the inert matter. The 50 mm screens remove 80 to 90 per cent of the inert material (SANDEC 2002). This mechanical separation is energy intensive and is physically demanding for the employees. On a regular basis employees need to manually unclog the sieves which accumulate un-segregated waste. Though safety equipment is encouraged, it is not insisted upon, leading to injuries from broken glass and sharp metal pieces. Due to failures in the enforcement and monitoring of the Bio-Medical (Management and Handling) Rules 1998, hospital waste is still often found here, further exposing employees to danger.

The KCDC is also the site for one of the largest vermi-composting sites in India, covering 4 acres (Basavaiah 2005). Pre-composted material, (compost after 25-30 days), is collected into bins to cool for 3-4 days. Once the temperature reaches a cooler 27-30 C°, worms are introduced. Four African worm species are used with three units. Vermi-composting takes an additional 30-40 days, making a total of 60-70 days. A top layer of vermi-casting is harvested every three days and sieved through the screens (SANDEC 2002). A specific density of worms is needed for proper and efficient degradation; 300 large robust worms per kg of waste or 1,000 small worms per kg of waste. Vermi-composting produces high quality organic manure.

A rag-picker family pays KCDC Rs. 5,000 per year for waste picking rights. The KCDC employs approximately 50 permanent staff at about Rs. 5,000-6,000 per month plus another 40 casual daily wages at Rs. 1,500 to 2,600 per month, all with social benefit packages (SANDEC 2002). In the coming years, the Advanced Bioresidue Energy Technology Society (ABETS) from the Indian Institute of Science will establish a 1 MW power generation unit at the KCDC premises by using the garbage rejects generated during processing.

Community Participation under Centralized System—Case of Ward 54

This section looks at one community under the '*Swachchha Bangalore*' programme, which has developed a participatory mechanism for monitoring civic services and ward planning. Ward 54 (Srinivasanagar Ward), one of the 10 wards in Bangalore participating with Janaagraha, (a citizen's movement towards improved public governance through strong community-based, participatory democracy) has been chosen as an example to illustrate this approach, used in other areas of the city. Under the Ward Vision Campaign by Janaagraha, Srinivasanagar has developed a system to promote participation in the decisions made concerning municipal services within the ward. In addition to addressing SWM this community initiative strives to expand participation in ward dynamics, which has meant that community members are demanding more from the municipal authorities, and this has resulted in a cleaner ward.

Srinivasanagar is located in the southwestern part of the BMP. It is a new ward, incorporated into the BMP only in 1995. The ward is divided into 14 blocks, and contains at least 16 RWAs that have formed a federation. It has an approximate area of 5 sq. km, a population of 82,072 and is predominantly a middle class area (Janaagraha 2004). There is a 79.3 per cent literacy rate, as compared to 76.7 per cent for the BMP average (Census of India 2001). There are two BMP recognized and seven undeclared slums in the ward. There are 8,751[14] properties, 609 registered as commercial and 8,142 as residential (Janaagraha 2004), indicating that this is a mostly residential ward. Of the registered properties in the ward, 7,408 (85 per cent) have water connections.

The Federation in Ward 54 and 55—Conducting Ward Sabha

After some time, a federation was formed to include the RWAs of Wards 54 and 55. The objective was to have unanimity of purpose and collective strength in contacting the public as well as officials. Having each RWA represent a block, each with a population of roughly 6,000 to 10,000, the federation has become a voice the BMP cannot ignore.

A monthly meeting or 'Ward Sabha'[15], facilitated by Janaagraha, is held in which members of the federation (RWA leaders) participate, and to which BMP officials are invited. Before the Ward Sabha is held the federation meets to discuss local issues, problems and possible solutions. Each RWA is asked to write down in an official document the concerns of their community. Once the federation has discussed all the issues from that meeting, a list of all the relevant BMP officials are asked to attend the 'Ward Sabha'. For example, if there is a solid waste problem in a particular street, the corresponding health officer in that area is invited. The BMP officials tend to be cooperative in attending the meetings, though remediation tends to be slow. To increase accountability with officials, follow-ups are performed after each meeting to check if the past month's issues were resolved. If not, the officials are asked again to attend the meeting until the issue is solved. Minutes from the meetings are produced and distributed to the entire community.

This approach is highly holistic in managing all the civic services within the area. According to the '*Swachchha Bangalore*' newsletter, "citizen participation is critical for a sustainable solution to a clean, garbage free city" [Janaagraha 2003(b)]. This requires that civic-minded people need to volunteer time and take on the responsibility of partnering with the BMP to ensure the cleanliness of the ward and to monitor local planning. Janaagraha plans on expanding this type of ward organization from only 10 wards to the rest of the city [Janaagraha 2003(b)].

Though positive results are produced in such a participatory system, this mechanism tends to be highly time consuming for BMP officials. Having 100 wards within Bangalore, Ward 54 and 55 may be receiving a significant amount of attention compared to others. This produces an inequitable division of services compared to other less organized wards in Bangalore. Also, RWAs, though highly participatory within the community, tend to attract the educated middle to upper class, while participation amongst the poor is low. In this regard, RWA's interests may only reflect the (relatively) privileged few rather than the needy and unorganized poor.

Janaagraha and the Ward Vision Campaign

Janaagraha, meaning 'the life force of the people', is a citizen's movement for improving public governance through participatory democracy. It runs three campaigns that engage stakeholders who have an interest in improving Bangalore's status as a national city, including citizens, government, local NGOs, and corporate institutions (Janaagraha Official Website 2005). The Ward Vision Campaign, in which Ward 54 is a participant, has the following objectives:

- To produce a plan that is articulated by the residents and other stakeholders of each ward, prioritized over 3 years.
- To make the plan a realistic one, and identify the sources of funds and their allocation.
- To make the process a representation of the demographic, economic and social profile of the participating ward.
- To make the process a collaborative one, with leadership from the councillor as the elected representative of the ward, and the administration of the BMP, as well as participation from all other concerned stakeholders.

The aim of the campaign is that high quality civic services through community participation will be achieved for the residents of the involved wards [Janaagraha 2003(b)]. Participating wards in the campaign have gone through various stages in their development to become a 'participatory community'. For this, civic matters are divided into 16 categories, including one on SWM, and also one for each of the 14 blocks within each ward. All of these categories have a corresponding citizen anchor[16] (Table 3.8). Information gathered on civic services by the citizen anchors, including issues on SWM, is presented to the councillors, RWA leaders, and other key community leaders and may be dealt with at the Ward Sabha. Then, the citizen anchors discuss participatory strategies and programmes with ward councillors and other key ward officials.

In addition to the primary talks, extensive surveying and mapping of the ward is done to get an up-to-date contextual base. The acquired information is also used to guide the BDA in developing the next Comprehensive Development Plan for Bangalore and which remains as part of the official city records.

Though the concept of citizen anchors is helpful in identifying ward level planning problems, it is quite difficult for them to gather information that is accurate and reliable. Also, convincing the corporators and other concerned officials to support the initiative is a great challenge, especially once a new political leader is elected [Janaagraha 2003(b)].

Table 3.8: Citizen Anchor Categories in Ward 54

Civic Matters	
Zoning & land use conversion	Storm water
Building violation & land encroachment	Public Toilets
Civic Amenities	Solid waste management
Heritage Preservation	Stray dog menace
Visual Pollution	Public transport, Parking and traffic
Tree Planting & felling	Street vendors
Rain water harvesting	Crime
Water and Sewerage	Roads, footpaths, drains, parks and street lights (Area wise)

Source: Janaagraha [2003(b)].

Solid Waste Management

By attending general meetings, doing field research and gaining public opinions on the matter, the citizen anchor summarizes the problems and possible solutions for municipal attention. Some of the problems that were recorded in-terms of SWM in Ward 54 include:

1. Lack of door-to-door collection from the '*Swachchha* Bangalore' programme for some of the residents.
2. Lack of awareness of the benefits in segregating solid waste into dry and wet biodegradable, inert and toxic categories.
3. Improper disposal of toxic and industrial waste by BMP officials resulting in the need for more stringent action against such illegal dumping.
4. Community waste bins are too low allowing for easy overflow of garbage and access by animals[17].
5. Bins are not being emptied quickly enough.
6. Waste transport trucks are being overfilled and are not being covered resulting in waste dropping on the roads and unpleasant odours. [Janaagraha 2003(b)]

Since Ward 54 is a fairly new addition to the BMP, the majority of the collection, transportation and disposal is performed by private contractors. There is no decentralized disposal scheme in the area, though the BMP is working closely with the community to develop composting sites in some local parks. The BMP has an interest in this due to the transportation costs that might be saved. In its early stages this development will have a strong dependency on community participation to achieve proper segregation at source throughout the participating areas. Strategically placed decentralized solid waste schemes throughout Bangalore, organized by the BMP, might provide the most efficient and viable solutions to SWM in the city and should be looked into further for future planning.

Ward Finances

The participatory movement was in part a response to scarce funding for projects in the community. Many development projects that were undertaken in Ward 54 in the past were not based on the community's needs but on the interests of the councillor at that point in time. This happened mostly because budgets were never scrutinized by the community. Through education and consultation, citizens in Ward 54 today have a strong sense of belonging and responsibility within their community and have a proper mechanism for inquiry and for demanding more accountability from their leaders. Instead of arbitrarily deciding the ward budget, it is important that both the community and the BMP work together to shape the budget to produce the most efficient results, rather than leaving an individual councillor to bargain for his/her ward.

Another issue with regards to finance addressed by the Janaagraha movement is revenue collection from the ward. A careful examination of the revenues in Ward 54 shows that there is a significant capacity for revenue increase. Currently there are 8,142 residential properties and 609 commercial ones that are under the tax radar, generating Rs. 1.79 crores in revenue (Janaagraha 2004). Community representatives performed a rough survey that estimated that there are in fact about 12,932 residential properties. According to these figures, roughly 37 per cent of residential properties were not taxed. And commercial properties were estimated at 1,507 compared to only 609 registered properties. In this case, only 40 per cent of commercial properties are taxed [Janaagraha 2003(b)].

If Ward 54 were to develop a property tax reform scheme, it is key that the municipal authorities work together with the community. First of all, there is a need to start a campaign to record and register the currently excluded properties. One way to do so is through updating locality maps to include correct street designations, sites and appropriate land use that can be compared to current revenue office records. Computerized mapping and associated data bases such as GIS are some options that will manage records, increase collection and further transparency. As well, a community awareness campaign that pushes for social responsibility can emphasize the idea that increased revenue means greater investment in civic amenities [Janaagraha 2003(b)].

There has already been collaboration in this matter between the BMP and citizen anchors. Revenue officers from Ward 54 have co-operated in sharing information regarding tax collection within the ward from previous fiscal years. Based on this information, the citizen anchors with Janaagraha plan to form a campaign aimed at educating the residents on social responsibility and augmenting ward revenue. Through the campaign, the citizen anchors propose that at least 70 per cent of the excess revenue, through tax reform, should go back into ward development. The anticipated annual revenue increase for the ward is 30 per cent for the first year, up to 75 per cent the second year, and 100 per cent by the third year [Janaagraha 2003(b)].

Decentralized SWM

As mentioned earlier, Bangalore has an active civil society that has resulted in a large number of decentralized SWM schemes throughout the city. Many NGOs and CBOs focussed on the environment became active during the 1990s when municipal services were at their worst. No longer trusting the municipal authorities to fulfil their duties, communities organized themselves and created their own waste management mechanisms, minimizing their dependency on the BMP. There are numerous decentralized schemes in Bangalore, but only two will be discussed here. The first, organized by the Residents' Initiative for a Safe Environment (RISE), illustrates a case when a community came together in developing a local SWM scheme but was abandoned when the '*Swachchha Bangalore*' programme was implemented in 2000, because the latter terminated many of decentralized schemes and handed over the areas to either municipal or contractual workers. The second decentralized scheme is organized by the Kalyananagara RWA, a group that works closely with the CEE, an NGO.

RISE and Its Take-over by 'Swachchha Bangalore Programme'

Founded in 1997 and headed by a retired community member, RISE organized citizens who were dissatisfied with the low quality of municipal services resulting in high health risks and environmental degradation. The main problem with the state of affairs was that the municipality was not emptying disposal bins on a regular basis leading to accumulated waste and overflow. In addition, waste would spread over the road due to rummaging rag-pickers and hungry dogs and cattle going through the heaps of garbage. Since garbage was not segregated, this became a very unhealthy environment for the rag-pickers and put the children playing near by at risk of heath problems.

RISE eventually gathered about 1,200 people for the initiative. Many households were included into the scheme even if they refused to pay for the service. Most of those refusing to pay did so simply because they were already paying their taxes for the municipal service and did not want to, or could not afford to, pay more—even if at a minimal fee. The monthly sum for those who did pay was Rs. 100 per household for worker salaries and for the operation and maintenance of equipment. Before starting, the University of Agricultural Science in Bangalore was consulted for the design of an appropriate composting pit that used simple materials and would be easy to maintain. A low cost build was chosen based on plastic aeration tubes and a concrete frame. The initial cost was Rs. 80,000 for the whole scheme and was borne by the first few members who joined RISE (SANDEC 2002). The compost site was donated to RISE by the BMP but was then destroyed by unknowing workers of the BDA. After much conflict with the BDA over property rights the plant was then rebuilt on its original site.

Rag-pickers who were already established in the neighbourhood undertook primary collection. They were paid to do door-to-door collection from the 1200 participating households. Waste was segregated at source into wet compostable waste, recyclable waste and inert waste. The waste collectors were provided with pushcarts, having four buckets for the segregated waste, and protective gear such as gloves and a uniform. The only secondary collection necessary for the entire SWM system was for the inert waste, which was placed in community bins for municipal collection[18]. Waste collectors were given name tags so that they could be well known and better integrated in the community. A monthly salary, health insurance and vocation days were also provided to the workers and their families. Waste collectors were allowed to keep any waste they found valuable to sell or reuse.

The compost site was set up at a nearby park. Approximately 300 kg per day or 108 MTs per year of biodegradable waste was disposed in the compost pit. Approximately 60 kg per day of compost output was produced that was sold to the community (SANDEC 2002). The University of Agricultural Science took compost samples on a regular basis for quality control.

By 2002, '*Swachchha Bangalore*' programme was well on its way in being implemented throughout the city. Without notice the contractual workers, later gaining the name of 'garbage mafia', showed up to collect household waste. Mr. Venkatramanan, the founder and manager of RISE, was told by these workers that he was to stop the community initiative at once. It was alleged that the group was not officially incorporated and thus illegal (Rosario *et al*). RISE had attempted to collaborate with the BMP, using RISE's own waste collectors and compost pit, but to no avail. The entire initiative was abandoned, including the compost pit, and the RISE waste collectors lost their jobs.

Today, the SWM system in place has some similarities to that of RISE, but also some stark differences. The door-to-door system is still in place, as is the informal mechanism of recycling valuable waste to itinerant buyers. The similarities end there. According to Mr. Venkatramanan, waste collectors are not provided with vacation days, health insurance, safety equipment, or name tags. Waste is no longer segregated at source; instead young boys are often found segregating remaining valuable recyclables in the bed of a moving truck for income. Safety equipment is not provided to any of these workers. Waste collectors are not being paid the mandatory Rs. 1,800 per month wage as defined by the municipal contract, but are earning maybe Rs. 500 per month. He suspects that municipal supervisors who are supposed to overlook payments to ensure fair salaries to workers are actually being paid by the contractors to look the other way as waste collectors are exploited. A probable explanation for RISE's failure is that the city scheme is cheaper for the households (since households were already paying for it in their taxes).

Kalyananagara Residents' Welfare Association

The Kalyananagara RWA is located within Ward 87, in Lingrajapuram. The Kalyananagara initiative started in 1998 and was one of four new layouts being developed by the BDA under the "Integrated Urban Environment Improvement Project (IUEIP) HBR Layout". Brindavan, Bhuvanagiri and Kasturi Nagar were the other three subdivisions. This was a partnership between the CEE and Technology Informatics

Design Endeavour (TIDE) where the BDA implemented the project with financial assistance from the Norwegian Agency for Development Cooperation (NORAD). The project's main components were: (1) Geographical Information System (GIS); (2) Solid Waste Management; (3) Open Space Management; (4) Project Secretariat. There are three main areas within Ward 87 that are not under '*Swachchha Bangalore*' that are being run under a similar manner as Kalyananagara. These have a total of eight decentralized composting sites within its system (SANDEC 2002).

In the SWM component of the project, the focus was on the development and implementation of a SWM system that would be operated at the community level. This included community mobilization and the development of an appropriate composting scheme.

The primary collection is done with handcarts door-to-door, and waste is segregated by the waste collectors. Biodegradable waste and "dry" non-biodegradables are loaded into separate bins on the pushcarts. Valuable recyclables are collected by the waste collectors to sell to itinerant buyers or taken home for reuse. The pushcarts are brought to the "recovery centre" that harbours the composting pits, a shed for the equipment and space for an office, a small garden and a metallic shed that holds all the community's inert waste. The pit structure is of low technology, mainly made from concrete bedding and a simple roof for temperature control. A small hedge surrounding the entirety of the vermi-pits containing water was needed to stop ants from crawling inside the pits and attacking the worms.

There are three composting pits; two wet pits, each with two tonnes capacity for the initial wet waste, and one pit for vermi-composting (SANDEC 2002). Wet waste is composted for approximately 25 days with manual aeration after the first 10-15 days. Once composted, the material is spread out on the ground to dry out excess moisture and get rid of large lumps. A couple of days later the compost is divided into fine and coarse compost where the fine portion is sold and the coarse portion goes into the vermi-compost pit. Approximately 300-400 kg of both regular and vermi-compost is produced per month. The compost is sold locally to the community at Rs. 5 per kg for the regular compost, (Rs. 1 of that goes to the individual selling the compost).

The Kalyananagara system employs 25 waste collectors, each of whom collects from approximately 100 households daily. Of the 2,500 households involved in the SWM system, each pay Rs. 15 per month for the service (Table 3.9). Waste collectors are currently being paid Rs. 1,000 per month, which is Rs. 800 less than what contractual collectors are meant to be paid and Rs. 4,000 less than what the municipal *Pourakarmikas* are being paid[19]. The association is hoping to increase their salary to Rs. 2,500 per month in the near future. Employees are not supplied with health insurance but doctors and other health workers within the community provide free services to the workers and their families.

The Kalyananagara SWM system internalizes about 80 per cent of its community waste within this scheme through composting and recycling. The last 20 per cent needs to be dealt with externally by the BMP for secondary collection. The inert waste is the only material from the entire SWM system that must be transported outside the community. A metallic shed has been built for storing all the inert waste from the community and has a capacity to hold up to three weeks worth. Since the community has such a low dependency on the BMP, three weeks is sufficient time for the BMP to do a pickup, meaning the community has a strong hold on what goes on in the system and has become the main player in the decision-making process.

It was always the intention to pass over all duties and responsibilities from the project into the hands of the community. A waste management committee was formed with community members who would volunteer their time and abilities to the community's waste system. The Waste Management Committee consists of 10 members including the President of the RWA, a VP, General Secretary, Joint Secretary and Finance Secretary. They work within the Kalyananagara RWA and are elected at a general body meeting every two years. The Committee meets every month on the second Saturday to discuss arising problems and implement on going programme adjustments.

The waste committee works with a budget of approximately Rs. 39,000 per month, which covers worker salaries, equipment maintenance and operation costs. After paying for salaries there is only Rs. 11,000 left for the operation and maintenance of the system[20].

Once '*Swachchha Bangalore*' was launched citywide, there was a similar scenario to the RISE example, in that contractual workers arrived to take over the SWM system in the neighbourhood. With so much work and progress within the project, there was a strong voice within the community to keep the existing system in place. Initially there was conflict between the BMP and the community, where the former wanted to put the Kalyananagara neighbourhood under '*Swachchha Bangalore*'. In the end, the Kalyananagara SWM system was not taken over by the contractual workers. The reasons for this include: the BDA's involvement in the project from the start; the development authorities recognition of the project; the fact that its start-up funds came from a foreign donor; and finally the fact that the Commissioner at the time had a close relation with the project and pushed its continuance.

Bangalor Municipal Corporation (BMP) Finances

As stated earlier, three main departments handle SWM within the BMP, the Health Department, which takes care of the collection and transportation; the Engineering Department, which is responsible for removal of silt from drains; and the Sanitary Engineering Department, responsible for the disposal of solid waste. The budget heads are made accordingly. In the last few years the BMP has spent, on average, around 8.4 per cent of its total budget on SWM related activities, namely collection, transportation and disposal of waste in the BMP. At the same time, the proportion of the total budget spent on SWM has decreased from 9.9 per cent in 2001-02 to 6.9 per cent in 2005-06 (Table 3.10). Actual expenditure made by the BMP in 2001-02 on SWM was Rs. 675 million, which has increased by 417 million (62 per cent) to Rs. 1092 million in 2005-06.

What is worth noting in Table 3.10 is that the amount going towards meeting the expenditures of the contractual system has increased steadily over the years. This is due to the growth of the city and the expansion of the areas under private contractors in SWM, as discussed earlier. Expenditure on the BMP managed system has also been on the rise even though the number of wards under its jurisdiction has remained at 114 during this period. This can be explained mainly by increased wages and social benefits for BMP workers. With changes in the BMP's property tax programme, the revenue from property taxes has dramatically increased. Therefore, even though the 2005-06 total budget for SWM has increased by over Rs. 150 million from 2004-05, relative to the total BMP budget, the percentage going towards SWM has gone down from previous years.

Table 3.9: Kalyananagara Waste Committee Monthly Budget

Particulars	*Rate*	*Expenditure Rs/month*	*Revenue Rs/month*	*Total for O&M Rs/month*
Salaries				
• Waste Collectors	• 25 × Rs 1,000/ month	25,000	/	(-)25,000
• Compost sales person	• 1 × Rs 1/kg (300kg/month)	300	/	(-)300
Waste management cess	2,500 x Rs 15/ month	/	37,500	(+)37,500
Compost revenue	300 kg/month × Rs 5/kg	/	1,500	(+)1,500
Total	/	28,000	39,000	(+)11,000

Source: Subbaiah (2005).

Further, the private contractors are managing solid waste in far more wards than the BMP, but they are doing so at a far lower cost. The BMP does the same job that the private contractors do at apporximately 2.5 to 3.5 times the cost per average ward as the private contractors. Contracting out the SWM to private contractors is one of the ways of keeping down the costs of SWM, so that budgetary funds are available for other purposes.

The BMP has been preparing a scientific landfill site in the city. Hence, expenditure on disposal has increased by 3,525 per cent between 2001-02 and 2005-06 (Table 3.10).

While the previous table (Table 3.10) looked at expenditures on collection, transportation and disposal, we turn now to the entire SWM expenditures by the BMP in greater detail (Table 3.11). Actual expenditure made by BMP in 2003-04 on all the SWM activities was Rs. 934.38 million, which has increased considerably by Rs. 276.61 millions (29.60%) in the proposed budget of 2005-06. This is mostly due to increased expenditure on infrastructure (86 per cent increase). Another reason for the increased total expenditure on SWM is the rise in capital investment in the landfill site, which constitutes Rs. 63 million of the 2005-2006 budget. More specifically, the capital investment includes facilities for dumping, small/medium scale composting, and a waste-to-energy plant of 8 MW capacity (as mentioned above). The discrepancy between the Rs. 800 million quoted by the BMP on annual

SWM spending and our estimate of over Rs. 900 million (arrived at by totalling the expenditures on each head of SWM), is likely due to the different accounting methods used.

Table 3.10: BMP Expenditure Incurred for Collection, Transportation and Disposal

(in Rs million)

Year	*Health Department (collection & transportation and other expenses)*		*SWM (disposal)*	*Total Budget*[21] *(for collection, transportation & disposal)*	*Total Budget of BMP*	*Percentage of Total Budget to SWM*
	*BMP System**	*Contractual System*				
2005/06	612.79	400.00 **	79.00	1091.79	15,717.42	6.9
2004/05	511.13	374.43 **	42.83	928.39	12,831.49	7.2
2003/04	518.51	346.53 **	25.00	889.73	9,979.21	8.9
2002/03	601.93	284.00***	21.75	907.68	9,721.42	9.3
2001/02	459.10	213.54***	2.24	674.88	6,793.95	9.9

Notes: * 114 wards (Ananthaswamy 2005).
** 183 wards (Ananthaswamy 2005).
*** 182 wards (Ananthaswamy 2005).
Source: BMP Statistical Department 2005.

Under '*Swachchha Bangalore*', the BMP has allocated Rs. 8.5 million for the year 2005-06 for public awareness, organizing workshops and capacity building. Increasing spending on this programme by the BMP is indicating its seriousness of achieving the objective of attaining total segregation of waste at source. This aim is highly citizen dependent, requiring immense participation from the people of Bangalore. In the past, it has happened that the revised budget estimates and then the actual expenditures made on '*Swachchha Bangalore*' project are far lower than the budgeted expenditures, as one can see from the data in Table 3.12. The main reason is BMP's inability to get citizen's support for carrying out a pilot project and laxity of the BMP in undertaking training workshops and even purchase of equipment for setting up a helpline. It is likely that for the last two activities also, there is no demand from the citizens.

The budget receipts for SWM are far lower than the expenditures. For example, while Rs. 934 million was actually spent on all the activities of the SWM in 2003-04, just Rs. 1.68 million was collected from this activity by the BMP. In light of this massive deficit, in the

Table 3.11: Expenditures on Different SWM Activities

(Rs in lakhs)*

Particulars	*2003-04*	*2004-05*	*Budget Estimates 2005-06*		
	Actual Expenditure	*Revised Estimates*	*Revenue*	*Capital*	*Total*
1. Health General					
Administrative	5282.91	5383.63	6127.91	/	6127.91
General Expenses (contingencies and uniform)	4.12	21.83	32.94	/	32.94
Infrastructure					
— M and R of push carts	1.26	20.18	26.00	/	26.00
— Procurement of vehicles	0.70	6.81	/	100.00	100.00
— Push carts, containers and bins	51.53	/	/	30.00	30.00
— Running & Maintenance charges	239.14	343.68	387.60	/	387.60
Public Health					
— Cleaning and transportation of garbage, toilets and urinals	3465.26	3744.34	4000.00	/	4000.00
— Decentralized composting	10.83	37.00	10.00	/	10.00
— Deputation of city cleaning workers	0.48	2.57	3.50	/	3.50
— Purchase of conservancy articles	6.41	100.00	100.00	/	100.00
— Storing and sorting area for secondary collection	/	/	/	100.00	100.00
Swachchha Bangalore Project (Campaigning and Capacity building)	24.18	10.00	85.00	/	85.00
TOTAL (Health General)	9086.82	9670.04	10772.95	230.00	11002.95
2. Landfill site development	249.57	428.33	456.89	630.00	1086.89
3. Others (GAD)					
Pourakarmikas' Day Celebration	7.39	20.00	20.00	/	20.00
Grand Total (1+2+3)	9343.78	10118.40	11249.80	860.00	12109.84

Source: BMP Budget 2005. * 1 lakh = 1 million

2004-05 BMP budget a SWM cess was introduced, which was supposed to collect Rs. 250 million for SWM activities provided by the BMP (that is covering one-fourth of the costs of entire SWM system). This cess could not be collected at all in this year as indicated by the revised estimates figure (Table 3.13). Thus, EMD collected from the private contractors was the only source of revenue receipt for the SWM activities. In 2005-06, debris collection fees (from the bulk producers of solid waste) have been enhanced and it is expected to bring in Rs. 15 million to the BMP in this financial year. The BMP has also estimated that Rs. 300 million would come from the SWM cess in this financial year. The cess was approved by the BMP Council in February 2005 (which was too late approval in the financial year 2004-05 for the collections in the year 2004-05). As per our discussions with the BMP officials, it was evident that this cess was not being collected even in the 2005-06 financial year.

Table 3.12: 'Swachchha Bangalore' Campaign Budget

(in Rs lakhs)

Particulars	*2003-04*	*2004-05*		*Budget Estimates 2005-06*		
	Actual expenditure	*Budget estimates*	*Revised estimates*	*Revenue*	*Capital*	*Total*
Campaigns, public awareness, rallies and IEC Materials	11.38	10.00	10.00	20.00		20.00
Equipments for health net and help line	4.18	80.00	/	25.00	/	25.00
Pilot project expenses for segregation of waste etc.	1.75	30.00	/	30.00	/	30.00
Training and workshops for capacity building HRD	6.87	8.00	/	10.00	/	10.00
Total	24.18	128.00	10.00	85.00	/	85.00

Source: BMP Budget 2005

Currently, property taxes are inclusive of SWM services but there are plans of levying a separate SWM cess to citizens. It was intended that through a SWM cess, the service would become a separate head altogether and revenues generated could be monitored and deployed separately, maybe even handed down to ward committees. The SWM cess rates are very low: Rs. 10 to Rs. 50 per residential building per month, Rs. 50 to Rs. 200 per month for commercial buildings, Rs. 100 to Rs. 300 per month for industrial buildings and Rs. 200 to Rs. 600 per month for hotels and nursing homes; the rates tied to the plinth area (Table 3.14).

Table 3.13: Receipts Pertaining to SWM

(in Rs lakhs)

Particulars	2003-04	2004-05		Budget Estimates 2005-06		
	Actual expenditure	Budget estimates	Revised estimates	Revenue	Capital	Total
SWM (EMD and others)	10.54	20.00	28.13	28.00	/	28.00
SWM Cess	/	2500.00	/	3000.00	/	3000.00
Debris Clearance Fees	6.28	0.33	0.35	150.00	/	150.00
Total	16.82	2520.33	28.48	3178.00	/	3178.00

Source: BMP Budget 2005

This means that the BMP itself would remain in charge of SWM, charging the citizens for the service provided and sub-contracting the task to private contractors to reduce costs. The BMP is not thinking of a model of community participation in the SWM, inspite of some success stories available at the city level. Such a model would deter decentralized waste recycling, for instance through vermi-composting. Why would such a model be considered? Is it because what was feasible with the middle class housing colonies, which have stable residency and residents with belongingness to the city, is not feasible with a highly mobile and uncommitted young population, as in the new economy sectors such as IT?

Table 3.14: Proposed Cess on SWM

(Rs/month)

Area (sq. ft)	Residential Building	Commercial Building	Industrial Building	Hotels, Kalyan Mantaps and Nursing Homes
Up-to 1,000 plinth area	10	50	100	300
1,001 to 3,001	30	100	200	500
3,001 and above	50	200	300	600

Source: FEDINA (2005).

Achievements

Bangalore is a typical middle-class city, its modern origin traced to the development of the public sector enterprises in the city in the period after independence. This is a congenial social group to organise into RWAs that can engage in civic affairs. And that is the reason for the development of active RWAs in some wards. These RWAs have taken up the cause of keeping their areas clean and to some extent are participating in local governance issues, including that of SWM.

At the same time, the BMP wants to remain in control of the entire city level SWM efforts. It has therefore developed an elaborate centralized system of SWM and has increased its investments in various city level systems such as the landfill site, awareness programmes such as '*Swachchha Bangalore*', the KCDC plant for composting and now an energy generating unit. To be able to make funds available for these city-level activities, the BMP has privatised collection and transport of solid waste in about 40 per cent of the city wards. If the BMP is carrying out these tasks, it is doing it at 2.5 to 3.5 times the cost than what is incurred by a private contractor.

Another feature of Bangalore's governance is the development of organisations like Janaagraha attempting to link decentralized governance institutions and the RWAs with the centralized system of the BMP for better service delivery. Moreover, BATF attempts to bring in corporate style accountability to the public organisations. This has just begun and has a long way to go. Critics, however, consider organisations such as Janaagraha and institutions such as BATF to be elitist, particularly the latter as it carries out the interest of the corporate sector.

Bangalore also has a good waste transportation system. Currently, waste from primary collection is brought directly to centrally located trucks throughout the city. This direct loading technique provides two very important benefits. It reduces the amount of time the waste is being handled. Today, Bangalore is predominantly a 'bin-free' city, which means no one needs to manually shovel waste into trucks as the waste collectors throughout the day fill up the truck directly. Direct loading decreases the spread of disease because in the past, community bins often overflowed, and garbage was spread around by rag-pickers and hungry animals. Bangalore plans on building three main transfer sites throughout the city at strategic locations to increase efficiency within the transportation system. Properly situated to maximize truck haul, these transfer sites will facilitate the use of larger capacity trucks to haul larger quantities of waste to the disposal facility, thereby decreasing transportation costs.

Concerns

The most important concern in Bangalore is the pre-eminence given to the IT sector in taking many of the city level decisions. Ghosh (2005) alleges that groups like BATF are to get the IT sector more concessions than are due to them. The citizens of Bangalore claim that the BMP is unfairly funnelling scarce financial resources towards supporting the IT businesses rather than its own citizens [*The Hindu* 2005(a)]. Since the state still holds executive powers within the BMP, infrastructural priorities give way to supporting the IT sector in the construction of costly flyovers and bypass roads connecting the industrial parks to the city, among other expensive developments. The disproportionate importance given to the IT sector is being noticed now and has created some resentment within the city. This is not an environment conducive to democratic self-governance.

With regards to SWM, there are some specific concerns. Since the BMP has taken charge of door-to-door solid waste collection, and privatized this operation in 60 per cent of the wards, it should provide a more transparent and equitable setting for waste collection tender bids. The current system prioritizes bids based more on lowest financial cost, rather than engineering feasibility and employment equity. Thus, the municipal employees within the SWM system are being paid up to five times the amount of an employee with the same job within the contractual system. Unequal salary, social benefits, and even a lack of basic protective gear set the differences between the municipal and private worker. This is not the best way to go about reducing costs of municipal government services. In fact, it would have been better if the households were made to participate in the door-to-door waste collection system, through some financial contribution and participation in various activities of the SWM. This would have been better than imposing the '*Swachchha Bangalore*' programme on the entire city, as has been done since 2001 and has resulted in the closure of earlier decentralised community-based programmes in practice. Further, the first set of contracts for the collection and transportation of waste given under the '*Swachchha Bangalore*' programme have not yet completed their term; their two year term being extended to first three and then two more, and hence have not yet been evaluated. There is a lack of transparency within this area of privatization, and a significant lack of critical evaluation of the contractual system. The loopholes within the contracts allow for exploitation and poor performance.

As discussed earlier, un-segregated waste from households contributes to high health risks for a segment of the population, notably the rag-pickers (often women and children), waste collectors, *Pourakarmikas*, people from the waste recovery chain, KCDC employees, and children in general. Whether privately or publicly run, segregating waste at the source will both increase the efficiency of composting and the amount of recoverable waste for recycling.

The bio-medical waste collection is not yet fully put in place. Citizens most at risk are rag-pickers, BMP conservancy employees, private waste collectors, street sweepers, and even the formal and informal workers who deal with recyclable material that has come in contact with infectious waste. The spread of disease can be extremely rapid as these high-risk workers come home and come in contact with family members.

Like most cities in India, Bangalore has not yet solved the problem of waste disposal as mandated under the MSW Rules, 2000. Solving the city's disposal site disputes, legally and financially, should be a priority. The city is moving in the direction of developing a sanitary landfill site.

There is a lack of public information concerning waste generation figures in the city. Bangalore does not monitor the amount of waste that is being collected. All figures on waste production from the city are based on population numbers and national daily per capita production figures. Examining and scrutinizing the amount of waste being brought into disposal sites is a great way to monitor the waste generation of the city over time, the performance of waste workers and planning for future space availability. Currently, the KCDC is the only legal dumping ground within the city. All other areas are considered illegal, and therefore are not monitored.

Bangalore's informal sector, as in other Indian cities, plays a vital role in SWM. According to the study by Beukering *et al* (1999), about half the city's waste is recovered by the informal sector working independently of municipal authorities. Understanding and integrating the informal sector within the system can encourage further recycling and might reduce fluctuations within the recovery system from economic influences.

Under the '*Swachchha Bangalore*' Programme, there is very little room for community participation. The top-down approach makes it difficult for the public to voice opinions. Janaagraha, a local NGO, is trying to setup a mechanism for community participation within the municipal government but has only mobilized a small part of the population. Questions also arise regarding equitable distribution of BMP attention between RWAs and non-organized areas of the city.

The 74th CA intends to bridge the gap between municipal officials and the public through the formation of ward committees and by passing down power to the people of the city. These committees have not been successfully formed to date. Of the ward committees that have been formed, many are not following through on their mandated purpose. The roles and responsibilities are not clear to either public officials, the citizens of Bangalore, or indeed to the committee members themselves. In spite of the Janaagraha initiative, the ward officials are often hostile towards the questions asked by the leading members of the RWAs.

Small-scale, decentralized SWM systems in Bangalore have varying degrees of public participation. These systems are mostly instituted by foreign donors after which roles and responsibilities are passed down to community groups. This type of system encompasses a larger degree of community participation than the centralized system. Participation from the community is vital for the system to succeed while communities become less dependent on the municipal authorities for service provision. Bangalore has many of these decentralized SWM systems throughout the city, with both positive and negative interference from the municipal authorities. There are cases of centralized workers, municipal or private, taking over areas under a previously smooth running decentralized system. However, there are examples where municipal and planning authorities are working together to develop an integrated system that is both centralized and decentralized.

Municipal authorities need to support existing systems as these may form a vital part of the centralized SWM system in the future. Having small-scale SWM systems in strategic branches of the city, according to geographical range, population, waste production, traffic, etc. may be a viable solution to some of the current and future problems and inefficiencies in the centralized system.

NOTES

1. The contents in this chapter are based on the interviews with: (i) Ananthaswamy, H.C. Executive Engineer (SWM); (ii) Basavaiah, C. Managing Director of the KCDC; (iii) Iyer, A., former Project Leader of '*Swabhimana*'; (iv) Roa, N., RWA member of Federation 54-55 and Janaagraha representative; (v) Subbaiah, C.M., President of the Kalyananagara RWA; (vi) Venkatramanan, P.V., former RISE President; (vii) Verghese, G., Project Coordinator FEDINA, former Project Coordinator for Center for Environment Education (CEE).
2. Solid Waste Management, in this study, will focus on municipal solid waste. The Municipal Solid Waste (Management and Handling) Rules 2000 defines municipal solid waste to include commercial and residential wastes generated in a municipality or notified areas in either solid or semi-solid form, excluding industrial hazardous wastes but including treated bio-medical wastes (CPCB 2005).
3. Figures on waste generation are scarce and unreliable and most estimations are based on population figures and generalized per capita waste generation averages.
4. Includes old clothes, gunny bags, coconut shells, dust etc.
5. The Ministry of Environment and Forest (MoEF), notified the Bio-Medical Waste (Management and Handling) Rules, 1998 under the provision of the Environment Act, 1986. The said rules extend to whole of India. The KPCB is the regulating and monitoring authority for bio-medical waste.
6. *Pourakarmikas* are BMP waste sweepers.
7. This figure excludes waste from industries as this waste is dealt with internally.
8. BATF webpage: http: //www.batf.org
9. The results of the sampling, in terms of the order of priorities may be explained by the strong middle-class profile of the city. Other studies that show how different sectors of the city prioritize their municipal problems reveal different concerns: availability of water, existence of job opportunities in poor neighbourhoods, claiming citizenship and voting rights to get on the voter's lists (Nair 2005: 336).
10. Since the inception of '*Swachchha Bangalore*' system, the '*Swabhimana*' initiative continues to be a important part in creating a participatory venue for Bangalore citizens in issues relating to urban planning. However, it has moved away from SWM issues and deals with other social and environmental matters.
11. Registered slums only.
12. BMP waste collectors are called *Pourakarmikas*.

13. Though there are designated landfill sites, mainly the Bettahalli and Nageshwara Nagawanahalli dumping yards, the BMP is not legally allowed to dump there due to court orders following opposition by the local public (Janaagraha 2003(a))
14. This figure does not include the seven undeclared slums.
15. 'Ward Sabha' meaning 'people coming together'.
16. Citizen anchors are community volunteers who are specially informed by Janaagraha staff on a civic amenity and who work closely with Janaagraha facilitators and coordinators. They then study and find out the issues involved in their corresponding area and develop solutions to the associated problems.
17. Bangalore is currently a bin-free city but when "Ward Vision Document for Ward 54" was written there were still some community bins within the city.
18. At this time Bangalore was still using community bins for waste collection.
19. Municipal *Pourakarmikas* are paid Rs. 2,500 per month in addition to health insurance and other social benefits, equating to approximately Rs. 5,000 per month.
20. BMP may or may not be contributing funds to this project. If so, that amount is not included here.
21. These numbers will be lower than actual expenditure as they exclude some hidden costs like the *Pourakarmikas* Day Celebration, silt removal from drains, and other SWM particulars. It provides a rough estimate of what is being spent on an annual basis as compared to previous years.

Photo 3.1: Doorstep Waste Collection by Pourkarmika - Indira Nagar, Bangalore

Photo 3.2 : Street Sweeping by PMC worker; No Dustbins on Streets and Footpath, Bangalore

Photo 4.1: Steel Bins Emptied into Trucks at Transfer Stations, Hyderabad

Photo 4.2 : Developed Landfill Site, Hyderabad

4

Local Government's Initiative: Hyderabad

DARSHINI MAHADEVIA and MANISH PATHAK[1]

City	Hyderabad—Secunderabad
Urban Local Body—1	Municipal Corporation of Hyderabad (MCH)
Urban Local Body—2	Secunderabad Cantonment Area (SCA)
Population—MCH, 2001	3,449,878
Population—SCA, 2001	204,182
Population—Hyderabad Urban Agglomeration (HUA), 2001	5.75 million
Population—Hyderabad Urban Development Authority (HUDA*) Area, 2001	6.38 million
Population growth rate, 1991-2001—MCH	1.21 % p.a.
Population growth rate, 1991-2001—HUA	2.85% p.a.
Population growth rate, 1991-2001—HUDA	3.17% p.a.
Population growth rate, 1991-2001—SCA	1.78% p.a.
Area—Hyderabad-Secunderabad	170 sq km (9% of HUDA area)
Area—HUDA	1865 sq km
Local authorities in HUDA	MCH, Cyberabad Development Authority (CDA), Secunderabad Cantonment Board (SCB), Buddha Purnima Project Authority (BPPA), other 10 Municipalities and 105 Gram Panchayats
Local authorities in HUA	MCH, SCB, Osmania University, 10 Municipalities and outgrowths
Water Supply and Sewerage in MCH	Hyderabad Metropolitan Water Supply and Sewerage Board (HMWSSB)[2]
Storm water drains	MCH
Total wards in Hyderabad-Secunderabad city	52 wards divided in 7 circles

Notes: *HUDA is the planning authority whereas the ULBs are the institutions of urban governance, thus taking up the task of SWM.
Source: Data based on Hyderabad Urban Development Authority (2003).

Introduction

Hyderabad (or Hyderabad-Secunderabad twin city), the sixth largest metropolis in India, with a population of 5.57 million in 2001 in the urban agglomeration is known as a historic city with some monuments such as Charminar and the Palace, and in recent years has become one of India's IT centres. The Musi river divides the Hyderabad city in two, the southern bank having the old city and the northern bank the new. Hussain Sagar lake, the largest among the city lakes, divides Hyderabad and Secunderabad. Hyderabad is among the first urban centres that have effectively experimented with a wide range of contracting systems, starting from leasing to management contracting. MCH has been awarded with Cleanest City Certification for two consecutive years by rating agency Crisil[3].

The innovative SWM programme of the MCH called 'Litter—free Hyderabad' has two components in it: (i) segregation of garbage into organic and recyclable waste at the source, and (ii) privatization of garbage collection and road sweeping in the entire city. The former is performed by the Resident Welfare Associations (RWAs) in the regular/ legal housing colonies and the Neighbourhood Committees (NCs) in the slums, through whom the member households participate in Voluntary Garbage Disposal Scheme (VGDS). These organisations convert the biodegradable waste into nutrient rich manure through vermi-composting in their respective localities. While the one time cost for setting up a vermi-composting centre is provided by the MCH, the recurring cost of maintenance is borne by the RWA/NC through collection of service charges and sale of vermi-compost and recyclable materials. A total of 120,000 households (17 per cent of the total households of MCH) are participating households in this innovative system.

Solid Waste Generated

In 2005 the city generated 2,200 Metric Ton (MT) of garbage daily (a figure that is little different from today). Of this, 60 per cent (1300 MT) is from the residential sector and the remaining 900 MTs from non-residential areas. Of the total solid waste generated in the city, 52 per cent is compostable organic waste (Table 4.1); 5.99 per cent leaves, 12.42 per cent fruit-waste, 18.08 per cent vegetable waste, 8.88 per cent coconut waste and 3.09 per cent cooked food waste.

The high percentage of coconut in the organic waste increases the latter's calorific value, which is good for power generation. Thus, power generation and palletisation companies have shown their interest in the city's solid waste management.

Table 4.1: Solid Waste Composition, MCH

Soild waste item	*Proportion in total (%)*
Leaves	5.99
Fruits	12.42
Hay and straw	3.67
Vegetables	18.08
Food Waste	3.09
Coconut	8.88
Paper	8.04
Plastics	8.36
Metals and Minerals	1.40
Glass	2.71
Ash and Fine earth	6.63
Fine organic Waste	15.57
Stones and Bricks	1.00
Others	0.42
Rubber/Leather	1.59
Racks	2.35
Coal	0.10
Wood and Metal	Nil
Crockery	Nil
Bones	Nil
Earthenware	Nil
Total	100.0

Source: MCH Handout.

Table 4.2: Price of Recyclable Materials, MCH

Item	*Rate (Rs/ MT)*	*Quantity (MT)*	*Total value per day (Rs.)*
Paper	3,750	176	660,000
Plastics	11,333	183	2,073,939
Metal	11,333	30	339,990
Glass	750	59	44,250
All recyclables		448	3,118,179

Source: MCH Handout.

The 21 per cent of the waste is total recyclable (Table 4.1), whose estimated value would be Rs. 3,118,179 per day (Table 4.2). This comes to an annual value of Rs. 1,138.14 million, which is 1.32 times the MCH's annual budget on SWM for the year 2005-06. In light of this a private company is trying to get the whole MCH waste-recycling contract.

The household sector, contributing the largest to the solid waste within the MCH, has registered the highest increase (64.6 per cent) in the six year period (1997-03) for which we have data (Table 4.3). The recreational places that generate the second largest amount of wastes, registered a growth rate of 51.75 per cent in the same period. Thus, fastest rate of increase in this six year period has been registered by the sectors having higher shares in the total waste generated. A point to keep in mind is that waste generated from the industries is not accounted for in Table 4.3 because the ULB does not consider managing wastes in industrial estates as its own responsibility.

Table 4.3: Change in Quantity of Solid Waste by Sources, 1997-03

	1997 (MT)	*2003 (MT)*	*% Increase (1997-2003)*
Households	1030.0	1695.0	64.60
Hotels	80.0	111.4	39.25
Markets	60.0	88.6	48.00
Slaughter	20.0	22.9	14.50
Hospitals	60.0	88.6	48.00
Recreation	80.0	121.4	51.75
Dairy	50.0	72.1	44.20

Source: Administrative Staff College of India (ASCI) (2005).

The System

The MCH has set-up the entire waste management system. The functions of this system covered herein are: sweeping, collection, vermi-composting case study, recycling system, transportation system, power generation and dumping. For this purpose, the MCH has divided the city into 7 circles (or zones).

Sweeping

As garbage on the roads is acknowledged as one of the main problems in SWM by the Supreme Court guidelines 2000, an efficient sweeping system is necessary. The MCH has therefore attempted to make this task more effective or thorough than before by deploying 8,000 strong conservancy workers consisting of the corporation staff (3,592) and the private contractors' employees (the rest), to sweep the city through the formation of units (day & night). Among the latter are also the workers of the 14 DWCUA (SHG) groups[4] given the contract to sweep the streets. Giving street sweeping contracts to the SHGs is an innovation introduced in Hyderabad and many other cities in Andhra Pradesh.

After sweeping the streets, the sweepers deposit the collected waste in the waste-bins located at the sites where the people were habitually throwing the waste and ones that had become de *facto* waste collection points. The entire city has such a waste collection system.

This private contracting system of MCH has evolved with time and experience. The first attempt at privatization was with regards to sanitation contracts for cleaning the streets and collection of garbage, initiated in 1995. Twenty-seven localities were identified and divided into two sectors for two separate one-year contracts. The contracts were awarded through competitive bidding to local firms and co-operatives that possessed the required manpower, vehicles, equipment and experience for the work. However, these contracts were not uniform in size in that the amount of work entailed was variable. This created an unhealthy competition among the contractors making monitoring the works difficult. An internal evaluation suggested that some improvements with regards to better monitoring of the contractors, a plan for movement of the vehicles, standardization of the payments by linking them to the quality of garbage removed, use of computerized weighing station at the dump sites, and guaranteeing of minimum wage payment to the workers were necessary. After bringing in these improvements in the system, the additional contracts for sweeping, collection and transportation of garbage to the official dumping sites were given out.

Again, in 1997 the system of private contracting was examined and subsequently modified. Then, the city was divided into uniform units of 8 km stretch road in housing colonies and 4 km stretch on main roads. Accordingly, a unit rate was formulated based on minimum wages, cost of tools, implements, disinfectants and contractors' profit. For lifting and transportation of garbage a stretch of road generating 7 MT of garbage was treated as one unit. The cost of transport unit was worked out based on the hire charges of the truck, minimum wages for labour, cost of tools and implements, disinfectants, and contractor's profit.

In 1998 MCH introduced a new unit-based system for private sector participation that integrated the two types of contracts—for road cleaning at night and for sweeping, collection and transportation by day. About 40 per cent of the day units were then reserved for the MCH staff and the rest were to be given out to the private contractors. 100 contractors were then awarded day units and around 25 contractors the night units. Provision was introduced to make payment to the contract subject to certification of their work by the local level community groups formed consisting of five men and five women. This certification is done on daily basis. According to the MCH, these latest reforms saw the collection increase to about 90 per cent.

Cleaning Unit (Day)	*Cleaning Unit (Night)*
• 7 to 9 km of road length • 15 female workers, 3 male workers and 1 supervisor • Fixed scale of tools and implements for cleaning • 1 tricycle for collection of sand and silt • 8 hours of work from 6 am to 2 pm	• 4 km of road length (double width road) • 15 female workers, 3 male workers and one supervisor • fixed scale of tools and implements for cleaning • one tricycle for collection of sand and silt • 8 hours of work from 10 pm to 6 am • 8 workers (50%) for day mopping work for each cleaning unit, to work from 7 am to 3 pm
Transport Unit (Day)	*Transport Unit (Night)*
• Lift 7 MT of solid waste in the cleaning unit area from the bins • Engage one truck to transport the waste in two trips • Engage one driver, 4 workers • Fix a scale of tools and implements • 8 hours, work from 12 pm to 3 am	• Lift 7 MT of SW in the cleaning unit area from all the bins • Engage one truck to transport the waste in two trips • Engage on driver, 4 workers • Fix a scale of tools and implements • 8 hours of work from 12 pm to 3 am

Fig. 4.1: Working of Units

Of the total units required to clean the city, 25 per cent are reserved for the MCH workers. Then, there is reservation for 14 DWCUA units. For the rest, bids are invited from the contractors through a notification. Contractor makes an Earnest Money Deposit (EMD) of Rs. 100,000 with the MCH at the time of applying. In response, the MCH receives around 7,000 to 8,000 applications of which only around 5 per cent are disqualified. Of these, 160 contractors are chosen at one go with a 15 per cent reservation for SC/ST contractors. Then the allocation of the units to the contractors is through a draw of lots. If a contractor does not want to work for the particular area then he/she can back out from the contract or mutually exchange the unit with another contractor. Each unit has equal area avoiding difference in work and payments. Each contractor is awarded 2 to 2.5 adjacent units to keep this activity profitable.

A day cleaning work unit sweeps 8 km of road length using a work force of 15 women, 3 men and 1 supervisor with prescribed number of tools and equipments. A night cleaning work unit covers 4 km of four-lane road (main roads) deploying 8 workers during the night followed with day mopping by 8 workers. A transport work unit (day/night) has one truck with 4 workers for lifting and transporting 7 MT of MSW in 2 trips from all the collection points in the cleaning area.

The contractor's contract is for 7 months. There is no assurance that the contractor may get the work next time. Therefore, costs and expenditure are planned for just the 7 month period. The unit rates are calculated on the basis of wages, equipments and consumables[5]. To the total cost of a unit, contractor's 8.5 per cent profit, including income tax, is added. The contractor is supposed to be on the site for the period of work. Also, cell-phone expenses are given to the contractor so that concerned MCH officials can contact him/her any time. One unit yields earning of Rs. 4,000 per month to the contractor taking his/her monthly income up to Rs. 10,000. The cost of a transport unit is arrived at by considering the vehicle hire charges, minimum wages for the driver and workers, cost of fuel, lubricants and contractors profit.

Table 4.4: Sweeping Logistics, MCH

	Day Sweeping Units	*Day Lifting Units*	*Night Sweeping Units*	*Night Lifting Units*
MCH	93.0	124.0	—	
Private	241.5	114.0	77.5	05
Total	334.5	238.0	77.5	05

Source: MCH Handout by Dy Ex-engineer, SWM

Table 4.5: Costs per Unit for Different Activities

	Rs. per each night unit	*Rs. per each day unit*
Lifting	41,334	46,834
Sweeping	69,250	48,853
Total	1,10,584	95,687

Source: From the MCH.

The DWCUA groups are given work on the lines of private contracting. These groups are registered under the community development department and come under the jurisdiction of state government. All members of the DWCUA groups are involved in the implementation and management of the contract. Unlike the private contractors, these groups are more or less permanent as their contracts are for longer period (1-3 years). These groups are mostly allotted the unit where they stay so that it is easier for them to work.

The MCH maintains a force of sweepers to be available for its work in the times of emergencies such as disasters as these are dedicated to do whatever the MCH directs them to do. This wisdom of the MCH indicates that it is not appropriate to outsource all the responsibilities of a public agency as the private actors may not be dependable in times of emergencies. On the whole, 75 per cent of the work is contracted out to either the private contractors/ DWCUA groups, who sweep around 2000 km length of roads daily.

Collection

The functioning voluntary garbage collection scheme was introduced by the MCH in 1994. Under the scheme, the voluntary agencies and the RWAs are provided a full subsidy, which includes one tricycle for every 100-150 households, for undertaking house-to-house garbage collection. For this, the RWA has to submit a signed list of households it will serve and then take the responsibility of appointing and supervising the staff and paying the tricycle puller at least Rs. 750 per month (at current prices) from the mandatory contribution of Rs. 10 by every household. The tricycle pullers are mostly in the age group of 10—16 years. About 600 RWAs and 350 NCs have been provided with the tricycles by the MCH. Although the NCs are supposed to be working in the slum areas, in effect, the scheme mainly functions in the middle and higher income areas covering 1.2 lakh households

(20 per cent of the city's population). These schemes are not concentrated in any specific area but are spread across the city. In other areas, residents themselves bring garbage to the MCH collection point. There are 3,850 notified collection points for garbage.

After segregating the waste collected from the households, the tricycle pullers who are mostly men, sell the recyclable waste in the recycling market and can make an additional income of up to Rs. 1,000 per month. This is his incentive to work at low wages. Also, after collecting the garbage, the tricycle pullers sometimes use their tricycle for other purposes. Each tricycle costs about Rs. 7,500 and can go on for longer period than two years (its expected life) by changing the base of the container when rusted.

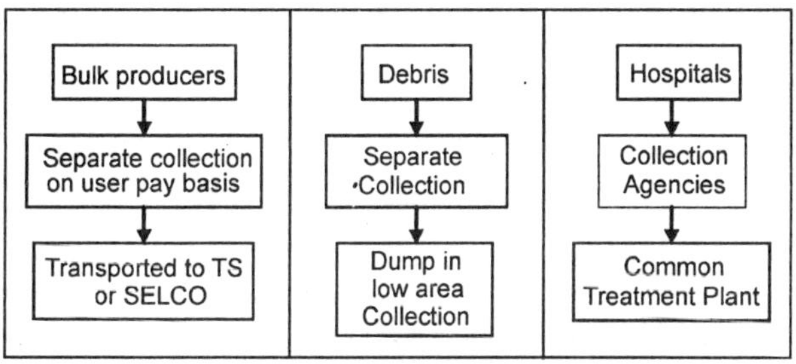

Figure 4.2: Collection System, MCH

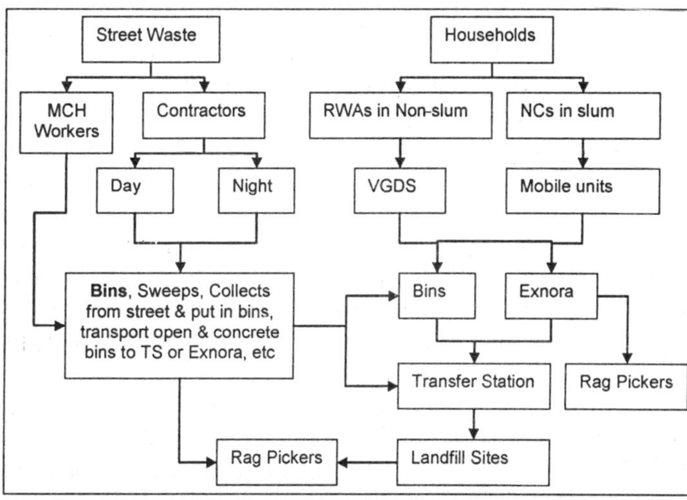

Fig. 4.3: Solid Waste Management System, MCH

Residents Welfare Associations (RWAs)	*Women's Neighbourhood Committees (WNCs)*
▪ The scheme is a MCH colony partnership programme ▪ MCH provides a tricycle free of cost to the colony association ▪ The association engages an operator for door to door collection and transportation to the notified collection point ▪ MCH will arrange lifting from this point to the landfill site by its truck ▪ The association pays honorarium to the operator raising fixed service charges from the residents ▪ Association maintains the tricycle	▪ The scheme is a MCH slum partnership programme ▪ MCH provides a tricycle free of cost to the NC ▪ NC engages an operator for door to door collection on MSW and transportation ▪ MCH will arrange lifting from this point to the landfill site by its truck ▪ MCH sanctions monthly honorarium for one year to the NC who will pay to the operator ▪ In one year time NC will enrol all the slum houses in the scheme and collect the service charges
▪ The scheme is under operation in 600 colonies	▪ The scheme is operational in 350 slums

Fig. 4.4: Logistics of Collection, MCH

There are 810 slums in the MCH area. All the slums are covered for cleaning services by the unit system. This is done through 2 systems, one through VGDS, where a community volunteer is paid minimum of Rs. 2,250 per month (for 300 households) to collect the garbage from the households and dump in the nearest municipal bin.

The other system of collection is run by the MCH itself, in which a mobile unit, such as a tricycle or a tractor (in case of large slum) collects the garbage from the households and the streets, at the frequency of once every four days and also on all special occasions such as festivals, holidays and dignitary visits. In some zones, the mobile units have been given to the private contractors. Even then, in several of the slums, optimal cleanliness standard is not being maintained, mainly because of infrastructure deficiencies and extreme overcrowding.

The non-recyclable waste is put in the special bins provided by the MCH at the collection points. The Bio-medical waste generated in the hospitals and the nursing homes is collected by specialised private agencies approved by the Andhra Pradesh State Pollution Control Board (APSPCB), to be transported to the hydroclaving/autoclaving plants. These agencies directly collect the charges from the hospitals and nursing homes, according to the rates fixed by the

APSPCB. Hospitals are required to segregate their waste as per the APSPCB norms and store in different bins. So far, not all hospitals have been covered under the bio-medical waste treatment norms and the APSPCB has sent them notices to be part of the system before a set deadline. Presently, there are two common treatment units in Hyderabad for the disposal of Bio-medical waste.

Under the principle of users'/polluters' pay, the MCH has introduced the scheme of collecting user charges from the bulk garbage generators in the city. 1700 establishments including hotels, restaurants, function halls, markets, commercial complexes, and other generators of bulk garbage, have been identified and classified into 12 categories for levying user charges[6] varying from Rs. 1,200 to Rs. 1,800 per month. A separate truck collects the bulk garbage from these establishments and transports it to the power generation plant, where the garbage with high calorific value goes to the landfill site. The slaughterhouses, which are mostly outside the MCH limits, are also part of the system. They pay the concerned local authority to collect their waste.

MCH has a separate collection system for debris from construction sites, once again on the users' pay basis. The collected debris is dumped onto the low-lying sites. The contractors themselves may collect the debris from their construction sites and dump it on the vacant sites if the owners of the latter demand.

Zero Garbage—Vermi-Composting: Jubilee Hills Civic Exnora (JHCE)

In 1998, few housewives from the Jubilee hill society initiated the programme of collecting garbage form households and vermi-composting it into organic manure to achieve a 'Zero Garbage' situation. For this, they formed an Exnora of the residents in Jubilee Hills, Kaveri Hills, Prashasan Nagar and Navnirman Nagar. Total of 2,000 households are participating in this programme now. The Jubilee Hills Exnora experience is discussed at length below.

Prior to constructing a vermi-composting plant, the household waste collected was transferred to three collection points of the MCH in the Jubilee Hills area, from where it was taken to the dumping sites. Now, the garbage collected daily from each household between 7 am and 12 noon by a team of cycle rickshaw boys is taken to the community level vermi-composting unit operated by the Exnora. The vermi-composting plant is a small-scale unit utilizing 7 MT of solid waste per day. The MCH gives the land for constructing this plant. The private contractors cleaning the roads in the four societies of the Jubilee Hill Exnora also dump the waste collected from the collection bins on to this site.

This Exnora has 13 tricycles, 4 of them given by the MCH and 16 workers including a supervisor and a collector. Each worker gets paid Rs. 2,000 and the supervisor Rs. 2,500 per month. There is a phone and a tube-well on the composting site for the efficient.

The entry road of the vermi-composting plant has been concreted for the purpose of maintaining cleanliness. The garbage is dumped in a corner/ shed of the plant. Here too, the rag-pickers rummage through the garbage every day and take away the remaining recyclables. They pay Rs. 100 per week per head to the Exnora for it. The segregation of the waste by source at the household is still incomplete and hence this arrangement has been made by the Exnora. The remaining segregated garbage is then put into the pits of 3 feet depth. In here, the worms convert the waste into manure in 3 months period. The worms are harvested in a separate pit using cow dung. Once the manure is ready, it is sieved to take out the worms which are then put back in the pit. Finally, the fine manure is sold at Rs. 4 per kg. Large packs of 40 kg per bag are prepared for bulk purchasers.

The profit earned through these activities by the Exnora is used for upgrading the system and provide workers' benefits such as on-the-job clothing, giving them bonuses, and supporting their education. Boys called 'Street Beautifiers' are employed for part of the day on the vermi-composting site and for the rest of the day either engaged in studies or other part time jobs[7].

The Exnora had carried out awareness campaigns and held workshops with the residents to encourage them to segregate the waste at the source. However, this campaign has not yet yielded much result. Even the members of this Exnora think that garbage handling is the responsibility of the MCH and they would not want to do more about it than what they are doing now.

Recycling

Even though the tricycle puller himself collects recyclable waste from the households for sale in the market, some recyclable materials remaining in the garbage get dumped in the collection bins. The rag-pickers rummage through the waste and collect whatever is saleable to be taken to the intermediaries.

The garbage collected from the households and the garbage brought by the private contractor is dumped in a corner of the Exnora site, if there is an Exnora vermi-composting site nearby, or else is carted to the landfill site. Here too, the rag-pickers are employed; 35 rag-picker families deployed to pick paper and plastics and another 35 families to pick only the metals. The latter pick up the waste during the day, using a magnet, and burn the waste in the night. In the process they are prone to inhaling smoke and toxic gases emitted from the waste. Besides that, since all the rag-pickers live in huts near the landfill site, their working and living conditions become quite hazardous. They earn about Rs. 2,000—Rs. 2,500 per month[8].

A private company had given a proposal to the MCH to buy up the entire recyclable waste of the city collected up by the tricycle pullers. They were willing to give strong and washable bags to the tricycle pullers for the purpose. In return they wanted the MCH to guarantee them a right to the entire recyclable waste of the city. The MCH refused to agree to this condition of contract knowing full well that it would have not have full control of the waste collected as there were many collectors and rag-pickers involved in the process who likely had their own channels through which their 'product' could be sold. The company was not willing to give any extra incentives to the waste collectors to ensure that the latter would sell the waste only to the company. In essence, this company wanted the MCH to police the waste collected from the households while it took away all the profits (to the tune of Rs. 31.41 lakhs per day) from recycling the waste. If this contract had gone through, the company would have acquired monopoly control over the city's recyclable waste—certainly not a good governance practice. The present waste collection system is highly decentralised with employment generated for a large number of informal sector workers.

Transportation

The MCH has kept the responsibility of transporting the entire waste of the city. At the garbage collection point two types of bins, steel (also called dumper placer bins) or concrete, are provided. For the purposes of transportation of waste, the steel bins are mechanically lifted by the dumper placer vehicles and carted to the nearest transfer station or landfill site. Waste from the concrete and open bin is loaded manually by the private contractors' workers in the small tipper trucks or mini compactor trucks of the corporation, which come at a designated time, to be carted to the Transfer Station (TS). The steel (dumper placer) are provided at 1,900 locations and the concrete bins at remaining 1950 locations. About 55 per cent of the garbage is handled manually and 45 per cent mechanically at the collection centre at present. The MCH has 102 dumper placer trucks with a carrying capacity of 2.5 MT each, 45 big Tippers (10 MT), and 79 small Tipper trucks (3.5 MT).

In the entire transport system, the TS is of immense importance, as it is from here that the garbage is taken to the final landfill site or the solid waste processing plant. It may be mentioned here that in Hyderabad, some waste is carted to a power plant for energy generation. Hence, the destination of the waste collected depends on the type of waste and whether there is vermi-composting unit in the collection area. After the recycled waste is picked by the rag-pickers from the local collection centre and after the compostable waste is composted, the non-recyclable waste reaches the local collection centre to be picked up for transport to the nearest TS or a dumping site. The garbage is brought in small dumper placer trucks and tippers to the TS, where the waste is directly transferred into large 10 ton tipper trucks. These larger vehicles transport the solid waste to the landfill site or the waste processing plant for final disposal.

By this arrangement, the small dumper placer vehicles are not required to travel long distances to the landfill sites which are, on average, located 20 to 25 km away from the centre of the city. Time and costs thus saved by these small vehicles can then be optimally used for frequent garbage collection trips, avoiding any overflowing of the garbage at the collection points. The transport fuel saved achieves the environmental goal of reducing air pollution. The landfill site is also better managed because the use of large trucks reduces vehicular traffic here.

The city currently has three transfer stations which are managed by the MCH. One is Tank Bund TS in the centre of the city, Imlibaug TS and Yousufguda TS in Ameerpet in the west of the city. The fourth one, at Jiaguda is under construction and will be completed shortly. The final plan is to have one TS in each circle. In spite of this arrangement, the MCH is spending 60 per cent of its SWM budget on transportation.

Disposal

At present, the solid waste is disposed of at 2 landfill sites in operation, one at Autonagar (45 acres) and the other at Gandhamguda (20 acres). Both these sites are equipped with the computerized weighbridges and other infrastructure like roads. These are sanitary landfill sites, with proper liners, leachate drains, treatment facilities, environmental monitoring equipment, computerised weighbridges, network of roads, high mast lighting, pumps and proper fencing. Mechanical equipment like front-end loaders and bulldozers are being deployed for spreading the solid waste and capping it with debris/gravel layers. It is contemplated that these landfill sites would be used for some recreational purpose on reaching their full capacity. One such site has already been developed at the Golconda landfill site. The Autonagar site, which was in use until recently, is being closed due to protest from the residents and ensuing political pressure.

Hyderabad requires 300 acres of land for a sanitary landfill site that would last 15 years. Its estimated cost is Rs. 260 crores. One 108 acre site was identified for the purpose by the MCH near the present airport. However, the SWM Rules 2000 states that permission from the airports authority is required if the landfill site is to be located within 20 km distance of any airport. This rule is to avoid the problem of bird-hits that can be caused by the birds hovering over a landfill site. The new site location therefore has not yet been finalised. Environment Protection Training and Research Institute (EPTRI), Hyderabad has been appointed to identify vacant government land around Hyderabad city, which is suitable for locating sanitary landfill sites and for carrying out environmental impact assessment study on these sites[9]. The MCH and HUDA are trying to identify a suitable landfill site in the master plan area, so that they can introduce a development control rule notifying a green belt area of 500 m. width. This problem of identifying potential landfill sites is complex because wherever such a site is developed, the builders move in the area and develop area around it.

Cost of landfill site is around Rs. 10 million per acre and the operation of it requires around Rs. 60 million per year. Further, every 100 MT of waste processed can reduce the land requirements for a sanitary landfill site by 14 acres. That is why the MCH wants the 'waste to energy' programme to increase so that the land requirement would not increase.

Power Generation

At present, 400 MT of solid waste generated per day is being converted into power by a private company, named M/s. SELCO International, which was commissioned in December 1999. This plant's full capacity is to generate 6.6 MW of power from 700 MT of garbage every day. First, the solid waste is converted into fuel pallets, for which there is a palletisation plant with a capacity of 105 MT per day. The MCH is not taking any money from the power generation company.

One more power plant is slated to come up with a capacity of utilising 700 MT of solid waste and generate 11 MW of power. A MoU has already been signed with a firm named M/s RDF Power Projects and project construction is to commence soon. The arrangement is that this power would be sold strictly to AP Transco[10] and not to any third party.

Institutional Mechanism

In this whole new system of the MCH initiated waste management, the policy level decisions are taken by the management committee of the MCH, that includes the Mayor, the MCH commissioner, additional commissioners, and people's representatives. The Municipal Commissioner has the signing powers and the Additional Commissioner (health and sanitation) oversees the working of the system (Fig. 4.5). Underneath the Assistant Commissioner (AC—health and sanitation) is the Chief Medical Officer, who is assisted in this work by 7 Medical Officers of Health (one each for the 7 circles of the city) and a Deputy Executive Engineer. While the latter assists the AC to manage the entire system, the former takes care of all the day to day operations in the 52 sanitary wards. Each of the sanitary ward has one Sanitary Supervisor who oversees the work of 310 sanitary workers of the MCH and the work of the contractors and the DWCUA units. In addition to the regular health staff, 52 officers of MCH belonging to other branches are assigned the duty of inspecting and supervising one sanitary ward each and every day. This arrangement is called "One Officer—One Ward Programme".

The MCH has introduced an enforcement system of levying fines for littering or dumping debris on the streets. Eleven mobile enforcement squads are deployed in the city to constantly track the

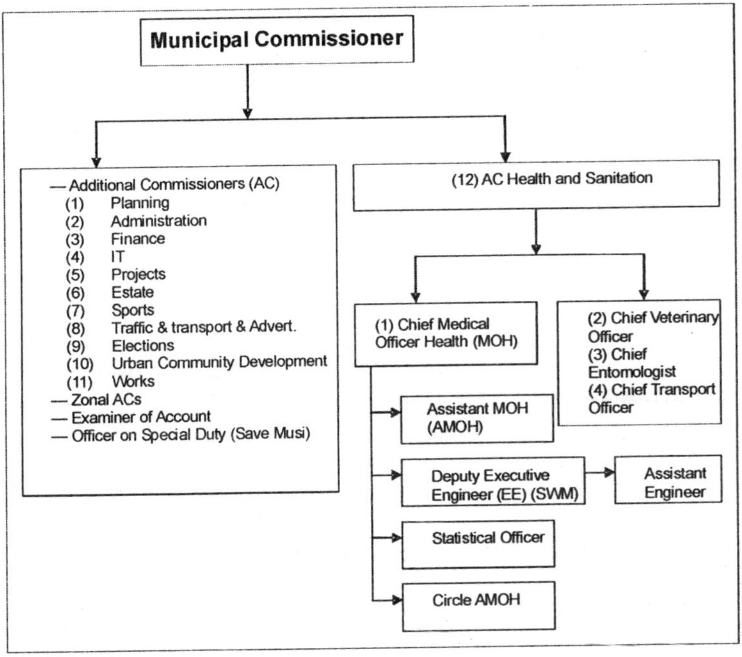

Fig. 4.5: Administrative Structure, MCH

offenders and fine them. In a period of 6 months in 2005, more than 5,000 offenders were caught and fined, the total fine collection coming to Rs. 4 million. Since this system has considerably reduced the road littering, it is being strengthened through regular monitoring.

All the officers, namely, the sanitary supervisors at the ward level, the traffic controllers of the TS, the vehicle inspectors at operational depots and the sanitary inspectors at the landfill sites have been given wireless walky-talky handsets and their offices have been give 'Base-station' sets to coordinate the work. These communication sets are the most useful for coordination of the primary collection of waste and movement of waste transport vehicles. Apart from this, all the officers are provided with cell phones to keep in constant touch with all other officers and be available to the public[11].

Also the ex-service people teams have been formed to assess the work of all the monitoring personnel and report in the writing to the MCH, particularly of any irregularities. Ten such teams have been formed by the MCH and each of them earn around Rs. 100,000 annually. Their work is to check the working of all the people involved and if there is any problem or irregularity then submit it in writing to the MCH. Thus, besides the RWA/ NC authorisation of work, the monitoring personnel also keep a check on private contractors' and DWCUAs' work. If the complaints of the latter's work increase, the MCH puts up an enquiry and takes necessary action such as keeping back the EMD, when required. However, so far such an action has rarely been taken.

Whenever, the official wants some information, give instructions or visit the site, the contractor is required to be available on the site. The officials don't talk to the labour because of various reasons, the important one being that the labour are the employees of the contractors and not of the corporation officials and hence they may not take directions or instructions from the latter.

Actors, Their Roles and Partnerships

Municipal Corportion of Hyderabad (MCH)

Rightly so, the MCH is in charge of the entire SWM system of the city. The MCH has divided the city into seven circles for this purpose and co-ordinates and manages the entire effort. It appoints the contractors for sweeping and garbage lifting and co-ordinates and monitors their activities. It supplies tricycles to the housing societies for door-to-door waste pick up. It provides land to the RWAs to set up the vermi-composting units. It forms NCs in the slums and provides salary for the garbage collection. The MCH is fully responsible for provision of bins at all the collection points, picking up the collected waste and transporting the same to the TS and from TS to the dumping/ landfill site or the power plant. To fulfil this entire responsibility that it has taken, the MCH has brought in different partners, as depicted in Fig. 4.6.

Contractors

The private contractors have been given the work of street sweeping and putting the waste into the collection bins, manually loading the collected waste into small tipper trucks or compactors of the MCH, applying disinfectants near waste bins, and day mopping. They manage the workers on the site, arrange the infrastructure required for cleaning the streets, pay the wages, instruct the supervisors overseeing cleaning operations and handling waste at the collection points. It is mandatory for the contractor to be present on the site during the working hours. The contractor is charged penalty, as per the contract stipulations, if the work contracted out is not done in time and to the satisfaction of the local residents. The contractor has to be accessible to the MCH officials all the time.

Partnership	*Tasks*
MCH—Contractors/ DWCUA	• MCH enters into a formal contract for a period of 7 months for the work of sweeping the city (Details in (Annexure A)).
MCH—RWAs & NCs	• Formal partnership • MCH provides equipments such as tricycles, gloves, dress, etc. • RWAs and the NCs manage the collection and composting of the waste.
MCH—Power Generators	• Formal Contractor • MCH provides waste and land at no cost to the power company and gives tax incentives to the power producing company • Power generating company supplies power to the state power distribution company.
RWAs & NCs—Volunteer	• The RWAs employ a volunteer who collects the waste from the households and transport it to the municipal bins. • The RWAs collect user fees from the households and pays the salary to the volunteer. • Volunteer's tricycle is maintained by the RWAs. • The RWAs monitors the tricycle puller and make sure that the households pay the user charges.
Rag-pickers—City	• This is an informal partnership. • This is unavoidable at the moment till the households fully segregate the waste. • The rag-picker collects the recyclable waste from the bin which reduces the quantity of waste to be managed by the MCH. • The rag pickers get some money out of it from the recycling industry.

Fig. 4.6: Partnerships

DWCUA Groups

DWCUA groups work like a contractor, except in so far as they function like a collective. The contractor works as an individual entrepreneur.

Households and Other Waste Generators

It is the duty of the households to get the garbage from their house to the municipal dustbins or the municipal collection point. They either do this task themselves or arrange someone on users-pay system to do so. Mostly, they engage recyclers, through their own RWA or NCs to do the task. Also, it is the duty of the households to monitor the work of the contractors/officers and report the irregularities to the concerned authority. Other wastes generators are largely bulk producers and they either dispose their garbage themselves or pay the MCH for the collection.

Resident Welfare Associations (RWAs)

In situations where the recyclable waste is locally composted, the RWAs arrange for the waste to be collected and transported to the vermi-composting site, compost it and then sell the recyclables and the composted manure to make some income. They charge the households for these tasks. They use the profit generated through the sale of compost and fees from the rag-pickers for the development of infrastructure and the workers' welfare. It needs to be mentioned that such RWAs working through the formation of a local Exnora are all of middle and high income groups and is limited to 2,000 households. RWA carries out awareness about the entire waste management system and also motivates households to segregate waste.

Volimory Garbage Disposal Scheme (VGDS)

A VGDS worker from a given community collects waste daily from the households for a monthly payment, as per the schedule decided by the housing societies. The collected waste has to be dumped by him in the nearest municipal dustbin or collection point. The recyclable waste is removed by him and sold to make addition income. Lastly, maintaining the tricycle used for collecting the waste is his responsibility.

Rag-pickers

Rag-pickers work at two levels in the MCH area, one at the local collection points and the vermi-composting sites wherever they are and the two at the landfill sites. They rummage through the waste to take away the recyclable waste to earn income. By this the waste to be dumped on landfill site reduces thereby reducing the land requirements for waste disposal.

Budget for SWM

Solid Waste Management is a very important activity for the MCH. This gets reflected in budgetary expenditures made by the MCH on this head. For example, the MCH has allocated Rs. 856.49 million for SWM; Rs. 735.49 million on the revenue head and Rs. 131.00 on the capital head in the budgeted expenditures (BE) of year 2005-06 (Table 4.6). This means that of the BE of this year, 10.9 per cent would be utilised for the SWM. On the revenue account, 17.0 per cent of the expenditures would be on the SWM, as compared to 11.6 per cent to be used for maintenance of roads and pavements.

Table 4.6: BE of Some Important Departments, MCH (Year 2005-06)

(Rs. Million)

Code	Department	Revenue Expenditure	Capital Expenditure	Total Expenditure	% to total MCH Expenditure
07	Budget and Accounts	1,454.75	0.00	1,454.75	18.5
35	Roads and Pavements	495.00	647.00	1,142.00	14.5
24	Solid Waste Management	725.49	131.00	856.49	10.9
34	Bridges and Flyovers	0.44	754.00	754.44	9.6
	Total MCH	4,250.86	3,629.20	7,880.06	—

Source: MCH (2005).

While 20.8 per cent of the capital budget is to be utilised for bridges and flyovers, just 9.6 per cent of it is to be used on SWM. The four departments mentioned in Table 4.6, get 50 per cent of the total budgetary allocations of the MCH, and importantly, the SWM is one of the four departments that gets high budgetary allocations. In terms of total budgetary allocation, SWM department is at third position, which indicates the importance the MCH gives to keeping the city clean. Further, the MCH has introduced a system of self-assessment for property taxes. The MCH is encouraging the citizens to pay the property tax so that 20 per cent of the taxes thus collected could be utilised for urban community development services, mainly in the slum development programme.

In the SWM department more than half the amount goes towards the salaries of the department staff (Table 4.7). Since the number of conservancy staff has been reduced and the responsibilities passed on to the contractors, whose staff would be working at one-fourth to one-third the salary of the municipal conservancy staff, the wage component is just over half the department budget. It would have been even higher if the sweeping and collection was not privatised. Another 40 to 50 per cent of the departmental expenses is on the operation activities. Over time there has been a gradual increase in the amount of expenditures on each of the heads of the SWM department.

Table 4.7: Revenue Expenditure by MCH Over Time

(Rs. million)

No.	Solid Waste Management Department	Actual expenditure 2003-04	Revised estimate 2004-05	Budget estimate 2005-06
A	Personnel Cost	396.11	399.53	406.84
	—Salaries	395.97	399.22	406.53
	—Benefits and other allowance	0.13	0.31	0.31
B	Operational Expenses	298.43	315.25	315.20
	Expenses on unclaimed dead bodies	0.06	0.25	0.25
	Sanitation/ conservancy expenses	287.88	298.00	300.00
	Maintenance of urinals and toilets	4.02	5.00	5.00
	Intensive/ special sanitation including that for fairs & festivals	0.25	0.50	0.50
	Conservancy tools	2.71	3.00	3.00
	Hiring charges for vehicles	3.11	5.00	5.00
	Hiring charges for equipments/machinery	0.40	3.50	1.50
C	Programme Expenses	1.29	2.50	2.30
	Environmental awareness programme	0.58	0.50	0.50
	VGDS	0.71	2.00	1.80
D	Administrative expenses	0.69	1.10	1.10
	Final total	696.52	718.38	725.49

Source: MCH (2005).

On the revenue collection side, in the year 2005-06, Rs. 18.16 million have been proposed to be collected by the SWM department, of which Rs. 8 million have been estimated to be collected in the form of penalty from the contractors (Table 4.8). Another Rs. 5 million would be collected as bulk garbage collection charges. On all other heads, current revenue collection is minuscule. On some of the heads such as sale of garbage and rubbish and charges for pay and use of toilets, collections can be improved. So can the collection charges for bulk garbage could be increased. These charges have already increased from 2003-04 to 2005-06. There is therefore a move in the direction of generating revenues from SWM activities.

Table 4.8: Revenue Receipts by the MCH, Year 2005-06

(in Rs. Lakhs)

No	Solid Waste Management Department	Actual expenditure 2003-04	Revised estimate 2004-05	Budget estimate 2005-06
1.	Registration of contractors	—	0.50	0.50
2.	Bulk garbage collection charges	19.48	50.00	50.00
3.	Special sanitation charges	28.11	30.00	30.00
4.	Charges for pay and use toilets	—	1.00	1.00
5.	Administrative charges for public nuisances	0.01	1.00	1.00
6.	Penalty on contractors	57.59	80.00	80.00
7.	Cost of tender schedules	0.01	5.00	5.00
8.	Sale of garbage	0.01	1.00	1.00
9.	Sale of rubbish	—	0.55	0.55
10.	Security deposit forfeited	—	12.50	12.50
	Total	105.21	181.55	181.55

Source: MCH (2005).

Table 4.9: Comparative Costs

	Area Covered (Sq km)	Workers Engaged	Street Sweeping cost (Rs million)	Qty. of MSW (MT/day)	Transportation Cost of MSW (Rs. Million)	Total Cost (Rs million)
MCH	42.5 (25%)	2200 (52/sq km)	428.3 (10 m/sq km) per annum	1500	111.0 (Rs 206/MT) per day	53.84 (12.6 m/sq km) per annum
Private	127.5 (75%)	6000 (47/sq km)	172.6 (1.3 m/sq km) per annum	700	54.2 (Rs 213/MT) per day	22.68 (1.8 m/sq km)/annum
Total	170	8200	600.9	2200	162.2	765.2

Source: MCH Hand out.

Costs

The transportation cost for solid waste in the MCH is Rs. 105 per MT, which is significantly lower than other municipal corporations in the likes of Chennai (Rs. 900 per MT) or Delhi (Rs. 10,000 per MT) (ASCI 2005). This is because the MCH has its own in-house fleet for transport and uses the contractors' fleet whenever necessary. In street sweeping, the MCH's annual cost works out to Rs. 10 million per sq km (Table 4.9), whereas that of private contractor works out to Rs. 1.3 million per sq km. On the whole, for the entire operation, that MCH annual costs work out to Rs. 12.6 million per sq km per year whereas that of a private contractor works out to Rs. 1.8 million for the same. Private contractors have higher costs as compared to MCH only for waste transportation. For all other activities, they have cost advantage over the MCH.

Achievements

The Hyderabad case study indicates that if the ULB itself takes the leadership build multiple partnerships, it can pursue the goal of reuse and recycle the city waste and thereby reduce the overall waste generated in the city. The goal then does not remain of just keeping the city clean but also leads to developing a model of a sustainable city. The city government has not sought to outsource its responsibility and stands firmly involved with planning, managing and monitoring all the processes of the SWM system. This is a quite refreshing approach among the rhetoric of downsizing local government by outsourcing all its responsibilities and privatizing public utilities or passing it on to what is called 'civil society organisations'. In urban policy discussions, it has almost become a taboo to say that the ULB need to be at the centre of all urban service delivery! This case also shows that both private and community sectors can be involved for their appropriate roles without the public sector abdicating its own responsibility.

This study also shows that ULBs can devote a noticeable proportion of its budget towards SWM and if done correctly it yields results. With financial contribution from the community sector, the small amounts available with the ULBs for all its functions can go a long way in improving services. This case also indicates the right type of incentives provided to the private and community sector to get their participation in the SWM. Small efforts and small steps have to be taken to improve the situation, and these are more cost-effective than going for expensive high-tech solutions to an activity such as SWM. In the MCH's case, it is able to manage the entire city garbage through its innovative system, inspite of its regular sizing down over the last two decades.

The city is also able to charge the bulk garbage generators for collecting and transporting their wastes on users' pay basis. We do not yet know whether the MCH is able to recover the entire costs of their service and if not how much more could be charged to the bulk generators to be able to do so.

This programme of 'Litter-free Hyderabad', that involves vermi-composting and collection of recyclable wastes, has enabled the MCH to reduce transportation of 500 MT of garbage per day and by that has saved transportation cost of Rs. 40 million per annum (of Rs. 120 million annual cost of the same) and reduced the need and the cost for development of landfill sites. When done through Exnora, it has led to cleaner local surroundings, more hygienic waste disposal, generation of organic manure and increased civic consciousness.

The Exnora approach and involvement of DWCUA as private contractors in SWM has added a livelihood dimension to the waste management activity in the city. Livelihood generation is an important aspect of a sustainable city model in the developing countries.

Independent monitoring and reporting system both from wards and the landfill sites has improved the efficiency of the MCH. Monitoring and evaluation by the RWAs and the NCs ensure proper execution of the works by the contractors. Regular surprise inspections by independent external agencies also ensure proper attendance and improved work culture among the contract agencies. A stringent fines regime is put in place to deal with possible malpractices by the private contractors, such as quality of sweeping, shortfall in weighing of garbage and for terminating and black listing non-performing contract agencies and littering of the streets by the citizens. All these factors have given the MCH a positive image nationally and goodwill among the citizens of Hyderabad. For example, 'Cleanliness is next to Godliness'; the main caption of the MCH is being used by other municipal corporations on their websites. However, this case study raises some important concerns in this new system of SWM.

Concerns

There are many concerns as well that emerge from this case study, some of which can be addressed and some not. MCH has been able to devote such high proportion of its budget to SWM because it does not take care of the water supply and sewerage systems, which is the responsibility of the HMWSSB, a special parastatal board that is directly funded by the state government. In states where the city government have water supply and sanitation as their mandatory responsibility, funds are not easily found for the SWM. This is a systemic problem, which cannot be answered by one city alone. Other specific concerns emerging from this city case study are discussed next.

Centralised Governance Structure

The entire system of SWM in MCH is centralised, through the administrative machinery of the ULB. The ward committees have not found any role in this system, although it can play an important function in decentralised waste management in a metro city as large as Hyderabad.

Lack of Participation of Households

At this juncture, the MCH is more active in the entire system of SWM than the households. This is how it should be. Nevertheless, households still need to take the responsibility of segregating the waste at the source and then vermi-composting it to assist the MCH to achieve the goal of sustainable city. Currently, only 2000 households are engaged in composting the waste through their RWAs. Further, the RWAs are predominantly middle and higher income group based. There is a big scope and need to increase the number of RWAs and NCs in the slum areas of the city. While it is easier for the MCH to fulfil its responsibility, efficiently organising communities for a collective task is difficult. The community-level has, as of now, become a weak link in the whole system. Formation of RWAs and NCs, and their functioning on a sustainable basis is a process that needs to be given time. At the same time, there is also a question as to whether that would be feasible given the fragmented social structure of Indian cites.

Non-functioning of Neighbourhood Committees

Although NCs in slums are given tricycles and one-year salary for the tricycle puller, after one year they are not able to maintain it. Most of the households do not want to pay for the services they use.

Source Segregation

There is no source segregation of waste in the MCH area. This increases the numbers of rag-pickers and also the quantity of waste to be transported. If the waste is segregated at source and vermi-composting units were set up then some of the burden may be removed and costs can be reduced as per the SWM rules 2000. Concerted campaigns among the citizens are required by the city government, NGOs and RWAs/NCs to segregate waste at source.

Collection

The collection system works efficiently in the middle and higher income groups but not in lower income groups especially in the slums.

Labour Conditions' Violation

From the discussions with the contract workers it came out that not all of them were getting the wages specified in contractors' contract with the MCH, indicating some extent of corruption. In such cases, the labour sign a receipt of the contract specified amount but get lesser amount than they have signed for. There were some instances of labour working longer time than that specified in the contract. The MCH officials refrained from interfering in such instances, considering this to be a matter between the contractor and his/her labour. Further, workers wages have to be paid through bank and for which the worker has to open an account, which they do not want to. They therefore agree to whatever payment the contractor makes.

Contracting

To escape the labour law provision that the workers employed for more than 210 days have to be treated as a permanent worker and that the public body is the principal employer even in the case of workers employed by the contractor working for it, the MCH awards the contract for only 7 months. As a result, large numbers of contractors are engaged in the system. The selection of the contractor is through a lottery system, which reduces the possibility for the contractor to get the work again. Last time, for 163 units around 8,000 applications were received by the MCH. The contractor therefore does not have an incentive to invest in this activity for long.

DWCUA Monopoly

DWCUA group creates a monopoly, as they are under the scheme of state government. The MCH is unable to take strict action against them if they do not fulfil their contractual obligations. Also, their contract, unlike private contractors, is for long period usually for 3 years and hence they are not covered under the system of accountability developed by the MCH.

Choice of Garbage Treatment System

Comparing the two waste treatment systems of vermi-composting and power generation, the former seems to be the better option. Still, the MCH wants power generation to be initiated. This is because of the low response of the people in vermi-composting. Power generation is a very quick and efficient system given that the calorific value of the waste of the city is good. Further, in case of power generation, the MCH has to deal with only one actor, the power generating company, whereas, for successful vermi-composting at the city level, number of actors at the level of RWAs and the NCs have to be partnered with, which may not be a guaranteed system of waste recycling. Other studies (Baud and Post 2003) have also found that the ULBs have found it easier to partner with the formal sectors such as a private sector company than informal actors and community organisations, which is also found in this case study.

Continuance of Rag-pickers

Most of the rag-pickers in the city are teenage boys and children who earn their living through scavenging. This is not a very desirable situation. Over time, it will be necessary to phase out the job of the rag-pickers; the city finding better segregation of waste and recycling system on one hand and the rag-pickers finding better jobs than what they are doing today.

Exposure to Environmental Hazards

The rag-pickers particularly are exposed to injuries while rummaging the waste. Those engaged on the landfill site and burning the waste are exposed to noxious fumes. Most women living on the landfill sites reported respiratory ailments.

Women

In all this system, women have benefited tremendously as most of the work is being given to them. But on the other side the question arises:

Why are women preferred over men for such jobs? Nonetheless, through the NCs in slums and the DWCUA groups, their financial status has improved somewhat.

Landfill Site Location

The problem of dumping yards is a major restriction in managing the waste. No body wants the landfill sites near their own residences—not in my back yard (NIMBY) and hence the MCH and even the HUDA is finding difficulty in identifying a landfill site that would last long.

NOTES

1. Acknowledgements due to Mr Rajiv Babu, Dy. Executive Engineer, SWM, MCH; Mr. Sharma, Engineer, EPTRI; Mr. Mustafa in-charge Tankbund transfer Station; Mrs. Indira Lingam, Jubilee Hills Civic Exnora; and Mr. Sunil Reddy, Sanitation Contractor, Hyderabad, for their valuable time for discussions and giving relevant materials. Our gratefulness due to the MCH for all the official documents such as development plan, city budgets and so on. Lastly, our friend Dr. C. Ramachandraiah of Centre for Economic and Social Studies (CESS) for logistic help.
2. The water supply is metered. Sewerage system covers about 70 per cent of the MCH area (as per discussion with Mr. RajivBabu, Deputy Executive Engineer).
3. MCH handout on solid waste management.
4. Under DWCUA scheme, women have made SHGs.
5. As per discussion with Dy Executive Engineer, SWM, MCH.
6. As per discussion with Mr Rajiv Babu, Dy. Executive engineer, SWM, MCH.
7. As per discussion with Mrs. Indira Lingam, JHCE.
8. As per discussion on the land fill site with the in-charge.
9. As per discussion with Mr. Sharma, engineer, EPTRI.
10. AP Transco came into existence as on 1.02.1999 as a consequence of the restructuring of the AP State Electricity Board to look after transmission and sale of electricity in the state of Andhra Pradesh (*Source*: http: //www.aptranscorp.com/).
11. As per discussion with the Dy Executive Engineer, SWM.

ANNEXURE A
CONTRACT BETWEEN THE MCH
AND PRIVATE CONTRACTOR

Scheme of Clean Hyderabad–2005 Works Package (Winter and Summer months)

Municipal Corporation of Hyderabad in Partnership with Resident Welfare Associations, and Neighbourhood Committees of slums introduced seasonal works for Winter and Summer under Clean Hyderabad–2005 works package with active citizenry participation in the twin cities of Hyderabad and Secunderabad. The scheme aims at involving the citizens in their respective areas for effective monitoring and day-to-day supervision in all the related activities of the proposed works and achieving the clean city. The works are purely seasonal in nature relating to Winter and Summer months and the agencies proposed to be selected for execution of works are free to carryout the works by engaging men or equipments and machines subject to the prescribed terms and conditions, work norms and the outputs. The agencies proposed to be selected shall themselves be personally present in the respective areas allotted to them for carrying out the works. No benamees and proxies shall be allowed on their behalf. Cases of such nature will be severely dealt with by cancellation of work, forfeiture of the security deposits and blacklisting of such agencies by debarring from works programme of any department under MCH.

Day Work

Each day work consists of one or more Cleaning units and Transport units.

Cleaning Unit

One cleaning unit will have a road length of 9 kms including streets, lanes/ bylanes etc., in the area of the unit. The cleaning work includes roads, central media, road dividers, footpaths, table drains and all the open places in the cleaning unit area.

Nature of work

- Surface cleaning of roads, streets including lanes and by lanes.
- Scraping and removal of road silt and sand.
- Cutting and removal of road side shrubs, tree stems, dry leaves etc., on the roads and foot paths.
- Collection and removal of litters, animal wastes, plastics, rubbles and all other wastes.
- Collection and removal of debris heaps, cutting wastes etc., from footpaths, road margins and garbage collection points.
- Cleaning of water stagnation points on the roads due to rains, leakages, breakdowns of pipes and man holes. (list not exhaustive)
- Disinfectant mixture of Bleaching powder + Lime shall be sprinkled at these points after lifting the garbage.

Time of Operations

- 6 a.m. to 11.00 a.m. (First Shift)
- 12.00 Noon to 3.00 p.m. (Second Shift)

Scale of Tools and Equipment

The agency shall carry out the work by engaging men and equipments or machines. The agency in case engages men shall provide the following tools and implements for each cleaning unit: (1) Three Wheel Barrows (2) Four Spades (3) Four Iron Baskets (4) Brooms @2 kg per worker per month (5) Scrapping Rakes one No. per worker (6) Date Baskets one No. per worker (7) One Crow Bar (8) One Axe.

Work Norms and Conditions

- The agency shall prepare and furnish a detailed cleaning plan for the first shift and second shift for each cleaning unit in the work area immediately after selection and before the allotment of work order.
- The work shall be carried out on all the days including Sundays and Public holidays.
- Attendance points should be fixed for verifying attendance

at 6 a.m., 11.00 a.m. for first shift and 12.00 Noon and 3.00 P.M. for the second shift.

- A full time supervisor for each cleaning unit has to be engaged and shall be physically present at the work site in uniform clothes (for easy identification) from 6.00 a.m. to 11.00 a.m. and 12.00 noon to 3.00 p.m. every day.
- The agency shall be personally present in the work site everyday from 6.00 a.m. to 11.00 a.m. and 12.00 Noon to 3.00 p.m. along with cell phone and execute the work.
- No benamees and proxies will be allowed. The agency should also keep a Photo identity card (for easy identification by the inspecting officer).
- The agency should submit daily progress report to the authorized officer of the MCH in the prescribed format after satisfactory completion of the work.
- The agency shall be prepared to accept additional responsibilities for attending the emergencies as directed by the MCH in keeping the area clean.

Transport Unit

Localities including streets, lanes by lanes, open places and collection points generating, 7 MTs. of solid waste per day constitutes one transport unit.

Nature of Work

- The Agency shall engage one tipper with capacity of not less than 3.5 MTs with 4 workers on it.
- The tipper shall make two trips per day. Single tipper for 7 Tons by one trip is not acceptable in view of the need for proper sanitation through out the day.
- All the Municipal solid waste including shrubs, tree stems, dry leaves, silt and sand, small heaps of debris from the garbage collection points like round RCC bins, garbage houses and open points shall be lifted. At the collection points it should always be kept clean by scraping and trimming.
- The vehicle loaded with Solid Waste should be covered with tarpaulin/plastic nets to avoid falling on the roads during movement.
- The vehicle has to proceed to the notified Transfer Station and unload there.

Timing

- 6 a.m. to 2 p.m.

Scale of Tools and Equipment

(i) 4 Danthis, (ii) 4 Iron Baskets, (iii) 1 Spade, (iv) 1 Crow Bar, (v) 1 Bucket, and (vi) 3 Mugs.

Work Norms and Conditions

- The Agency shall have to furnish a detailed lifting plan for each trip.
- The work shall be carried out on all the days including Sundays and Public holidays.
- The Agency should be personally present in the work site to ensure completion of work satisfactorily everyday.
- The Supervisor shall monitor and supervise the lifting operations.
- The Agency should furnish daily progress report to the authorised officer of the MCH on the performance.
- The Agency should also be prepared to accept additional responsibility for attending emergencies by engaging additional vehicles.

The Scheme for Night Work

The work package consists of cleaning units and lifting units.

Cleaning Unit

A total road length of 3.5 to 4 kms of main roads makes one cleaning unit. The roads, footpaths, table drains, central media, road dividers and open places on the road side are part of the cleaning unit.

Nature of Work

- Removal of litters plastic rubble and other wastes.
- Scraping of road silt and sand and removal of the same from the entire road, dividers, road margins etc.,
- Surface cleaning of the entire road stretch including foot paths, road margins etc.,
- Removal of dust and other wastes from inside the tree guards on the foot-path.

- Removal of roadside shrubs, tree stems, dry leaves etc., on the roads and footpaths.
- Removal of posters and wall writings from notified sites.
- Removal of heaps of debris on the roads, foot-paths and the garbage collection points.
- Attending to water stagnation points due to rains etc., on the roads.
- Mopping up of work by litter collection, scrapping and collection of sand, and silt and removal from foot-paths and road margins in the following fore noon. (list not exhaustive)
- Disinfectant mixture of Bleaching powder + Lime shall be sprinkled at these points after lifting the garbage.

Timings

- 11 p.m. to 5 a.m. (in the main cleaning work)
- 7 a.m. to 3 p.m. (Mopping operation and maintenance of litter free roads)

Scale of Tools and Equipment

The work may be done by men or machines. If the contractor chooses to engage labour the following Tools and Implements shall be used: (1) Three Wheel Barrows (2) Three Spades (3) Three Iron Baskets (4) Brooms—3 kgs per worker per month (5) Scrapping Rakes One per worker (6) Date Baskets One per worker (7) One Crow Bar (8) One Axe.

Work Norms and Conditions

- The Agency shall have to furnish a detailed cleaning plan for the main cleaning operation in the night and for mopping up operation in the forenoon separately.
- The work shall be carried out on all the days including Sundays and Public holidays.
- Attendance points should be fixed separately as
  - Reporting attendance point(s)—11 a.m.
  - Departure attendance point(s)—5 a.m.
- A full time supervisor for each unit has to be engaged by the Agency. He shall be available in uniform clothes for easy identification by the inspecting officers.

- The Agency should be personally present in the field with a cell phone to ensure completion of work satisfactorily. The agency should carry a photo identity card (for easy identification by the inspecting officers) in the work site.
- The Agency should furnish daily progress report to the authorized officer of the MCH on the performance.

(The above are indicative but not exhaustive)

Transport Unit

Continuous stretches of main roads including footpaths, table drain, central media, road dividers, and open places on the road sides generating 7 MTs. of solid waste per night constitute one transport unit.

Nature of Work

- The Agency shall engage one Tipper with capacity of not less than 3.5 MTs. with 4 workers on it.
- The vehicle shall make two trips per night.
- All the Municipal solid waste from the garbage collection points like round RCC bins, garbage houses and the wastes around steel dumper bins in the area/open points shall be lifted.
- All the sweepings, scraped silt and sand resulting for the cleaning work unit shall be lifted every night.
- The vehicle has to proceed to the notified land fill site and unload there.

Timing

- 11 p.m. to 5 a.m.

Scale of Tools and Equipment (Total)

(i) 4 Danthis, (ii) 4 Iron Baskets, (iii) 1 Spade, (iv) 1 Crow Bar, (v) 1 Bucket and (vi) 3 Mugs

Work Norms and Conditions

- The Agency shall have to furnish a lifting plan and route for each trip.
- The work shall be carried out on all the nights including Sundays and Public holidays.

- The Agency should be personally present in the work site to ensure completion of work satisfactorily everyday.
- The Supervisor shall monitor and supervise the lifting operations. He shall be available in uniform clothes for easy identification by the inspecting officers.
- The Agency should furnish daily progress report to the authorized officer of the MCH on the performance.
- The Agency should be prepared to accept additional responsibilities to attend the emergencies like engagement of workers and transport vehicles.

  (The above are not exhaustive)

Security Deposit Towards Performance Guarantee

The individuals/ firms/ registered labour cooperative societies willing to apply for pre-qualification shall pay a security deposit of Rs. 1 lakh (Rupees One Lakh Only) along with the application. This applies to the existing contractors as well as the new entrants equally. The existing contractors should also pay a security deposit of Rs. 1 Lakh (Rupees One Lakh Only) if they apply for the new package. The security deposits of the contractors held by the MCH under the existing works package cannot be considered as security deposits for the new package.

After selection the agency shall deposit one month value of the work towards performance guarantee to receive the work orders. The security deposit amount of Rs.1.00 Lakh paid along with the application will be adjusted towards the performance guarantee amount and the agency will have to remit the balance amount only to receive the work order.

Applications received from the agencies without security deposits will be summarily rejected.

The security deposit amount of Rs. 1 Lakh (Rupees One Lakh Only) shall be paid by Demand Draft drawn on any Scheduled Commercial Bank in favour of The Commissioner, Municipal Corporation of Hyderabad. Cheques will not be accepted.

The performance guarantee of work value will be released on production of clearance certificate to the extent from E.P.F. Organization and E.S.I. Corporation for the entire period along with the no due certificate issued by the concerned AMOH. Further the first month bill will be released only after code numbers is taken from E.P.F. Organization and E.S.I. Corporation.

Unit Rates for Cleaning and Transport Units

For Day Works

Sl. No.	*Unit*	*Circles*	*Amount per Month in Rs.*
1.	(One) Cleaning Unit	All Circle Offices	Rs. 56,488
2.	(One) Transport Unit	I, II, III, V, VI & VII	Rs. 38,796
		IV	Rs. 41,660

For Night Works

Sl. No.	*Unit*	*Circles*	*Amount per Month in Rs.*
1.	(One) Cleaning Unit	All Circle Offices	Rs.80,054
2.	(One) Transport Unit	I, II, III, V, VI & VII	Rs.38,796
		IV	Rs. 41,660

The payment of wages to the workers employed by the contractors shall be by way of crossed cheque/demand draft and in accordance with the rates notified by the Commissioner of labour in the Andhra Pradesh Gazette Part—I Extraordinary No.524 dt.26.11.2002. A certificate that the wages have been paid in accordance with the said notification should invariably furnished by the contractor every month to MCH.

Fines Regime

Since the work is of very important nature for keeping the city clean and protect the health of the citizens, a detailed Fines regime is introduced by the MCH for regulating the works. The Agency if fails to carryout the work as per the norms and conditions laid down by the MCH, the authorized officers will impose fines and their recurrence may lead to termination of the contract and forfeiture of the performance guarantee deposits.

Commissioner Municipal Corporation of Hyderabad.

5

Modernizing Municipal Management
A Case Study of Kolkata Municipal Corporation

NATALIE KEHLE, DARSHINI MAHADEVIA and JEANNE WOLFE

City	Kolkata
Urban Local Body	Kolkata Municipal Corporation
Population — KMC, 2001	4.58 million
Population — Kolktata Urban Agglomeration (KUA), 2001	13.22 million
Population — Kolkata Metropolitan Area (KMA), 2001	14.50 million
Population Growth Rate, 1991-2001—KMC	0.20 % per annum
Population Growth Rate, 1991-2001—KUA	1.93 % per annum
Population Growth Rate, 1991-2001—KMA	2.03 % per annum
Other ULBs important in KIMA	Kolkata Municipal Corporation (KMC) Howrah Municipal Corporation (HMC) Chandannagore Corporation
Local Authorities in KMA	KMC, HMC, Chandan-nagore, 38 municipalities, 23 panchayat samities, 58 census towns and 16 out-growths including a number of rural mouzas
Number of Wards in KMC	141 grouped into 15 Boroughs
Slum (*bustee*[1]) population in KMC	1.5 million (33% of the population) (KMC 2005b)

Introduction

Kolkata is located on a series of wetlands in the delta of the river Ganga. It is situated on the east bank of the Hugli River, a distributary of the Ganga, which reaches the Bay of Bengal about 180 km downstream. On the opposite bank lies Howrah. Kolkata is the largest urban centre in the Eastern Zone of India and is located less than 100 km from the Bangladeshi border. Kolkata Municipal Corporation (KMC) is ranked 7th largest in India covering an area of 187 square km with a population of more than 4.6 million. Kolkata Urban Agglomeration (KUA), with a population of 13.22 million in 2001, is the second largest after Mumbai Urban Agglomeration. Once referred to as "the jewel in the crown of the British Raj", today, the city is an industrial, cultural and historical hub, and is the home of many educational institutions, museums, and festivals (KMC Website 2005).

Kolkata makes an interesting case for study because in the state in which the city is located (West Bengal), the Left Front Coalition has been in continuous power since 1977, the Communist Party of India (Marxist) [CPI (M)] being the leading party in association with the Forward Bloc (FB), Communist Party of India (CPI), and the Revolutionary Socialist Party of India (RSP). In the early years the left front state government focussed its economic policy on small and medium sized farms and businesses in the rural areas, a move which is thought to have contributed to the economic stagnation of Kolkata. When the economic reforms were introduced in India in 1991 by the central government, the Government of West Bengal (GoWB) took the opportunity to reform its economic policy to become more urban focussed and thus provide a more open climate for private sector investment.

In order to promote Kolkata and West Bengal as desirable in the new economic regime, the state first changed its industrial policy. It issued a policy statement on Environmental Protection and Conservation of Natural Resources, which aimed to reconcile pro-development and pro-environmental goals. This led to an extensive research program on the overwhelming environmental problems in Kolkata, identifying improvements needed to make it a better place to live, work and importantly, to invest. In other words, to promote Kolkata on the national and international stage as a centre for growth, the city's 'image' had to be improved, including social welfare such as poverty alleviation and raising the standard of living for urban poor, in order to promote at least a sense of social justice (Dutt 2000). Today, both the local and state governments plan to make Kolkata an Information Technology (IT) hub. Part of this plan is to lower property taxes, lower trade license fees and lower building permit fees for IT and Small Scale Industries (SSI) thereby providing a financial incentive to establish businesses in the area (KMC 2005a).

Commitment to Decentralization

The GoWB has a long history of passing down responsibilities to local authorities, particularly in the rural areas. With limited success during its initial inception, West Bengal nonetheless was able to pave the way for urban decentralization policies, years before the GoI passed the 74th Constitutional Amendment *Act* (CAA) in the early 1990s. Most notable are the Constitution of Borough Committees in the Kolkata Municipal Corporation (KMC) Act 1980, and the rapid constitution of Ward Committees in 1994 following the passage of the 74th CA, the first state to do so. Not only that, but Committees were created in each municipal ward, and have since been running for almost a decade. The composition, nomination and powers of the Ward Committees are clearly laid out in the West Bengal Municipal (Ward Committee) Rules 2001, later amended in 2003. This level of commitment is rarely seen in India (*see Annexure* A).

The decentralization commitments from the state and local governments are worthy of further comment. In 1983, the newly created Kolkata Municipal Corporation (KMC) Act 1980, was a significant step towards decentralization. The Act constituted three important elements that provided more decision making powers to the municipality, namely: a Mayor-in-Council (MIC) system, Borough Committees, and finally, enlarged obligatory responsibilities for the local government.

The first element of decentralization is the MIC system, unique at the time for municipal administration in India. This system gives more decision-making power to the Municipal body rather than the State government. This was introduced in the amended KMC Act 1980 in 1984 that aimed at making the Corporation more efficient and effective. In this system, the decisional power rests with the elected

Mayor and Council, who thus have a superior position to the Municipal Commissioner, an appointed state bureaucrat and traditionally the most powerful administrator in a city (Jalal 2005: 103). Today, the Municipal Commissioner is the principal executive officer in the Corporation but is subject to the supervision and control of the Mayor (KMC Act 1980).

When first constituted, Kolkata (KMC) was divided into 100 wards, each electing a City councillor. The state also appoints council members with expert professional knowledge, but who have no voting rights within the Corporation. The Mayor and Chairperson are elected by the councillors at their first meeting. The MIC consists of the Mayor, Deputy Mayor, and no more than 10 members of the Corporation (Deputy Mayor and members are among the elected Councillors, nominated by the Mayor), and function as a cabinet. Thus, there are three levels of municipal authority in this system: the Corporation, the MIC, and the Mayor.

The second element in the KMC Act 1980 which promotes decentralization is the creation of a two-tier governing system which promotes decentralized service delivery. The wards are grouped into Boroughs, each organized with a Borough Committee. Each consists of elected Ward Councillors (from the included wards) and a chairperson (elected from those ward councillors). The Borough Committees are subject to the general supervision of the MIC, and look after functions such as water supply, drainage, collection and removal of solid waste, disinfection and health immunization, *bustee* services, lighting, repairs of certain categories of roads, and maintenance of parks, drains and gullies. Borough Committees can override decisions made by the Corporation on some matters, related to service delivery. The government remains fully committed to the functioning of the Borough Committees and recently has developed the Integrated Borough Scheme, which ensures their empowerment and also aims at developing their capacity to undertake proper local planning processes (KMC 2005a and KMC website 2005).

The third element in the KMC Act 1980, outlines in detail obligatory and discretionary functions of the KMC, passing down much of the planning, developmental and service responsibilities to the Borough Committees. To appreciate this, of the eighteen municipal functions named in the 74th CAA (which the GoI had left to the discretion of the states to delegate which ones were appropriate to implement in municipal law), twelve had already been incorporated into the KMC Act 1980.

From West Bengal's already strong commitment to decentralization, the implementation of the 74th Constitutional Amendment Act 1992 is seen as a success as compared to many other states. The implementation of the act introduced some other elements that furthered the decentralization process in the state. The first is the provision for appropriate and adequate representation of all elements of society. The amendment provided for the reservation of seats for the scheduled castes and scheuled tribes and for women. The second is the increase of obligatory functions for ULBs, specifically the endowment to the KMC of the power to plan for economic development and for improving, upgrading and the promotion of the social, cultural, educational and aesthetic aspects of life for the citizens of the city. Finally, the 74th CAA constituted Ward Committees, creating a three-tier system of governance in the KMC. It was hoped that this feature would both enhance the proximity of elected representatives for better accountability and performance and ensure an increase of participation from citizens in the delivery of services and governance at the local level.

Recent Environmental Initiatives

In 1994, the GoWB with the aid of the British Government Overseas Development Administration (ODA) prepared a Calcutta Environmental Management Strategy and Action Plan (CEMSAP). The objective of CEMSAP was to provide a framework for enabling sustained environmental improvements to reconcile pro-development and pro-environmental goals and to develop the means to introduce stronger environmental decision-making processes at all levels. The Environmental Management Strategy (EMS) aimed at strengthening the formation and effectiveness of public-private partnerships (PPP) for making sustained environmental improvements. The EMS is founded on Kolkata's defining environmental problems: poverty and poor resource management (Dutt 2000).

More recently, the Kolkata Municipal Corporation (KMC) with foreign donors and lenders, and in conjunction with the GoWB has developed an environmental improvement scheme with a serious emphasis on results, and a long term commitment to building on municipal management capacities in the city. The Kolkata

Environmental Improvement Project (KEIP), started in 2003 and projected to be complete in 2007, aims at improving the urban environment of Kolkata and, more specifically, improving the living conditions of slum dwellers (Box 5).

A feasibility study was undertaken for the KEIP, which identified the need for capacity building in the KMC. The United Kingdom's (UK's) Department for International Development (DFID) has provided £28.3m million, spread over six years, to implement the Capacity Building Programme (CBP). The prime objective of the CBP is to transform the KMC into an institution representing good governance, providing sustainable and equitable civic services, with a focus on the rights of the poor. Approximately Rs. 12,450 millions loaned by the Asian Development Bank (ADB) for infrastructure development requires that the KMC get its financial accountability in order while making management decisions more public (Table 5.1).

Table 5.1: Sources of Funding under the KEIP

Sources of Funding	*Amount (US$ in millions)*
ADB	220.00
GoWB	54.60
KMC	55.40
DFID	40.00
Total	370.00

Source: KEIP Website 2005

The main environmental improvement components are sewerage and drainage, Solid Waste Management (SWM), slum improvement and canal improvement under stakeholder consultation processes and capacity building (Table 5.2). The project affects all 141 Wards within

Table 5.2: Spending for Environmental Improvements under the KEIP

Environmental Components	*Fund Allocation (US$ in millions)*
SWM	31.00
Slum Improvement	15.00
Sewerage and drainage	135.00
Canal Improvement	22.00
Total	203.00

Source: KEIP Website 2005

the KMC, except for the sewerage and drainage component, which concentrates on Wards 1-6, and Wards 101-141. However, the KEIP is focusing on the added areas within the KMC, which are in desperate need of infrastructure updating, and 100 slums that have a similar need for attention.

Box 5.1

SWM Under the Kolkata Environment Improvement Project (KEIP)

The KEIP started in 2003 (projected to end in 2007) and is based on the collaboration of foreign aid and the KMC. It aims to improve environmental management within the KMC, focusing on improving the living conditions of slum dwellers while increasing the management capacity of the KMC. The SWM component of the project is aimed at improving the management of municipal waste, bio-medical waste, and hazardous waste, including the management system, treatment and disposal facilities. Approximately US$31 million is being allocated to the SWM improvement component for the five-year duration of the project.

The SWM component aims at enhancing stakeholder awareness through informal education and training trash sorters, extending the already existent door-to-door collection system to include the entire KMC area, and to encourage waste generators to segregate recyclable and non-recyclable waste at the source. In addition, the project is replacing open containers with cleaner, closed equivalents, optimizing container locations, and replacing open-back trucks with compactor trucks. The municipal disposal site in *Dhapa* is being upgraded to include new bulldozers and other equipment for easy management, the introduction of mechanical street sweepers and phasing out of night soil tankers with the expansion of the sewerage system. Bio-medical waste management is also being assisted through improved handling and transportation system in conjunction with a cost-effective treatment and disposal system. Hazardous waste management is being developed through the introduction of specific transport vehicles for hazardous waste and the introduction of an integrated hazardous waste management facility at *Dhapa* landfill site for proper disposal.

Source: KEIP website 2005.

Community mobilization and participation under the project was essential for the planning, implementation, monitoring and success of the project, undertaken by the Stakeholders Consultation Process (SCP). Citizen interventions, particularly in the selected Nagarpalika area, occurred in consultation with community-based organizations, local clubs and NGOs. Some of these innovative events included friendly football matches, health camps, *safai abhiyans*[2], community level meetings to break the ice, disseminating information on the project and receiving information and feedback from the community on their needs.

Innovative Practices by the KMC

Even though the KMC council has often been of an opposing political power to the state, the management of the city has undergone some impressive reforms. Most importantly, the KMC has developed a management approach that is focused on efficiency and transparency.

The KMC has developed a more corporative strategy that focuses on quality service delivery while increasing municipal revenue. It has reduced staff from 41,000 in 1999, to 37,000 in 2005, without a compromise of services (or so they say) (KMC 2005b). Upgrading the 'incentives' structure has increased productivity and is a morale boost for staff at the Corporation and the other service providers, as is the case for SWM in the following section (Jalal 2005: 103). It has introduced modern management and e-governance practices, through the computerization of municipal systems and information system using the Geographical Information System (GIS).

Perhaps the most striking outcome of the modernization of the municipal system is the change in financial management of the KMC. By digitizing the tax and revenue programmes while upgrading the property mapping data system, the KMC has been able to increase its own revenue base significantly. With municipal tax reforms and improved non-tax revenues such as fees and levies, the municipality has been able to increase its own revenue by an estimated 355 per cent since 1992-93. In the 2005-06 budget, the KMC projects earnings of Rs 4917.70 million, excluding state government grants. (Table 5.3). In the 2003-04 budget, tax revenues increased 25.40 per cent from the previous year and non-tax revenue grew 52.60 per cent (KMC 2005b). Between the 1992-93 budget, and that of 2005-06, the proportion of total revenue received from the state government decreased by about 10 per cent (Figure 5.1). This decreased dependency on the state government also means that the KMC has more decision making power in development and improvement projects. The KMC also makes all records open to the public by providing annual publications of both the finances and the status of services.

Table 5.3: Internal Revenue Changes from 1992-93 and 2005-06

(in Rs. million)

Budget	*Tax revenue*	*Non-tax revenue*	*Total revenue (excluding state grants)*	*Government grants*
1992-93	847.95	233.21	1,081.16	1,040.85
2005-06	2,925.01 (245%)	1,992.69 (755%)	4,917.70 (355%)	3,172,72 (205%)

Source: KMC (2005a).
Note: Figures in parenthesis show the percentage increase from the 1992-93 budget.

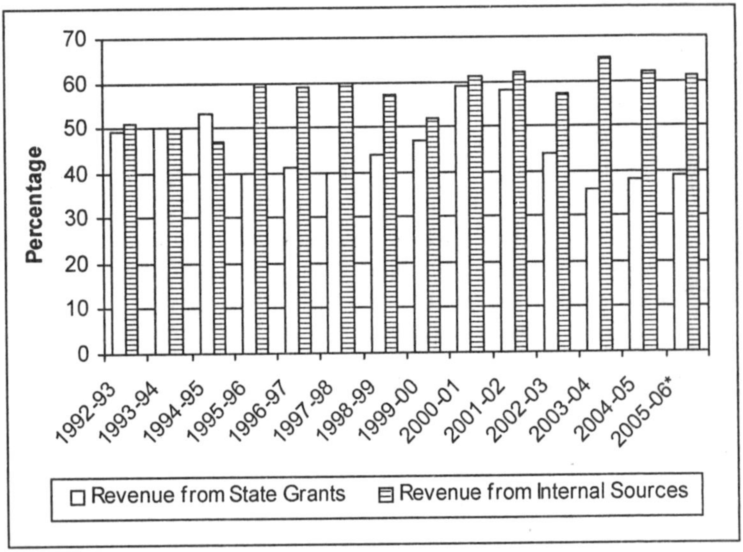

Fig. 5.1: Changes in Revenue Sources

Source: KMC (2005a).
Note: * 2005-06 figures are based on estimates.

Other examples of the innovative practices by the municipal government that both increase efficiency, transparency and accountability are:

- Reforming the tender process by creating a three-tier clearance process involving Works Committees, Controller of Municipal Finance and Accounts (CMFA) and the MIC, for Projects worth Rs one lakh or more, rather than having only the Chief Engineer making the decisions. This has made the process more selective in the interests of getting more value for KMC money.
- Making public all financial accounts
- Publishing a bi-weekly newsletter, '*Poura Tender Barta*', that supplies all the names of bidders, making a clearer process as to who is applying and being awarded projects.

Solid Waste Generation

The KMC area generates approximately 4,000 MT per day, 450 gm per capita waste, which consists of municipal solid waste, debris and silt removed from drains and open drains, night soil, animal carcasses, construction waste and non-hazardous industrial waste. About 2,500 MT is collected daily under the municipal solid waste category, which includes commercial and residential waste generated in a municipality or notified areas in either solid or semi-solid form, excluding industrial hazardous wastes but including treated bio-medical wastes (CPCB 2005). Like most Indian cities, Kolkata generates a high percentage of biodegradable waste, at about 45-50 per cent of the total (Table 5.4). Interestingly, the proportion of this waste in the total has increased from 45 per cent to 50 per cent in the seven years data presented in Table 5.4. although recyclables remain the highest proportion.

Table 5.4: Composition of KMC Waste over Time

(MT per day)

Year	Inert (ash, earth)	Bio-Medical waste	Biodegradable waste	Other (plastic, coconut shells, paper, metal, glass etc.)	Total municipal waste generated
1995-96	136	2.0	817	1,140	2,095.0
1996-97	189	2.1	872	1,119	2,182.1
1997-98	142	1.9	861	1,200	2,204.9
1998-99	184	2.4	972	1,213	2,371.4
1999-00	159	2.6	918	1,186	2,265.6
2000-01	161	2.7	968	1,259	2,390.7
2001-02	202	2.9	1,051	1,218	2,473.9

Source: Bhattacharjya and Shylajan (2004).

SWM System

Adminstrative System

In the wake of the economic reforms on one hand and the decentralization of ULB management under the 74th CA on the other, KMC too has undertaken serious management reforms including employing the private sector in the delivery of services. This section looks at Kolkata's administrative organization, the players involved in SWM, the workings of SWM from collection to disposal, slum or *bustee*[3] initiatives, bio-medical and market waste management, and associated sewage and drainage management.

The KMC has a distinctive administrative structure, as described earlier. To briefly recapitulate, from the top down the three tiered structure consists of, the Corporation governed through a Mayor-in-Council system, the fifteen long-standing Borough Committees, and the fairly newly constituted Ward Committees, as per the 74th Constitutional Amendment Act (CAA).

Actors Involved

The KMC is the key player in the delivery of SWM services in the city. However, in recent years, its role is mutating to become one of facilitator and of regulator rather than just service provider. The key players in the delivery of SWM services in the city are the KMC Solid Waste Management (SWM) Department and the Kolkata Metropolitan Development Authority (KMDA). Chapter IV (a) "Powers and functions of the Corporation" in the Kolkata Municipal Corporation Act (amended to 2005), lays out the obligatory functions of the Corporation relating to SWM, which are essentially:

"(b) the construction, maintenance and cleansing of sewers and drains, sewage and drainage works, and public latrines, urinals and similar conveniences;

(c) the scavenging, removal and disposal of filth, rubbish and other obnoxious or polluted matters;

(d) the reclamation of unhealthy localities, the removal of noxious vegetation and generally the abatement of all nuisance;

(f) the lighting, watering and cleansing of public streets and other public places;

(g) the removal of obstructions and projections in or upon streets, bridges and other public places;

(q) the regulation of places for the disposal of the dead and the provision and maintenance of places for the said purpose;

(r) measures for preventing and checking the spread of dangerous diseases;

(u) the construction and maintenance of municipal markets and slaughter-houses;"

(*Kolkata Municipal Corporation Act 1980* (amended to 2005)

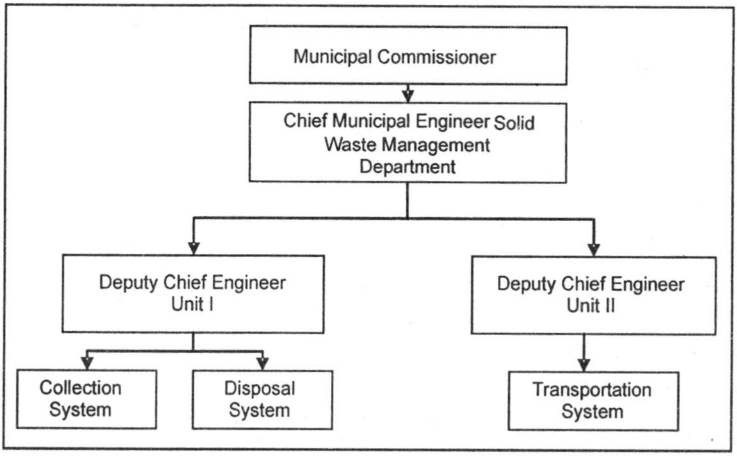

Fig. 5.2: SWM Department of the KMC

Source: KMC 2005c

The entire SWM services are under the SWM Department of the KMC, which is subdivided into two units. Unit I takes care of the collection and disposal of solid waste in the city and is headed by a Deputy Chief Engineer. Unit II takes care of the transportation of solid waste and is also headed by a Deputy Chief Engineer. Both Deputies are under supervisors of the Chief Municipal Engineer for the SWM Department (Figure 5.2). In 2004, there were over 16,000 people working within the two main SWM Units (I and II) which includes all supervision, engineers, officers, conservancy *mazdoors*, support staff, etc. (Table 5.5). This number does not include the workers providing SWM services through the private or informal sectors. The number of employees in SWM services accounts for 42 per cent of the 37,829 total KMC employees (KMC 2005b).

The KMC is divided into two territorial areas, 'added' and 'proper'. Wards 1-100 are the 'proper' area, or the core city, and the 'added' areas are wards 101-141 which are those that have most recently been added to the KMC jurisdiction. The KMC mostly takes care of the SWM services under the 'proper' areas, while for the most part SWM services are outsourced to the private sector for the 'added' areas of the city. By gradually privatizing its solid waste services, the KMC hopes to move its role more towards facilitating, monitoring and supervising, while private operators perform the services.

Table 5.5: KMC: Employees in SWM 2005 (Excluding Private and Informal Sectors)

Position	*Collection*	*Disposal*	*Transportation*	*Administra-tion Head Office*	*Total*
Exe. Eng	—	1	3	5	9
Dy. Director	—	—	—	4	4
Ass. Eng.	—	1	9	17	27
Ass. Director	15	—	—	1	16
Special Officer	—	—	—	3	3
Sub-Ass. Eng.	—	2	9	19	30
Supervisor	21	—	—	6	27
Overseer	258	2	1	2	263
Sub-Overseer	870	2	—	6	878
Driver	34	—	285	—	319
Loader	207	—	—	—	207
Cons. *mazdoor*	10,414	89	1,002	42	11,547
*Methors	1,286	—	—	—	1,286
**Domes	107	—	—	—	107
Mechanics	—	—	450	515	965
Lorry Driver	—	—	—	5	5
Worker	—	—	—	147	147
Computer Op.	—	3	—	—	3
Admin. Staff	235	3	56	191	485
Total	13,447	103	1,815	963	16,328

Source: KMC 2005c

* *Methors* are those who remove silt from sewers

** *Domes* are those who remove carcasses from the street

Private Sector

The main areas of privatization have been for the collection and transportation of municipal waste, bio-medical waste management, composting services and plastic waste recovery, all of which will be described here.

Under the amended Kolkata Municipal Corporation Act 1980, the outsourcing of some SWM activities is a clear option for the KMC. Section (327) under the provision of means for processing of solid waste states that:

> "the Corporation may, for the purpose of receiving, storing, treating, processing and disposing solid wastes or converting such wastes into compost or other matters, construct, acquire, operate, maintain, develop, and manage any work within or outside Kolkata and run it on a commercial basis."

Municipal Waste Collection and Transportation

There are 53 private operators that collect, transport and dispose municipal waste in Kolkata. Of these contractors, 40 transport waste, and 13 undertake street sweeping and door-to-door collection. As the KMC jurisdiction expands, the 'added' areas, are mostly relegated to a private contractor, consequently contracting out has been gradually increasing. In 1999, about 54 per cent of total municipal waste collected was serviced by the KMC, with the remaining 46 per cent by contractors. By 2005, these proportions were reversed (Table 5.6). There are approximately 232 private vehicles collecting waste on an average day compared to 108 KMC owned trucks.

Contractors are paid on a weight basis, a unique strategy that is incentive driven. The entire system is computerized at the disposal site and contractors are paid between Rs. 130 and 170 per metric tonne (MT), according to the distance from the disposal site. There are four zones (A–D) for payment purposes. Zone A (the nearest) pays Rs. 130 per MT, Zone B, Rs. 140 per MT, Zone C, Rs. 150 per MT and Zone D, Rs. 160 per MT. These payments include transportation costs, operation and maintenance, and salaries. Each contractor is paid on a monthly basis.

The collection of silt and rubbish from sewers, open drains, and ditches is completely contracted out to the private sector, with approximately 120 trips made on an average day, amounting to an average of 825 MT per day. This waste is also disposed of at *Dhapa*. The KMC does not perform this service itself (KMC 2005d).

Table 5.6: The Gradual Privatization of SWM Services

Year	Total (MT/day)	Waste collected (%)		No. of Vehicles per day		No. of Trips per day to Dhapa	
		KMC	Contractors	KMC	Contractors	KMC	Contractors
1999	2,340	54	46	96	96	281	174
2000	2,408	54	46	106	102	315	174
2001	2,387	52	48	106	114	328	193
2002	2,341	49	51	105	128	186	325
2003	2,619	46	54	101	143	321	215
2005[3]	3,349	46	54	108	232	357	391

Source: KMC (2005b) and KMC (2005d).

Bio-Medical Waste Management

The KMC has completely privatized the collection, transportation, treatment and disposal of bio-medical wastes. Medicare Incin. Pvt. Limited has been hired to provide these services to the Kolkata (KMC) and the Howrah (HMC) areas. Bio-medical waste management is monitored by the West Bengal Pollution Control Board (WBPCB), a state run agency. The incineration facility is located in HMC which treats and disposes all the waste from the two municipalities. The Common Bio-Medical Waste Treatment & Disposal Facility (CBMWTDF) which started operating in 2003, is the largest of its kind in India. The HMC provided the company with two acres of land on lease for 30 years, on a public-private partnership basis (PPP). The total capital investment for the facility was Rs. 20 million. It has the capacity to treat bio-medical waste generated by 30,000 hospital beds. The treatment facility consists of two incinerators, an autoclave and two standbys, and a specialized Effluent Treatment Plant (ETP).

The collection and transportation system uses 15 vehicles. Two types of colour coded bags are used by the health units for waste segregation purposes; yellow for incineration and red for non-incineration. Services provided by the operator (Medicare Incin. Pvt. Limited) of the facility are:

- Doorstep collection and transportation of untreated bio-medical waste (BMW),
- Treatment and disposal of BMW,
- Supply of bags and containers to health care units,

- Supply of needle cutters and chemicals to health care units, and
- Workshop and training on BMW management.

Treatment costs are based on the number of beds per day: Rs. 2.70 per day per bed for private hospitals and nursing homes, and Rs. 2.45 per day per bed for state government hospitals (excluding costs associated with supplying materials such as coloured bags, containers and needle cutters, etc.) (WBPCB 2005).

Currently, the quality of service is good for some users, but low or non-existent for facilities that either disregard the need, or are not serviced for various reasons. One cause for this is the inability for the company to develop an adequate business plan. The privatization of bio-medical waste management is fairly new in India, and developing a profit-making scheme is difficult in this sector. The WBPCB and Medicare Incin. Pvt. Limited are working closely together to develop a plan that will adequately satisfy the financial needs of the company while abiding by the guidelines under the Bio-Medical Waste (Management and Handling) Rules, 1998.

Composting

The KMC has partnered up with Excel, a composting firm that converts urban solid waste into environmentally safe fertilizer. The scheme expects to reduce the demand on urban land for landfills.

In early 2000, Excel's compost plant, with a daily capacity of 700 MT per day was commissioned at the *Dhapa* waste disposal site. The KMC has leased 30 acres of land for the purpose of composting to the firm, while the KMC receives 2.5 per cent of the sale proceeds. The KMC provides urban waste for free. The initial investment for the facility was Rs. 70 million. The site uses the aerobic digestion method making the final product a bio-fertilizer, with the commercial name '*Bhu-Dhan*' meaning 'gift to earth' (Bhattachariya and Shylajan 2004). Around 300-400 tonnes per day is supposed to be produced. Today, the amount of input the plant receives is only 70 MT daily, as opposed to its 700 MT daily capacity. The enterprise encompasses the operation and maintenance, as well as the marketing of the final product. The composting plant employs 50 permanent staff.

Like most composting endeavours, marketing the urban compost is a major challenge for the company. Excel sells the fertilizer at Rs. 2,500 per 50 kg package. The high price is because the plant is not working at its full capacity and also due to the low demand for the product. Tea farmers are the biggest customers of this fertilizer. Currently however, the company is no longer receiving waste from the city because of differences between the Commissioner and the company over payments to the KMC.

Plastic Waste Recovery

The KMC is developing a partnership with K.K. Plastic Waste Management where the company will provide the KMC with the KK polyblend (bitumen/tar and plastic) which will be used for paving Kolkata's roads. The polyblend uses molten plastic bags and other disposable plastics that would otherwise go to the dumping grounds. In addition, it is claimed that the polyblend will save the KMC approximately Rs. 10,000 per km. of road because the polyblend increases surface longevity. A one km. road with a width of 3.75 m and 25 mm thickness, requires around 1 million plastic bags. This method of road work has been successful in other cities (KMC 2005b).

Civil Society

Civil society groups have an increasingly important role in both decision-making and the delivery of services in the new economic and political environment. Most importantly, with regard to SWM, civil society may be able to ensure equitable and environmentally sustainable service delivery in Kolkata. With the devolution of power to lower levels of government, there is an impetus on seeking community participation at the institutional level. Under West Bengal's relatively more developed decentralization the governance system in Kolkata is increasingly more dependent on lower levels of social organization such as Ward Committees (WCs), Neighbourhood Groups (NHGs), Neighbourhood Committees (NHCs), Residents Welfare Associations (RWAs) and the like. According to Jalal (2005: 108), because of Kolkata's social structure and history there are relatively few formally registered NGOs as compared to other cities: however there are over 2,000 informal associations within Kolkata's civil society, such as caste and mutual benefit societies. The urban-NGOs that do exist have been created mostly by pressure from donor agencies and supported by larger NGOs from other parts of India, or outside the country.

Public United for a Better Calcutta (PUBLIC) was formed under pressure from donor agencies and other organizations in India in the

Natalie Kehle (2005)

Photo 5.1 : Compost heap from the plant at Dhapa, Kolkata

Natalie Kehle 2005

Photo 5.2 : A waste collector depositing the day's collection, Kolkata

Natalie Kehle 2005

Photo 5.3 : Waste truck at Dhapa weighing station, Kolkata

Natalie Kehle 2005

Photo 5.4 : Large concreted transfer station, Kolkata

Natalie Kehle 2005

Photo 5.5 : Informal waste pickers are abundant at the disposal site, Kolkata

early 1990s. Though not directly involved in SWM practices in the city, PUBLIC organizes publicity campaigns to raise civic consciousness through activities, poster campaigns and public service announcements. This NGO had protested against the encroachment on the Kolkata wetlands by private estate developers and the misuse of public property such as heritage sites, by launching public interest litigations in the High Court. Even though its initiatives are well founded, there are claims that their scope is restricted to mainly the middle-class Bengalis (Jalal 2005: 108).

There is a unique institution in Kolkata called *Nagarik Sammelan*, which is a citizens committee, or 'local' committee that is a quasi-political organization. They function as a parallel welfare organization which recruit political party members, represent the voices of its members and undertake periodic welfare activities, such as a campaign for improving the living conditions in slum areas. These committees openly reveal their biases against those that support other parties or hold different political beliefs (Jalal 2005: 109).

MASS Education is one locally run NGO that is involved in social and environmental sustainability advocacy in the city and West Bengal. This NGO is a good example of how the KMC has collaborated with the civil society for equitable urban services but also shows how far the KMC is from fully utilizing this resource. MASS Education began its work in 1978 among Kolkata's slum population to upgrade basic education, health and hygiene awareness, by running clinics and providing other services. In 1997 it initiated a SWM program in conjunction with a poverty alleviation programme in Kolkata to remove bio-medical wastes from domestic houses, institutions, hospitals, clinics and industrial sites while providing employment opportunities to Kolkata's poor. At this time, segregation and special handling of bio-medical waste was not practiced in Kolkata, or most of India. The objectives of the programme were:

- To ensure protection at all levels of handling of hazardous wastes,
- To prevent unauthorized disposal of hazardous wastes,
- To generate employment opportunities with protection against health risks, and
- To create awareness for waste segregation and proper waste disposal methods.

By 1999, MASS Education had organized and educated 200 rag-pickers who were already established in the area, to be *Paribesh Bandhu* or waste collectors. They were paid a daily wage to collect recyclable waste from the doorsteps of households rather than from picking through bins, street gullies or other disposal sites. Each waste collector earned Rs. 1,500 to Rs. 3,000 per month. Approximately 50 per cent of the workers were women. The workers were given uniforms, identification badges, and protective gear, and were supplied with regular vaccinations and health care services. They provided service to households, hospitals, markets, shops, hotels, restaurants, street vendors, construction sites and the like. They also collected night soil from households and surface drains. Some slum areas were also provided with services free of cost. Residents were asked to place waste in community bins which would then be mechanically lifted by MASS Education trucks (MASS Education 1999).

MASS Education was collecting daily about 400 kg of domestic waste, 200 kg of industrial waste and 300 kg of bio-medical waste from 5,000 households, 20 industries and 400 hospitals, nursing homes and clinics. Collection for domestic waste and clinical waste began at 7.00 am and ended at 2.00 pm everyday while industrial waste was collected on a weekly basis. Participating households were asked to segregate their waste into two bins; biodegradable and non-biodegradable waste. Service costs were between Rs. 30 to Rs. 50 per month, or Rs. 20 to Rs. 30 per household if waste was put into fixed bins in front of a residential complex. Shops and other establishments were charged a minimum Rs. 40 per month per bin, and waste was put into fixed bins outside with a covering. Ten specially designed covered vehicles were used for collection, as well as auto trailers, 3-5 tonne trucks, and some push cycle vans. All waste collected went to the city's dump site and the clinical waste went to the local incineration plant (MASS Education 1999).

Payments for industrial and clinical waste were based on weight and frequency of collection. The NGO was predominately run under foreign aid such as from the World Bank. At this time, MASS Education had little to no association with the KMC since most of SWM services were just not being done, especially the collection of bio-medical waste.

Today, the KMC provides citywide door-to-door SWM services, though still not at 100 per cent coverage. The NGO plays a different role now than it did in the past. First, bio-medical waste management and handling has been taken up by the WBPCB and, in part, has been contracted out to a private company. As already discussed, MASS Education no longer plays a part in this type of waste management. The privatization of the waste service has created the problem of profit-making. As discussed earlier, the company, Medicare Incin. Pvt. Limited, who is supposed to provide the service to all of the KMC and HMC is unable to develop a business plan which produces adequate profits. As a result, services are inconsistent if not unavailable. Secondly, the relationship between the KMC and MASS Education is like any other private contractor for SWM tenders. Even though the organization is a non-profit and has already developed an equitable SWM system in some areas, they have to go through the same bidding process as private contractors, (submitting technical and financial bids). However, the payment of MASS Education for the services is not the same as those to private contractors, though the service is the same. Instead of being paid on a monthly basis, as are the private contractors, MASS Education is paid every six months or payments are skipped altogether. The NGO claims that the KMC is exploiting the organization because it views it as affluent due to foreign ties. Also, it alleges that the KMC is only allowing MASS Education to win bids so that the city can keep an image of 'association' with socially and environmentally equitable activities.

MASS Education is one of few known cases of an organized group, that is neither private or governmental, involved in the provision of SWM in Kolkata. It is clear that the KMC is not taking advantage of the benefits of socially equitable and safe SWM practices that NGOs and some other community groups might offer.

Informal Sector

Kolkata, like most Indian cities, has a sophisticated informal system for managing and trading solid waste. What is unique to Kolkata compared to other cites in the country is the consistent involvement of KMC *mazdoors* in waste trading. Almost everything that is thrown out, especially from affluent households, is useful to someone else. Bose (1996) estimated that the private recovery system in Kolkata reclaims between 20 and 25 per cent of waste and that the 'total turnover of the recycling operation' was well over Rs. 60 million.

Waste trading starts at the household where *mazdoors* are the first people to claim waste rights. Due to higher wages paid to *mazdoors* now for the collection of door-to-door waste, collecting and selling valuable waste is no longer worth it for them. In place informal waste collectors, such as rag-pickers have stepped in. *Mazdoors* have changed their role from formal waste collectors, to facilitators in the informal system as contractors, rentiers and auctioneers of waste bins and other waste property rights to rag-pickers (though this might have changed since the development and expansion of the city-wide door-to-door waste collection scheme by the KMC) (Bose 1996).

After much negotiation with *mazdoors* and others seeking rights to the waste, rag-pickers got the right to sell valuable recyclables like paper, glass, metal and plastic to itinerant buyers, and other waste traders. Popular places for collection are; directly from householders, household bins, community bins, transfer sites, industrial waste bins, construction sites, hospitals, factories, disposal sites such as at *Dhapa*. Recycling units that process the waste are privately run and are located at the Recycling Region located along the Eastern Metropolitan By-Pass, partially within the KMC and the district of 24 Parganas North and South.

Rag-pickers are usually seen without any protective gear for rummaging and use, at best, poking sticks. Rag-pickers are usually women, members of the backward castes and children. Besides the multitude of health problems that rag-pickers face, the problem of scattered waste on streets makes for increased health risks to other members of society and becomes an expensive task to clean up afterwards. During the monsoon season, scattered waste easily clogs up open drains and sewers causing extensive flooding in the city.

From Generation to Disposal of Solid Waste

The solid waste management practices in Kolkata operates in two different systems. The first is the public or municipal system based on service delivery by the KMC and the private contractors. The second is a complex private stream based on waste picking and trading. The two systems work in tandem with each other: however the informal private system is substantially dependent on the municipal system.

Collection and Transportation

The door-to-door collection of waste is performed by both the KMC waste collectors and the private contractors. *Mazdoors* collect un-segregated waste from households with the help of handcarts and alert households at collection time with a whistle. The waste is then carried out to community vats throughout the city, which await secondary collection by trucks. There are some areas in the KMC that have direct loading, specially fitted trucks, thereby avoiding double handling. KMC serviced areas mostly use mechanically fitted trucks to load waste, while in privately serviced areas the waste is usually manually scooped from the community bins into the trucks with shovels. Private operators must provide at least five *mazdoors* for garbage truck loading and off-loading. Trucks must report at the garbage collection site by 7.00 am each day. Late arrivals are penalized to the extent of 0.5-1.0 MT of the garbage collection fee. Truck operators must clear and disinfect community vats and surrounding areas by 10.30 am.

In addition to door-to-door collection, *mazdoors* also sweep roads. There are about 12,000 KMC *mazdoors* in addition to over 1,000 *domes* (carcass collectors) and *methors* (night soil collectors), compared to approximately 800 private *mazdoors* (KMC 2005c).

Mazdoors are predominantly drawn from the caste known as *Bhangi*, traditionally having the occupation of night soil removers. In the KMC, the *mazdoors* mainly come from the very poor and underdeveloped area of the Chhotanagpur plateau of Bihar (Bose 1996). The KMC and contractual *mazdoors* are the first set of actors to lay claim over the waste, once it has left the household. This is the first step in waste recovery by the informal sector system, as described earlier.

The KMC has set-up an incentive driven collection scheme in an attempt to persuade workers to do a thorough job and to avoid illegal dumping. Each truck, both KMC and privately owned, is weighed by a machine at the *Dhapa* site. Payments to contractors are based on per MT basis, as described earlier. The KMC claims that 95 per cent of the city's solid waste is collected daily, up 25 per cent from 1996 (KMC 2005b and Bose 1996).

The KMC has set up an incentive driven collection scheme in an attempt to persuade workers to do a thorough job and to avoid illegal dumping. Each truck, both KMC and privately owned, is weighed by a machine at the Dhapa site. Payments to contractors are based on per MT basis, as described earlier. The KMC claims that 95 per cent of the city's solid waste is collected daily, up 25 per cent from 1996 (KMC 2005b and Bose 1996).

The KMC is slowly starting to remove community vats in the city, based on an experimental initiative in selected wards. Direct loading, from handcarts into trucks takes place in Jodhpur Park by three trucks and in Jadavpur by one truck (KMC 2005a). By removing the vats, the KMC is reducing health risks to conservancy workers by avoiding unnecessary contact with the waste. Also, direct loading makes for cleaner streets since rag-pickers are not scattering garbage outside the bins, animals are not being attracted, and the putrid smell of decomposing waste is eliminated.

Transfer sites are used as a secondary point of collection, which are mostly found in the higher density 'proper' areas. Like community vats, the transfer sites are the secondary point in which waste from the door-to-door collection awaits pickup before going to *Dhapa*. Instead of a bin for a small area, these transfer sites are for a greater area and are housed in a concreted corner of the neighbourhood. The site is large enough to allow conservancy trucks in, but is often filled with rag-pickers and animals going through the heaps. Waste from here is manually scooped by shovels into the trucks.

Disposal

The waste is disposed of mostly by uncontrolled dumping at Dhapa and Bantala, but also at Noapara and Garden Reach, all within the KMC. The majority of the disposal occurs at Dhapa, which is located in a system of wetlands. The naturally depressed marshes have a thick clay layer that supposedly protects the groundwater system from the waste and its effluent. The present Dhapa disposal area is approximately 8-10 hectares and is located south east of Salt Lake City.

All trucks, whether owned by the municipalilty or the contractors, must reach Dhapa by 12.30 pm. A collection incentive was developed so that garbage collectors are encouraged and rewarded for extra loads that are brought into the dumpsite. A computerized system at the dumping grounds weighs the extra amount and adds a bonus sum to the daily wages of the waste workers. Any amount exceeding 7.0 MT is not remunerated. This is to discourage those that take advantage of the system by filling trucks with heavier material rather than with municipal waste. Having a computerized system allows engineers to plan space availability, while decreasing the chances of manipulating the system for bonus wages. All vehicles must be covered by tarpaulins, as per the *Municipal Solid Waste (Handling and Management)* Rules 2000. Non-compliance is penalized with a fee deduction equivalent to 0.5-1.0 MT of waste. Strict monitoring of payments also allows for detailed study of the fluctuations in waste generation in the city, which is a rarity in India.

Table 5.7 shows the detailed report for a day (October 10, 2005) of waste disposal at the Dhapa site, where each truck is weighed on entering the site. A report of all the data collected (with hourly summaries) is produced at the end of the day and given to the Deputy Chief Engineers and the Chief Municipal Engineer (SWM) for analysis in addition to the Commissioner and Mayor. For any given day, the KMC authorities know the exact number of vehicles dispatched, total weight of waste collected and disposed of and the number of trips made with the name of each officer responsible, increasing accountability.

Table 5.7: Vehicle Report at Dhapa for October 10th 2005

	No. of vehicles	*No. of Trips per Day*	*Total Weight of Collection (MT/day)**	*Average Truck Capacity (MT/truck)*	*Weight of Silt Collected*
KMC	108	357	1,541.8	4.3	0.0
Contractor	232	391 (+120 from silt/ rubbish collection)	2,523.9	6.5	825.3
Total	340	748 (+120)	4,065.7	—	825.3

Source: KMC (2005d).

* Excluding silt and rubbish

Once weighed, trucks drive up the waste mound to dump the waste. The landfilled waste has an average density of 500 kg. per m^3, and rises 25 m. above ground level, a height equivalent to a seven-storey building. Rag-pickers eagerly wait for the trucks to unload the waste so that they can pick out the valuables.

Currently, the vast wetlands of Dhapa also house two controversial agricultural sectors. The first is the waste based agricultural fields. The municipal solid waste heaps from previous years are spread out flat and converted into vegetable plots, known as 'Garbage Farms'. Up to 11 crops are produced here on approximately 700 hectares of land. The plots produce 150 tonnes of vegetables per day, about 30 per cent of the demand of fresh vegetables in the city. They are then sold at city markets and along the highway.

Secondly, the Dhapa wetlands are also used for disposing untreated sewage and surface water drainage of the city while supporting an aquaculture sector. About 680 million litres of sewerage waste per day run through the system. About 3,000 hectares of the wetlands are 'naturally' treating the city's sewage. Dhapa supports 3,000-4,000 hectares of aquaculture and 7,300 hectares of paddy cultivation. The aquaculture activities based on sewage-fed fishponds produce 2.5 tonnes of fish per hectare. Through the process of oxidation, radiation and biological breakdown of organic waste, the whole process is said to improve the water quality as well as well-managed stabilizing tanks.

These waste-based activities generate about 20,000 jobs. The KMC owns the land and rents out the fields at Rs. 4,500—Rs. 6,000 per hectare (Hasan and Khan 1999). A recent study however, by the Geological Survey of India (GSI), at the Dhapa-Chowbagha region found that soil and water samples contain an alarmingly high level of toxic elements including chromium, cadmium, chlorides, nitrates, and fluorides. Soil and water pollution levels in these areas have increased significantly over the last few years, well over the permissible limits. The main reason for the rise has been the indiscriminate dumping of toxic industrial effluents, such as waste generated from illegal tanneries, small-scale electroplating and casting industries, and from human waste. Elements like lead, chromium and arsenic are carcinogenic and when consumed above permissible limits, are linked to cardio-myopathy, irregular heart beats and heart blocks (*Calcutta Times* February 2004).

The KMC is in the process of recovering around 250 acres of land at Dhapa from the encroachers living on the site. Residents here are mostly rag-pickers or workers in the waste-based agricultural sector. The KMC plans on having a Scientific Sanitary Landfill Site at Dhapa by 2007-2008, and construction is being funded by the KEIP for Rs. 400 million. Under the KEIP, it is being proposed to increase the size of the Dhapa site to 113 hectares, an added area of around 90 hectares. Two pay loaders and two tractors have been purchased under the KEIP project to facilitate disposal (KMC 2005b).

Slum, or *Bustee,* Improvement Initiatives

Historically, the KMC has had extensive problems with providing minimum living standards for slum, or *bustee*, residents. Such individuals must manage with poor basic infrastructure (water and sanitary toilets), few services, high levels of unemployment and an absence of coordinated improvement initiatives. Recently, the KMC has conducted a comprehensive survey of the slums, studying location of dwellings, number of residents in each, income profiles, and other social characteristics, preparatory to adopting a programme of social and economic development for the least advantaged under the requirements of the 74th CA. Under the newly formed Nagarpally Unnayan (slum improvement) department in the KMC, the focus is on adequate supply of drinking water, proper drainage and sanitation, concrete roads and passages, access to proper health services and the availability of electricity. There are presently seven slum improvement initiatives within the KMC each of which has a bearing on solid waste management (KMC 2005b). These are: (i) KEIP, (ii) Pay-and-Use Toilets, (iii) Community Taps, (iv) Swarna Jayanti Shahari Rozgar Yojana (SJSRY) and DWCUA (self employment opportunities for women and youth), (v) Thrift and Credit groups and (vi) the National Assistance Programme for the old.

Of these initiatives, only KEIP (Kolkata Environmental Improvement Programme) has a direct SWM component, details of which were given in Box 5.1. About 100 slums in Kolkata are being targeted for improvement at a cost of US$ 15 million. Among the various components to improve physical and socio-economic aspects of the *bustees*, the important ones for this SWM case study are: (i) Needs assessment with regard to water supply, sanitation, hygiene, solid waste collection, sewage, drainage, lighting and overall health, (ii) Long-term development to improve living conditions by widening and realignment of neighbourhood roads, lighting and construction of sewage and water lines and installation of solid waste containers, (iii) Conversion of over 160 km of open slum drains to an underground sewage system and (iv) Awareness campaigns for wise resource use.

The Swarna Jayanti Shahari Rozgar Yojana (SJSRY), the national level self-employment programme encourages formation of DWCUA groups, which are Self Help Groups (SHGs) or Neighbourhood Groups (NHGs) of 10-40 *bustee* women who work collectively to improve opportunities, manage micro-loans, generate income and support planning, implementation and activity monitoring at neighbourhood levels. There are 650 NHGs across Kolkata. Further under the SJSRY, formation of Neighbourhood Committees (NHC), coalitions of NHGs, is encouraged. Some 65 NHCs in Kolkata have been formed which identify local problems and priorities, make suggestions for group involvement in meeting community goals and needs, support local action in partnership with responsible agencies, and facilitate timely loan recovery. Lastly, there is also the formation of Community Development Society (CDS) in the city, which regroup NHC representatives and one member of the executive committee, with principal responsibilities to represent the broader community needs, coordinate with agencies and departments to promote need fulfilment through projects and identify and implement necessary training. There are six CDSs in Kolkata. This is an important institutional structure, which could be utilized where it exists for decentralised SWM work, an as yet unexploited opportunity.

Bio-Medical and Market Waste

The clinical waste is collected from different healthcare units in the city by a private company, Medicare Incin. Pvt. Limited, as discussed earlier. There are around 602 healthcare units including government hospitals, private hospitals, nursing homes, clinical laboratories, and health centres throughout the city. Ten vehicles are used daily to collect bio-medical waste in the KMC area.

The KMC and contractual workers collect waste from markets. There are 23 registered KMC markets and 252 markets in total. There are no formal partnerships that have been developed by KMC or private markets to provide highly valuable market waste to local farmers.

Sewage and Drainage

Kolkata is known for its annual flooding during the two monsoon seasons that hit the region. In the month of October 2005, the city faced unusually heavy rainfall during the returning monsoon. The city was at a standstill for two days since most roads entering the city were inaccessible. More importantly, millions of people, mostly the poor and slum dwellers, were exposed to unhygienic living conditions and were at risk from possible contaminated drinking water. There are a few factors to blame for this crisis, but most importantly are the inadequacies of the sewage system in the city and the poor solid waste collection.

Kolkata's sewer system dates back to 1876, installed in the 'proper' areas of the city. This leaves most of the newer 'added' areas of the city without a system, or with open drains at best. There is 953 km of open drains. If the SW is not picked up, it is likely to clog the open drains. Even if it is picked up, more often than not, the waste dumped at the community vats is not collected immediately—it usually takes some time between waste being dumped and the trucks' arrival, and during this lag the scavenging rag-pickers and animals scatter it all over the road. In times of heavy rains, such waste quickly gets into open and closed drains and sewers and clogs them reducing their capacity to carry rainwater. This is what caused flooding in the city in October 2005[4].

Budget Spending on SWM

All SWM expenses are recorded under the SWM Department at the KMC. In 2005-06, the KMC expected to spend Rs. 1,481.43 millions for SWM (Table 5.8). Since 2002-03, the city has decreased the proportion of funds going towards SWM by about 6 per cent, but actual spending has gone up by about Rs. 200 million.

Table 5.8: Total Spending for SWM from 2002-03 to 2005-06

(Rs in million)

	2002-03	*2003-04*	*2004-05*	*2005-06*
SWM	1278.23	1,305.42	1,477.83	1,481.43
	(22.9 %)	(21.5 %)	(17.3 %)	(16.6%)
Total KMC expenditure	5,581.17	6,083.45	8,563.98	8,942.20

Source: KMC 2005a

Note: Figures in parenthesis indicate percentage of total spending on SWM

The most costly aspect of SWM is operation and maintenance (O & M) which includes the salaries of all the workers (KMC and any outsourcing). As mentioned, there are over 16,000 SWM employees under the KMC, plus 53 contractors with all their employees. The next table describes in more detail the amount being spent on SWM, subdivided into collection, transportation and disposal costs. In this, collection gets the highest proportion of expenditure allocation on O & M, followed by transportation (Table 5.9). The O & M expenditure on the former is about 3 times that on the latter. Capital costs have jumped in recent years because the KMC is building a scientifically engineered landfill facility which is supposed to become operational soon.

Table 5.9: Breakdown of SWM Expenses from 2002-03 to 2004-05

(Rs. in million)

Task	*Capital Cost*			*Operation & maintenance Cost (including salaries and outsourcing)*		
	2002-03	*2003-04*	*2004-05*	*2002-03*	*2003-04*	*2004-05*
Collection	8.31	9.38	10.31	921.39	938.00	1,025.95
Transportation	2.62	2.66	4.60	283.38	290.11	360.18
Disposal	9.56	14.47	26.00	52.97	50.80	50.80
Total Cost	20.49	26.51	40.91	1,257.75	1,278.91	1,436.92

Source: KMC (2005c).

Solid Waste Management (SWM) costs are covered by revenue from property taxes or other sources, such as state grants. Spending for SWM represented over 17 per cent of total city expenditures in 2004-05. The KMC raises a tiny proportion of SWM costs through non-tax collection, about 0.4 per cent of the total, mainly collected by the boroughs (Table 5.10). Since the KMC has not privatized collection, or contracted it out to SHGs, the financial burden on the KMC for collection is relatively high.

Table 5.10: Revenue from SWM Receipts compared to SWM Spending

(Estimated for 2004-05, in Rs. Million)

Particular	*Borough*				*Total*
	I-X	*XI-XII*	*XIII-XIV*	*XV*	*(all of KMC)*
Non-Tax Revenue-Receipt from SWM	21.85 *(0.33%)	2.39 *(0.83%)	1.84 *(0.91%)	0.41 *(1.19%)	26.4.80 *(0.37%)
SWM Expenditure	1,204.31 **(15.97%)	51.92 **(20.49%)	77.95 **(21.97%)	52.15 **(28.42%)	1,386.33@ **(16.64%)

Source: KMC 2004

* As percentage of Total KMC Revenue

** As percentage of Total Expenditure

@ This total figure does not tally with total figure in Table 5.8 and 5.9 because of head-office costs

For the year 2004-05, it was estimated that the SWM department in Boroughs I-X would raise about Rs. 21.85 million from SWM receipts (see *Annexure B*). The most significant revenue from SWM receipts come from the removal of encroachment, for scaffolding and for depositing building materials and other heavy debris. Spending for SWM in those same boroughs (I-X) was estimated at Rs 1,204.31 million. The major expenses here are the housing-Harijan Employees/ Street Cleansing & Solid Waste Collection, the collection of solid waste, and the removal of night soil, respectively costing 71, 12 and 8 per cent of the total expense for SWM (*Annexure* C). Under the newly formed Nagarpally Unnayan (slum improvement) Department, the KMC aims at improving the environment of slum dwellers through increasing the number of sanitary toilets in *bustees*. Approximately Rs. 52 million was spent under the Nagarpally Unnayan Department between 2002-03 and 2004-05.

Achievements

Solid Waste Management (SWM) in Kolkata has undergone significant changes in the last decade since the implementation of the Municipal Solid Waste (Management and Handling) Rules 2000, with the continued decentralization practices of urban management under the 74th CAA in West Bengal, and with the utilization of the private sector for some service delivery. Though not an exemplary case of municipal SWM in urban India, the Kolkata Municipal Corporation (KMC) is an interesting case because of the enormous improvements in recent years through the adoption of modern management practices. Most notably are the upgrading of financial management, the use of the private sector in service delivery and the development of an incentive-driven solid waste collection scheme. The most important achievement in the city is that new practices have been introduced by the city government itself, unlike other cases in this book, where NGOs have taken the initiative. Given that this is a KMC initiative, the entire city comes under the purview of the new efforts. The other interesting aspect of the city is that waste management has been integrated with urban agriculture—horticulture, fisheries and aquaculture, in an effort to create a sustainable city.

In some neighbourhoods, source segregation has been introduced that allows for more recyclables to be retrieved from the total waste. This gives income to the waste collectors, who in case of Kolkata are the municipal workers themselves. In recent years however, they have taken to sub-contracting this to the rag-pickers. The wet waste of the city is converted into bio-fertilizer by a private sector firm, Excel. Currently, the plant is not functioning. But, when and if it starts again, it would divert 700 MT of waste per day to the plant, thereby reducing landfill dumping. This is very important in the Kolkata region as there is a large number of wetlands and other water bodies. In the neighbourhoods with waste at segregation at source, community vats are eliminated, reducing environmental and health risks.

The KMC claims that of the 650,000 households in the KMC, about 95 per cent or 617,500 households are served by door-to-door waste collection, up from 70 per cent in 1996. Currently, the KMC is working towards 100 per cent service delivery under the KEIP.

KMC has developed an incentive driven scheme for the collection of municipal waste. At the disposal site at Dhapa, the KMC has constructed a large weighing scale for all the trucks that enter the site. Contractors are paid on a per metric tonne (MT) basis at the end of the month. This method has two important benefits. The first is that waste workers are encouraged and rewarded according to the amount of waste they collect. It also decreases the chances of illegal dumping outside the municipal dumping grounds. The second benefit is that the municipal government has a clear measurement of the amount of waste that is being generated in the city and may plan for this accordingly. These kinds of figures are rarely available for governments in India. In this case, engineers and planners can organize the dumping site and can make decisions based on solid empirical data as opposed to mere estimates. At the end of each day, the Municipal Commissioner, Chief Municipal Engineer and Deputy Chief Engineers receive a data summary from Dhapa. This includes the amount of solid waste that was collected, the number of trucks that have entered, the number of trips each truck made, and the names of all the workers who have been rewarded bonus for the excess waste they collected. This system is much more accountable, transparent and efficient compared to other methods of contracting between the private sector and government.

Further Action Areas

There are many areas of action necessary for achieving even better results in the city. The first one is that segregation at source of wet and dry wastes should be introduced in all parts of Kolkata, not just the few selected neighbourhoods. This is a fundamental need in order to reduce the amount of waste going to the dump, to properly supply the compost plant already set up, and to avoid the contamination of recyclables. Local environmental organizations should be encouraged to collaborate in educating citizens through environmental and social awareness campaigns so that the importance of segregation becomes widely known.

The KMC has developed a partnership with a local non-governmental organization, MASS Education, for the delivery of some SWM services. This organization has a history of being involved primarily with organizing and training rag-pickers for the collection, transportation and disposal of wastes, especially bio-medical waste in the KMC, before it was mandatory under Indian law. However, today, MASS Education is treated like a private enterprise: the organization has to put out a financial and technical bid like any other contractor, but is not rewarded in the same way. The role of such NGOs needs to be appropriately redefined. KMC should review its relationship with MASS Education and other environmental NGOs, perhaps working with them for SWM education and capacity-building for more environmentally and socially friendly practices, instead of relying on relatively low wages paid by NGOs to their workers for service delivery. Further, NGOs have to depend on their own funds for carrying out local level activities, such as training rag-pickers to work effectively in neighbourhoods, as MASS Education did in past times. KMC has not paid some amounts due to this NGO for its contracting services, assuming that it is an affluent organization with foreign funding.

Two things are important here. First, the multi-facetted role of NGOs makes them different from the private sector. Not only can they perform services, at the same time they train and nurture unskilled labour, and teach social competency, time management and financial responsibility. These life skills are essential for the disadvantaged to make a living. Now that municipalities are responsible for their own economic and social development, such instruction is fundamental for cities to progress. Human development must become more than just the responsibility of departments of economic and social welfare: thought must be given to how all city departments can contribute, and SWM is an ideal candidate for this since so many slum-dwellers already work in the field. Second, as the Indian GDP continues its amazing climb, foreign funding for NGOs will wither in favour of less favoured nations. To make the assumption that NGOs can and will be able to materially aid the poor for ever is a delusion. For all these reasons, Kolkata and the Kolkata environmental NGOs would be astute to make mutually satisfying, transparent and equitable written agreements about their partnerships in solid waste management and any other city services where there is an NGO component.

The waste collection transport system from the community vats to the final disposal site needs to be made more efficient through the deployment of better machinery. For example, the KMC serviced areas mostly use fitted trucks that efficiently lift vats with mechanical arms, while the private sector usually scoops the waste manually from the vats into trucks, a time consuming method that exposes workers to danger, leads to unnecessary littering, causes local environmental and health problems, and also clogs storm water drains. Better vat design might discourage rag-pickers and animal scavengers. In short, better design and management at the collection points would help reduce litter.

With Kolkata's geographical location (proximity to wetlands and two monsoon seasons), coupled with the city's ancient sewer system, the city is extremely vulnerable to flooding. For this reason, it is urgent that the city decreases the time solid waste is waiting to be picked up from primary collection points since it soon spreads and rapidly blocks the sewage and drainage system of the city. The most obvious solution is to remove selected community vats and use direct loading methods to decrease the chances of solid waste going into drains. Through prevention methods, the city can save repair and water pumping costs.

Today, citizens pay for SWM services through their property taxes. There is no direct contribution of households towards waste collection. Experience in other cities shows that households can be charged a small fee for refuse collection. This could be considered if KMC needs additional resources for other purposes such as upgrading equipment.

Bio-medical waste management has been privatized and given to one company, Medicare Incin. Pvt. Ltd. for their collection, transportation, treatment and disposal. Health care units are responsible for the proper segregation of bio-medical waste as per the Bio-Medical Waste (Management and Handling) Rules, 1998. Under the privatized system, some areas are not being serviced, and for those that have the service, proper management is wanting: segregation is not being practiced as per the rules, workers are being exposed to contaminants, bio-medical waste is still being found in the municipal waste system and at the disposal site. Used needles, tissues, body parts, and a slew of other medical by-products are found in waste heaps putting thousand of waste workers, rag-pickers and children at risk. Strict implementation of the current regulations must become a priority.

The current system of SWM is a top-down approach to management, with little input from the community, despite all the efforts at decentralization to the fifteen Boroughs and the 141 Ward Committees. Recently, the KMC has created the Integrated Borough Scheme to aid in the devolution of powers to this level of government through capacity-building initiatives. This could be extended down to the Ward level, wherein, local citizen participation can well be enhanced, but in a non-partisan manner. A good beginning has been made with decentralization of administrative structure; it has to be furthered by enhancing citizens' participation in this structure.

Acknowledgements

Interviewees

Adhikari, S.K. (2005). Senior Environmental Engineer, Waste Management Cell, West Bengal Pollution Control Board (WBPCB), Department of Environment. Salt Lake City, Kolkata.

Bhattacharjee, S. (2005). Deputy Chief Engineer of the Department of Solid Waste Management (SWM), Kolkata Municipal Corporation (KMC). Market Street, Kolkata.

Bhowmick, S.K. (2005). Municipal Secretary, Kolkata Municipal Corporation (KMC). Banerjee Road Kolkata.

Ghosh, A.K. (2005). Director of Environment and Development and President of the Endev Society for Environment and Development. Jodhpur Park, Kolkata.

Manna, D. (2005). Environmental Management Specialist, SENES Consultants India Pvt. Ldt. Sector I, Salt Lake City, Kolkata.

Sarkar, A.K. (2005). Chief Municipal Engineer of Solid Waste Management (SWM), Kolkata Municipal Corporation (KMC). Market Street, Kolkata.

Sarkar, S.K. (2005). Member Secretary, Indian Administrative Service (IAS), West Bengal Pollution Control Board (WBPCB), Department of Environment. Sector III, Salt Lake City, Kolkata.

Singh, S. (2005). Secretary General for MASS Education. October 2005, Garia, Kolkata.

Shaw, A. (2005). Professor at the Indian Institute of Management Calcutta, Regional Development Group. October 2005, Joka, Kolkata.

NOTES

1. *Bustee* is the local term for slum.
2. Cleanliness drives.
3. Data for 2004 are not available.
4. City coming to standstill for a few days in the monsoon period is quite usual for the city and happens nearly every year.

ANNEXURE A
EFFECTIVE WARD COMMITTEES

As mentioned, the KMC is one of the few cities in India with operational Ward Committees. Not only are the roles and functions of the Ward Committees clearly defined under the West Bengal Municipal (Ward Committee) Rules 2001, but Ward Committees in the KMC are constituted for every ward, providing for a greater participation by the citizens in the management of urban planning and development in the KMC. This unique constitution will be looked at here.

Highlights from the West Bengal Municipal (Ward Committee) Rules 2001 (Ghosh and Mitra 2003)

Formation of Ward Committees

The municipal authority should form Ward Committees within one month from the date of the first general meeting of the Board of Councillors after the general election. Ward Committee member positions are the same as the Board of Councillors at the municipal authority.

Size and Composition

Ward Committee members include elected councillors and members that are nominated by the Ward Councillor, in specific proportions (see Table A). The number of nominated members depends on the population of the ward. Generally, for a population more than 5,000, there should be 14 members, excluding the councillor of that ward.

Selection Criteria of Members

The Ward Councillor and the local body are the sole nominating authorities. The elected Ward Councillor acts as the chairperson and presides over the Ward Committee meetings. The Ward Councillor has the right to remove any members, if necessary. To increase transparency and objectivity in the selection of members, the councillor must place the names of the members in the first general meeting. If there are any objections from the citizens about a member, it will be considered by the councillor and municipal authorities as to remove that member.

Table A: Composition of Ward Committee (Amended in 2003)

Total No. of Members	*No. to be Nominated by the Ward Councillor*	*No. to be Nominated by the Municipality*
7	5	2
8	5	3
9	6	3
10	7	3
11	8	3
12	9	3
13	10	3
14	11	3
15	11	4
16	12	4
17	13	4

Source: Ghosh and Mitra (2003).

All nominated members should be residents of the ward. During nomination, proper representation should be kept in mind so that all sections of the population are covered. Specifications are given to ensure access of the poor in the participation in the civic affairs. Ward Committees with nominated members up to nine must have at least two members from the CDS, constituted from implementing SJSRY. Ward Committees with more than nine members must have at least three representatives from CDS, but none of the CDS members should be nominated for more than one Ward Committee. In the case of unavailable CDS members, women members of the Below Poverty Level (BPL) families from the concerned ward may be nominated instead.

Powers and Function

The Ward Committees are responsible for most of the key functions that a ULB has at the ward level. According to the Ward Committee Rules, under the supervision of ULBs, Ward Committees are supposed to perform the following functions;

- Supervision and Monitoring,

- Financial,
- Planning, and
- Execution of Development Schemes.

The Kolkata government, like other local governments in West Bengal today, still faces problems with the implementation and management of the Ward Committees. The following is a list of some of the challenges Ward Committees face (Ghosh and Mitra 2003):

- Local people are generally not interested in participating at the Ward level on a regular basis.
- The system of Ward Committee member nomination is often violated (regardless of the Ward Committee Rules) with ward councillors nominating people with similar political backgrounds, making the Ward Committee a political entity (which is illegal).
- Ward Committees have no financial power to either generate resources, nor do they have the financial resources to take up development work.
- Ward Committees functions are mostly constraint to supervisory and monitoring, due to the above.
- Ward Committees face discrimination from the Corporation if councillors are from the opposition party, often affecting financial allocation, provision of services and redressal of complaints in their wards.

ANNEXURE B
BUDGET ESTIMATES OF REVENUE FROM SWM IN BOROUGHS I-X

(in Rs. Lakhs[5])

	2004-05 (estimated)	2003-04	2002-03
A—SWM District Administration—Recovery from			
Seizure of animals	0.20	0.17	0.40
For filling low land	0.10	0.04	0.01
Notification charges	3.00	2.24	3.09
For removal of encroachment	80.00	64.79	64.38
For temporary occupation of public thoroughfare	5.00	4.43	3.26
Receipt from ceremonial functions	0.10	0.09	0.07
From removal of sullage	5.00	0.02	0.06
Miscellaneous receipts	95.00	91.24	92.26
B—Recovery for Removal of Solid Waste			
From petrol pumps	2.00	0.71	1.38
From nursing homes	0.01	23.42	18.32
From trade refuse and solid waste	8.00	21.86	18.77
From markets	2.00	1.89	2.13
From Fort William	—	0.03	—
For scaffolding and depositing of building materials	10.00	14.52	10.49
From hotels and lodgings	8.00	6.84	6.22
C—Recovery from Carcass Service			
Removal of carcass	0.10	0.25	0.06
Total Conservancy (A+B+C)	218.51	232.54	220.90

Source: KMC 2004-05 Budget 2004.

ANNEXURE C

TABLE 16: BREAKDOWN OF SWM EXPENDITURES ESTIMATED FOR BOROUGHS I-X

(in Rs. Lakhs)

Expenditure	*2004-05 (estimated)*	*2003-04*	*2002-03*
SWM	24.25	14.67	12.65
SWM Office Administration	72.63	55.03	54.02
SWM District Administration	45.10	32.82	25.83
District SWM Office	216.02	180.87	186.46
Housing-Harijan Employees/Street Cleansing & Solid Waste Collection	*8,560.00*	*7,790.12*	*8,227.46*
Removal of Solid Waste	1,415.00	1,236.53	1,106.35
Ward Depot	12.25	4.61	1.57
Cleansing Public Latrine	7.60	4.53	2.72
Removal of Night Soil	*920.50*	*782.65*	*808.27*
Carcass Bin Service	0.18	—	—
Carcass Removal Service	27.00	22.94	20.13
Street Watering Service	11.00	8.33	11.27
Ground Hydrant Service	—	—	1.13
Dumping Service	545.50	479.52	468.40
Composting Service	—	—	—
Incineration	—	—	—
Conservancy Head Quarters	117.02	90.51	90.80
Bio-Medical Waste	30.00	10.84	8.19
Contributions to P.F.	39.00	38.49	41.41
Total	12,043.05	10,752.46	11,066.66
Under the Nagarpally Unnayan (Slum Improvement)			
Conservancy	200.00	150.74	20.59
Pay-and-Use Toilet	150.00	—	—
Total	350.00	150.74	20.59

Source: KMC 2004-05 Budget 2004.

6

Multiple Partnerships Mumbai

DARSHINI MAHADEVIA and BELA PHARATE[1, 2]

City	*Mumbai*
Urban Local Body	Municipal Corporation of Greater Mumbai (MCGM) or Brihanmumbai Mahanagar Palika (BMC)
Population—MCGM, 2001	11.92 million
Population—Mumbai Urban Agglomeration (MUA), 2001	16.37 million
Population Growth Rate, 1991-01—MCGM	1.83% per annum
Population Growth Rate, 1991-01—MUA	2.65% per annum
Area—MCGM	437.71 sq km.
Area—Mumbai Metropolitan Regional Development Authority (MMRDA)	4207.90 sq km*
Local Authorities in MMRDA	MCGM, Thane Municipal Corporation (TMC), Kalyan Municipal Corporation (KMC), Ulhasnagar Municipal Corporation (UMC), Navi Mumbai Municipal Corporation (NVMC), 12 municipalities, 5 census towns and rural areas
Local Authorities in MUA	MCGM, TMC, KMC, NVMC, UMC and 3 municipalities
Water Supply and Sewerage in MCGM	MCGM
Storm water drains	MCGM
Total wards in MCGM	24
% population living in slums	60

Introduction

Mumbai is the largest metropolis of India, with a population of 16.37 million within the urban agglomeration in 2001 (Mumbai Urban Agglomeration—MUA). Mumbai is the capital of Maharashtra State and the financial capital of the country. The city, once a leading industrial centre, has become a tertiary economy, with just one-third of the male workers and one-fifth of the female workers employed in manufacturing and construction (the secondary sector) in 1999-00 as per the National Sample Survey data. The rest of the workers are in the tertiary sector. Less than 20 per cent of the male workers are employed in higher end tertiary sector jobs, indicating a predominance of low-end jobs in this sector, and hence the informalization of employment at low income levels (Mahadevia 2005).

The MCGM area is divided into two major geographic divisions—the island city, a strip of 24 sq km, and the suburban area, north of the island city. While separating the two sections, Mahim creek is gradually filling up because of indiscriminate land development by the builders and city planning agencies such as Mumbai Metropolitan Regional Development Authority (MMRDA). The River Mithi which runs north-south in the suburbs discharges its waters in this creek, and it too has been choked up due to such encroachments.

The city district has one of the highest population densities in the country, with 46,000 persons per sq. km in Mumbai and 20,000 persons per sq. km in suburban Mumbai (SWM Dept. 2004). For the sake of comparison consider the densities of Kolkata with 24,760 people per sq. km, New Delhi with 4,909 people per sq. km, and Bangalore with just 2,979 people per sq. km.

As is often the case, the fact that Mumbai is the financial capital of the country has not translated into the economic uplifting of the city's entire population. Indeed, while the population census of 2001 states that 6,475,440 people or 54.06 per cent of the population live in slums (Office of the Registrar General & Census Commissioner, India 2005: 22), the MCGM website puts this figure at 60[3]. The population density in some of the slum enclaves can reach as high as 400,000 per sq. km (Mumbai pages 1997). To make matters worse, according to an independent estimate, just 12 per cent of the total SWM budget is utilised in the slum areas (Davis n.d.).

Administrative System

The three geographic divisions of Greater Mumbai shown in Fig. 6.1

are further divided into six zones and 24 wards for the purpose of administration (Table 6.1). Each administrative ward constitutes of 10 to 12 electoral wards (Table 6.2) and in total the MCGM has 227 Councillors. Ward (s) Committee (WC), as per the requirement of the 74th CA, is formed as per the jurisdiction of the administrative ward. Even then, the city has just 16 WCs, indicating that in some WCs, there are two or more administrative wards (Table 6.2). All WCs represent a population far exceeding the norm of 300,000 population per WC stated in the 74th CA and the WC number 13, consisting of wards M East and M West, represents about 1 million population! Such large wards committee in a way mocks at the spirit of the 74th CA, which is that of decentralised urban governance.

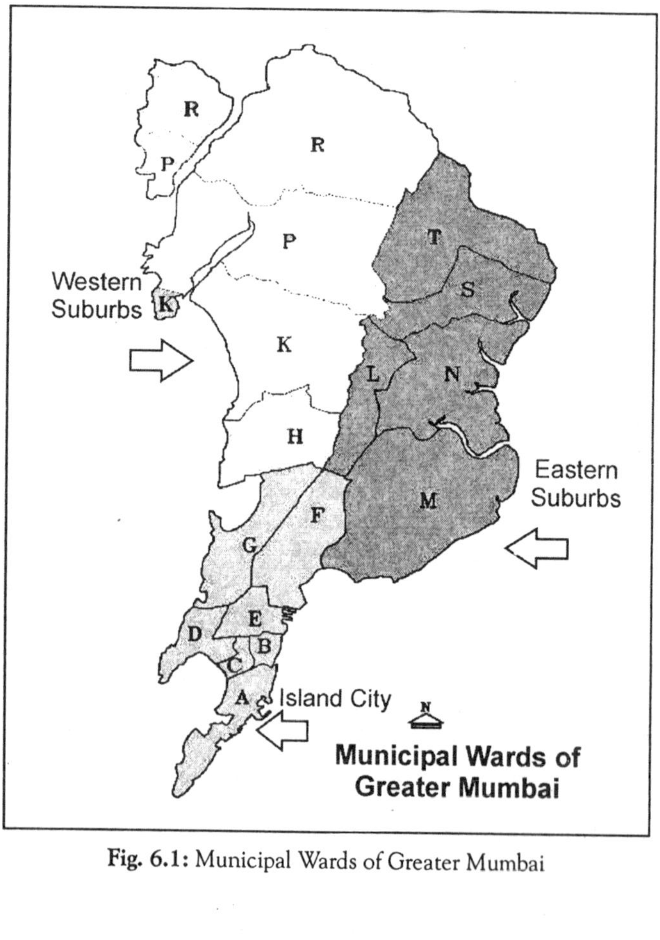

Fig. 6.1: Municipal Wards of Greater Mumbai

Table 6.1: Zones and Wards of MCGM

Island City		*Western Suburb*		*Eastern Suburb*	
Zone 1	*Zone 2*	*Zone 3*	*Zone 4*	*Zone 5*	*Zone 6*
Ward A	Ward F/North	Ward H/East	Ward P/North	Ward L	Ward N
Ward B	Ward F/South	Ward H/West	Ward P/South	Ward M/East	Ward S
Ward C	Ward G/North	Ward K/East	Ward R/North	Ward M/West	Ward T
Ward D	Ward G/South	Ward K/West	Ward R/South		
Ward E			Ward R/Central		

Source: *http: //www.mcgm.gov.in/forms/Wards.aspx?slmno=MTE%3d-7nR%2f8MPTOxg%3d* (accessed on March 14, 2007).

Table 6.2: Wards Committees Details

Ward committee no.	*Ward name*	*No. of electoral wards*	*Total councillors*	*Total population*
1.	A, B & E	16	16	776,066
2.	C & D	13	13	529,564
3.	F South & F North	19	19	922,333
4.	G South	10	10	380,340
5.	G North	13	13	666,776
6.	H West & H East	18	18	893,144
7.	K East	15	15	798,116
8.	K West	13	13	694,151
9.	P South	8	7	421,006
10.	P North	14	14	778,237
11.	R South	8	8	579,954
12.	R North & R Central	15	15	740,722
13.	M East & M West	19	19	1,009,551
14.	S & T	20	20	805,195
15.	L	14	14	744,812
16.	N	12	12	829,640
Total	24 wards	227	226	11,569,607

Source: Mukhopadhyay (2005).

The MCGM has two wings; the deliberative wing and the executive wing. It has nine statuary authorities, out of whom six are deliberative namely (i) The Corporation itself, (ii) The Standing Committee, (iii) The Brihanmumbai Electric Supply and Transport (BEST) Committee, (iv) The Education Committee, (v) Wards Committees and (vi) The Improvement Committee. Of these six, all except the last one have powers to sanction the funds. The remaining three authorities, which are executive in nature, are: (vii) The Mayor, (viii) The Municipal Commissioner (MC), and (ix) The General Manager (Bombay Electricity Supply and Transport—BEST Undertaking). There are two sets of parallel bodies, the elected body formed of the Councillors that is led by the Mayor and the administrative body formed of bureaucracy, that is led by the Municipal Commissioner. In absence of a Mayor-in-Council system, the Municipal Commissioner is the most powerful executive power in the city. In essence, the Municipal Commissioner, who is an employee of the state government, takes decisions on behalf of the state government. This person can be transferred out as per the decisions of the state government and hence cannot be expected to have total dedication to the city. Such an arrangement becomes particularly problematic in case of such mega cities as Mumbai.

The Standing Committee is the most important decision-making authority in the MCGM. It is in effect the bridge between the elected body and the Municipal Commissioner. The main functions of the committee are to sanction contracts, scrutinise estimates of income and expenditure prepared by the Municipal Commissioner and prepare the budget from those estimates, frame service regulations, sanction investment of funds, prescribe the form of keeping accounts and conduct their scrutiny. The Municipal Commissioner can attend the meetings but do not have the right to vote in the Standing Committee. While the Municipal Commissioner prepares and presents the budget, it is approved by the Standing Committee whose tenure is just one year.

A Ward Committee is formed of councillors representing the electoral wards within the territory of the WC, the concerned Assistant Municipal Commissioner, and not more than 3 members of the NGOs/ CBOs engaged in the ward, as nominated/ elected by the WC (Pinto 2005). The functions and powers of the WC's are mainly to:

(i) Redress the grievances connected with water supply, drainage, sanitation and storm water disposal.
(ii) Consider estimates of expenditure pertaining to the wards under different heads of account of the budget.
(iii) Grant administrative approval and financial sanction for municipal works up to Rs. 500,000 each, provided there is a

specific provision in the budget, sanctioned by the corporation. With respect to the planning and implementation of larger works in the ward area, WCs have only advisory role.

The head of the WC is generally a locally powerful person. Often, this deters common people from participating in the discussions in the committee. To further decentralise the system another set of institutional arrangements are required. Towards this end, the MCGM has issued orders for the creation of Local Area Citizens' Committees (LACC) in each of the 227 electoral wards, in an attempt to bring together bureaucrats, citizens and the Councillors closer and have them on the 'same page' (Pinto 2005). Still, how these members will be selected is not known.

Functions of MCGM

Under the Mumbai Municipal Corporation Act of 1988, it is the mandatory duty of the Corporation to maintain the area falling under its jurisdiction in clean and hygienic conditions in order to ensure a good and healthy environment. What follows are the obligatory duties of the Municipal Corporation under section 61 (A), 61(C) and 61 (N) of the Mumbai Municipal Corporation Act of 1988:

- Cleaning of public streets
- Collection of solid wastes including temporary storage
- Removal and transportation of solid wastes
- Disposal of solid wastes
- Disposal of dead bodies of animals
- Construction, maintenance and cleansing of urinals and public sanitary conveniences.

Solid Waste Generation

There are varying estimates of total solid waste generated in the city. As per the CPCB data (Chapter 2, Table 2.2), 450 gms of per capita or 5,320 MTs of solid waste is generated in the city per day. The MCGM website gives a figure of 450 gms per capita and a total of 6,000 MT of solid waste generated per day of which 4,000 MT is the garbage and another 2,000 MT is the debris, silt, cow dung, etc[4].

Another estimate states that 630 gm per capita and total of 7,800 MT of solid waste is generated in Mumbai per day (Table 6.3)[5]. The waste generated figures for the city are estimates and are quite different in the various sources. For this article, the figures given by the MCGM have been accepted as the total waste generated in the city. The island city, housing about 28 per cent of city's population generates as high as 48 per cent of the total solid wastes (Table 6.3). This is because the island city is a major employment centre with very large day time population.

Table 6.3: Garbage Generation by Geographic Locations, Mumbai

City region	*Population (%)*	*Garbage generation*		*Garbage density (MT/ sq. Km)*
		MT per day	*%*	
Island city	27.92	3,700	48.00	53.85
Western suburbs	42.77	2,500	32.00	12.07
Eastern suburbs	29.31	1,600	20.00	10.17
Total	100.00	7,800	100.00	16.80

Source: TERI data as in MCGM website: *http: //www.demographia.com/db-mumbaiward91.htm*

Table 6.4: Physical Characteristics of Municipal Solid Wastes by Region (% Weight)

Category	*Island city*	*Eastern suburbs*	*Western suburbs*	*Whole city***
Biodegradable	42.29	35.72	39.52	39.60
Paper and cardboard	6.16	10.93	6.61	7.78
Plastics	4.23	4.87	5.47	4.69
Metals (ferrous)	0.85	0.65	1.42	0.90
Glass	1.28	0.87	3.48	1.60
Bio resistant*	4.15	11.81	11.07	8.02
Inert materials (debris & silt) & others	41.09	35.20	32.43	37.41
Total	100.00	100.00	100.00	100.00

* Includes leather, rubber, bones, and synthetic materials.

** Calculated using the total waste generated by city region in Table 6.3 and the waste composition by city region in this table.

Source: Salvi (1997).

The composition of solid waste differs across the city regions. There is a higher proportion of biodegradables and inert materials in the solid waste generated in the island city as compared to their proportion in the solid waste generated in the western and the eastern suburbs. In the latter two regions, the proportion of paper and cardboard, and bio-resistant materials in the total solid waste is higher than their proportion in the island city. This might be because of location of manufacturing units, often small ones in the residential areas, in the suburbs.

On the whole, 40 per cent of the total solid waste of the city is biodegradable and another 37 per cent is inert materials consisting of debris and silt and other materials (Table 6.4). Thus, just 40 per cent of the waste of the city is compostable and another 37 per cent can be used for landfill purposes. About 15 per cent of the city waste is recyclable. The composition of waste stated in Table 6.4 does not tally with that in Table 2.2 of Chapter 2. Nor does it tally with the statistics of waste composition within the MCGM website[6]; which is 37.5 per cent compostable matter, 35 per cent inert matter, 15 per cent paper and cardboard, 0.75 per cent plastics, 0.80 per cent metals (ferrous), 0.40 per cent glass and 10.55 per cent others.

In short, this wide difference in two sources of information on the composition of solid waste in Mumbai indicates that more systematic data collection and recording is required for proper management of solid wastes and that even a rich ULB like MCGM is unable to do so. It further indicates the need for decentralised management, which would generate decentralised data with regards to city garbage.

Table 6.5: Constituents of Recyclable Wastes at Source and Disposal Sites (in %), 1994

Zone	Paper		Plastics		Glass	
	Source	DS	Source	DS	Source	DS
Island city	6.16	5.38	4.23	4.10	1.48	1.30
Eastern suburb	10.93	7.08	4.87	3.54	2.07	1.02
Western suburb	6.61	3.98	5.47	3.85	4.23	3.64

DS = Disposal Site

Source: Data from SWM Department, MCGM.

A part of the recyclable waste generated is sold by the households themselves to the itinerant waste collectors and the small remains in the solid waste dumped for removal is scavenged by the rag-pickers. The former does not get accounted in the total waste generated figures given by the city government, but the latter do. Hence, before the waste reaches the final dumping (disposal site), part of it is already taken away by the recyclers. There is therefore a gap between the percentage of recyclable waste generated and the percentage of recyclable waste that reaches the disposal site. For example, 6.16 per cent of the total waste generated in the island city is paper, but only 5.38 per cent of the total waste from the island city that reaches the disposal site is paper (Table 6.5). For the eastern suburbs the two figures are 10.93 per cent and 7.08 per cent respectively and for the western suburbs these figures are 6.61 per cent and 3.98 per cent respectively. Similarly, for plastics and glass, part of the waste generated is picked up for recycling by the rag-pickers. This is happening in all the three regions of Mumbai. Thus, the waste that reaches the disposal site has a lesser proportion of recyclables than that generated at the source.

Overall Waste Management System

Organizational Structure for SWM

The MCGM has an elaborate structure for the purpose of SWM. The Chief Engineer (SWM) is the head of the department, who works under the Deputy Municipal Commissioner. The Municipal Corporation is headed by a Municipal Commissioner, below whom is Additional Municipal Commissioner and then the Deputy Municipal Commissioner. The SWM Department is divided into two wings, namely, the Conservancy Wing and the Transport Wing. The Chief Engineer heads both the wings. For the conservancy works, the Chief Engineer is assisted by the Head Supervisor and three Assistant Engineers below the Head Supervisor. The Head Supervisor with one Assistant Engineer in each region manages the street sweeping and primary garbage collection. For this purpose, there are 53 Supervisors, 433 Jr. Overseers, 1772 *Mukadams* (labour contractors) and workers for sweeping, waste collection and loading of collected waste on the waste transport vehicles.

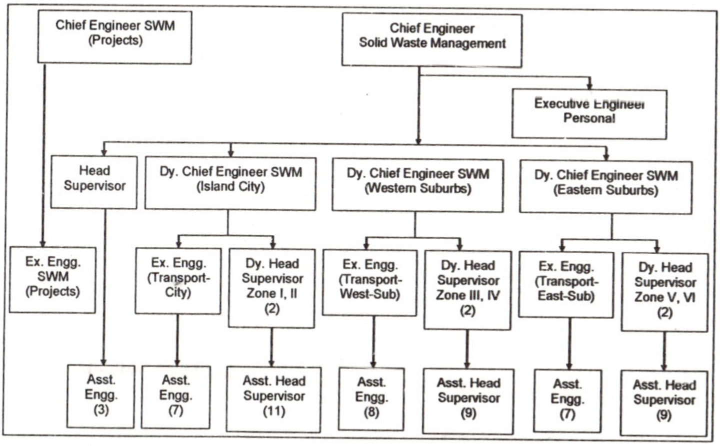

Fig. 6.2: Organisation Chart of SWM Department of MCGM

The transport wing functions by regions and there are 3 Deputy Chief Engineers who head the transport operation of each of the three regions (Fig. 6.2). Under the Deputy Engineer are one Executive Engineer and two Deputy Head Supervisors (DHS) in each region. Since there are two zones in each region, there is one DHS for one zone, who functions as the zonal coordinator. The Executive Engineer is assisted by an Assistant Engineer, who is attached to the ward office. The DHS is assisted by the Assistant Head Supervisor (AHS). Each ward has atleast one AHS. The total staff in this wing consists of 13 Asst. Engineers, 25 Sub-Engineers, 23 Junior Engineers and the workers for waste transport. At the ward level, the Assistant Head Supervisor (AHS) heads the SWM works.

For special projects there is a Deputy Municipal Chief Engineer (Projects) who handles special projects related to solid waste management. The SWM department has total staff of 38,000, who works in 3 shifts round the clock. The MCGM has 7,900 refuse vehicles for transporting the waste. Sweepers sweep road length of approximately 1,800 km every day[7].

The day to day functions are organized at the ward level in each of Mumbai's 24 wards (Fig. 6.3). These functions encompass cleaning of roads, waterways and drains; collection of garbage and debris and its transfer to disposal sites. The SWM is organised at the administrative ward level and the ward(s) committee does not have any role in it.

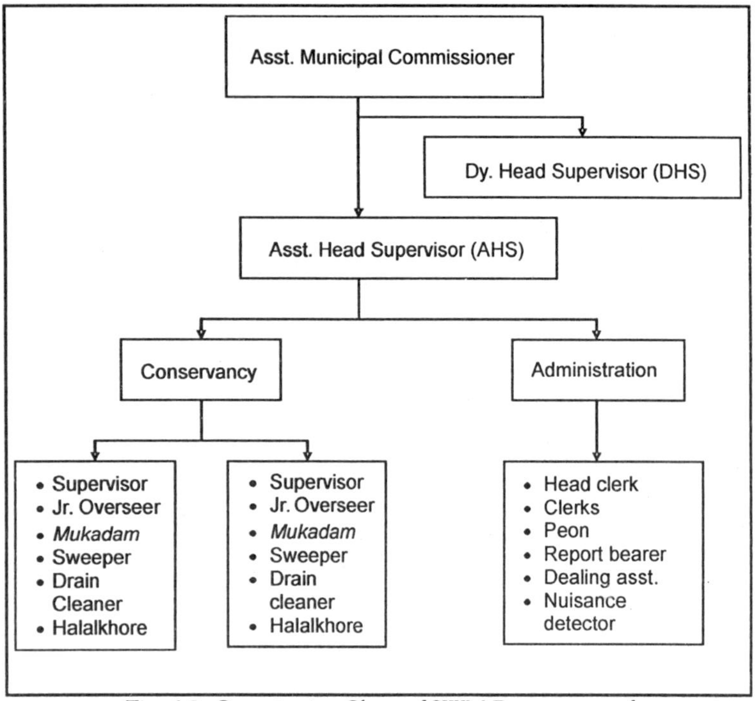

Fig. 6.3: Organisation Chart of SWM Department of MCGM at Ward Level

The AHS, who is the head of the ward, is responsible for day to day sweeping of the streets within the wards as well as collection and transportation of garbage from residential areas in the ward area. He is responsible for all activities of the SWM in the ward under the direction and control of the Assistant Municipal Commissioner (formerly called Ward Officer).

The Conservancy Wing and the Transport Wing are involved in the secondary stage of garbage collection. The responsibilities of the supervisory staff of the Conservancy Wing are to ensure provision of adequate number of community bins at the collection points, sweeping of streets to maintain overall cleanliness and hygiene and to plan its transportation without allowing any backlog. The basic duty of the officers of the Transport Wing is to provide an optimum number of refuse vehicles for safe, timely collection of garbage and its disposal to the land fill site.

Street Sweeping and Collection of Waste

Street sweeping is organized by dividing the city into 'beats', where one beat covers an area of about 4,000—5,000 sq.m. in the island city area and 8,000—10,000 sq.m. in the suburbs. A pair of sweepers is assigned one beat. In all internal areas, the sweeping is from 6: 30 a.m. to 1: 30 p.m whereas on the arterial roads and the main station roads, the sweeping is during the night. There are around 4,200 beats for Greater Mumbai and about 8,400 workers employed for this activity alone. A pair of sweepers use one handcart with 2 containers and brooms. The waste collected is deposited in nearby community dustbins provided by the MCGM.

The MCGM sweepers do not collect waste from individual households. The citizens and institutions are instructed to store their garbage in their own bins, which are to be kept at the gates of their complex, from where the municipal vehicles pick them up mechanically at a specified time. The citizens and the institutions have been instructed by the MCGM to segregate wet and dry waste and store them separately, so that the former be collected daily and the latter one or two times a week depending on the amount generated. The individual households of the housing societies employ waste collectors/sweepers to collect the waste and dump it in the specified community dustbins. In places where local level institutions have been set-up, as in the case of the 'Advance Locality Management' (ALM) scheme (discussed in more detail below), garbage is collected from door-to-door. Of the total population of Greater Mumbai, 90 per cent is served by the community bin collection system and 9 per cent by door-to-door collection[8].

The MCGM has made it mandatory for the households to segregate waste. Firstly, wide notification u/s 368 of MMC Act, it mandated this segregation and also stated that there would be fines levied in case of violation of the rule. In effect since March 1, 2006, the 'Municipal Solid Waste (Prohibition of Littering and Regulation of Segregation, Storage, Delivery and Collection) Rules 2006' have been enforced in the MCGM area. These Rules state the obligatory responsibilities of the MCGM as well as citizens in managing the solid waste and lay out a penalty structure for the violation of the same (See *Annexure A*).

There are about 5,600 community dustbins of different designs provided through out the city for temporary storage of all waste other than the debris, silt and the like. Many of these dustbins are open and bottomless, and require manual and multiple handling of waste. Collection of waste from community dustbins is carried out once a day. The waste collection by the MCGM from hotels, restaurants, office complexes and commercial areas, is partial at present and is separately done. It has been proposed to charge these bulk producers of solid waste for collection and disposal.

Removal and Transportation of Waste

The waste is transported in conventional trucks, compactor vehicles, dumper vehicles and tipping vehicles. The MCGM does not have adequate vehicles and hence hires vehicles from the private contractors. All municipal vehicles are of the hydraulic tipping type but the contractor vehicles are open body and tipping type. The contractors are hired on 3 years term and paid on the basis of rate per trip, or per vehicle shift or per cubic metre.

The removal and transportation of garbage is done in 2 shifts i.e. morning shift from 6: 30 a.m. to 2 p.m. and afternoon shift from 2 p.m. to 10 p.m. As well, depending on the situation, the vehicles are deployed during the night shift too. The removal and transportation of debris and silt is done in one general shift i.e. 8 a.m. to 5 p.m.

For removal and transportation of garbage, 6 municipal workers and 1 Mukadam are deployed on each refuse vehicle, for municipal as well as contractor's vehicles. The workers uses 2 baskets and 2 iron rakes per each vehicle for loading the garbage from the collection point on to the truck. Only the municipal workers are used for loading the transport vehicles. For debris and silt, however, only the contractors' vehicles and their labours are used. 45 per cent of the transportation is through municipal transport and 55 per cent is contracted out[9].

The municipal vehicles and the drivers are under the control of the transport wing, whereas the staff for loading is provided by the conservancy wing. Every vehicle is given the route to serve at a check post in the ward and then it reports back there to register the total waste collected. The vehicle then proceeds to the TS or the dumping site, as the case may be and where again the quantum of waste collected is noted.

Mumbai is a linear city (See Fig. 6.1). The waste dumping sites are in the northern parts of the city, at a distance of 30-40 kms from the southern most tip. Thus, Transfer Stations (TSs) have been set-up, one at Mahalaxmi in the western part of the island city and the other at Kurla just to the north of the island city. Both the TSs combined can handle about 600 MT of garbage everyday. A separate transport is arranged for transferring the garbage from the TS to the dumping grounds located in northern Mumbai. From the suburbs, garbage collected from the collection dustbins is directly transported to the dumping grounds. Given heavy congestion on the city roads during the day, a refuse collection truck can make only one journey to the disposal site in the north in one shift of the transport workers. If roads permit, as is the case in the suburbs, large collection trucks are deployed to collect waste from the collection points to be taken to the dumping sites.

For primary collection, transportation and disposal, MCGM deploys 141 refuse vehicles for the city region and 120 for the suburbs. The Corporation does the operation and maintenance of these refuse-vehicles through a network of transport garages. A total of 13 garages are provided at different locations within Greater Mumbai area (Coad 1997). The maximum trips made by the municipal vehicles in a day are 425 and that by the Contractor vehicles are 660 and the minimum number is 395 and 651 respectively (MCGM 2004a).

At the Mahalaxmi TS, the municipal vehicles carrying waste from south Mumbai directly discharge their contents in the bulk refuse carrier because the latter are parked at a lower level than the former. But, the Kurla TS is at a single level and here the waste from south Mumbai is dumped on the ground and then loaded onto the bulk refuse carrier for transport to the final dumping site. This loading is done by the front end loaders. The TS at Mahalaxmi is underutilized and is capable of handling at least twice the load it is handling (Coad 1997).

Disposal of Municipal Solid Wastes

There are two methods through which solid waste is disposed off by the MCGM. One is through the usual dumping of the waste at landfill sites. The other method is through decentralized systems, where the waste is recycled. We deal with each method in turn.

At present the 4 dumping sites in operation are: Deonar and Mulund in the eastern suburbs, and Malad and Gorai in the western. These were originally creeks and swamps that have been reclaimed through continuous dumping of solid waste. The dumping sites are open thereby providing free access to any one who wants. Slums have come up on these sites and rag-pickers scavenge the waste here.

Refuse and debris are levelled at these sites by means of bulldozers and landfill compactors. Two more landfill sites have been proposed: one 82 ha at Kanjurmarg and other one of 40 ha at Mulund (AIILSG 2003).

Details of three of the four sites are available (Table 6.6). The largest one among them is Deonar with an area of 111 ha and is the closest to south Mumbai where 48 per cent of the total waste of the city is generated. The Deonar site needs to be upgraded and the waste appropriately treated because estimated time it would serve is only 5 more years (AIILSG 2004). The MCGM has started to seek decentralised solutions such as manufacture of organic manure and electric power from the city garbage, which is discussed below.

Table 6.6: Amount of Waste Disposed at Three Dumping Sites

Location	*Area (hectares)*	*Quantity (in MT) of solid waste received per day (maximum)*
Deonar	111.00	6,826
Mulund	25.30	598
Gorai	14.50	2,200
Total	150.80	9,624

Source: MCGM, Dec. 2004

Budget Allocation on SWM

The city is allocating and spending between 7 to 8 per cent of its total budget, which includes revenue (recurring) and capital (development) budgets, on SWM (Table 6.7). The table has Revised Estimates (RE) of the expenditures incurred for 2002-03 and 2003-04 financial years and the budgetary allocation for the financial year 2004-05. The city government is making far more expenditures on the revenue account than on the capital account. For example, in 2002-03, while Rs. 3,884 million were spent on the revenue account, just Rs. 58 million were spent on the capital account on the SWM. Thus, of the total revenue expenditure for the year, 8.70 per cent was expended on the SWM, where on the capital account just 0.55 per cent was expended on the same. It is obvious that most of the expenditures are on personnel costs. Jain (2004) states that in 2001-02, 62.5 per cent of the total budgetary expenditures on the SWM went towards the salaries of the municipal workers and staff. This is nearly the case for all the three years represented in Table 6.7.

Table 6.7: Annual Budget on SWM

	Item	*On SWM (Rs. Million)*	*Total (Rs. Million)*	*SWM as % of total*
2002-03 RE	Revenue	3,884	44,668	8.70
	Capital	58	10,493	0.55
	Total	3,942	55,161	7.15
2003-04 RE	Revenue	4,175	49,880	8.37
	Capital	144	11,081	1.30
	Total	4,319	60,961	7.08
2004-05 BE	Revenue	5,149	54,301	9.48
	Capital	427	16,126	2.65
	Total	5,576	70,427	7.92

RE = Revised Estimates of expenditure;
BE = Budget Estimates indicating allocation.
Source: Data from MCGM (2003, 2004b).

A slight increase in the proportion of expenditures/allocation on the SWM has been noted on the capital account. In 2002-03, this figure was 0.55 per cent, which increased to 1.30 per cent in 2003-04 and to 2.65 per cent in the proposed budget for 2004-05. This is an indication of purchase of better equipments for SWM. The SWM budgetary expenditure increased by 9.56 per cent from 2002-03 to 2003-04 and by 23.33 per cent in the next year. The leap in budgetary allocation in the last year is explained by the leap in allocation on the capital side.

Of the total expenditure on the SWM incurred by the MCGM, in the same year, 38 per cent is on sweeping activities, 57 per cent is on collection and transportation and the remaining 5 per cent is on disposal activities (AIILSG 2003). If part of the sweeping and collection was contracted out, the same 38 per cent can give better results than now.

The New Efforts in Solid Waste Management (SWM)

Three new efforts discussed here are:

(i) The Advance Locality Management (ALM),
(ii) The Slum Adoption Scheme (SAS),
(iii) The Parisar Vikas programme by the Stree Mukti Sanghatana (SMS), and
(iv) The partnerships for disposal of solid waste.

The first three efforts have an element of sustainable disposal and along with these, the MCGM has worked out some other partnerships to take care of the waste disposal problem, given that the present waste disposal sites are not to last much longer.

Advanced Locality Management (ALM) Programme

Genesis

The Advanced Locality Management (ALM) Programme, initiated by the MCGM in 1997 with the main objective of mobilizing citizens to participate in local SWM in an environmentally friendly manner, focuses on 'waste minimization' and the 'segregation of waste at source'. It started in Joshi Lane, Ghatkopar, a North-eastern suburb of the city, where the citizens were concerned about the piling up of garbage on the streets, emanating a stench and blocking the traffic. The MCGM officials diagnosed the cause of this problem as citizen indifference. To solve the problem, the MCGM held meetings and discussions with the representatives of different housing societies of Joshi Lane and convinced them to participate in the applying the principle of three 'R's ('reduce-reuse-recycle'), along with segregation of waste at the source. The residents of the lane then formed a 'Street Committee', which engaged rag-pickers in door-to-door waste collection and the speedy removal of segregated waste and then maintained vigilance to prevent littering. The MCGM then formalized its partnership with the 'Street Committee', wherein the latter's task was to ensure the three 'R's, waste segregation and keep vigil on street littering and that of the latter to regularly clean the streets, pick up the collected garbage and transport it to the final disposal site. The scheme was then voluntarily adopted by the surrounding lanes and six Street Committees were formed in Ghatkopar. In November 1997, it was accepted by the Additional Municipal Commissioner, MCGM for the whole city [Yashada 2004(a)].

Process of Formation and Functioning

An ALM is formed either with members of one housing society or a group of housing societies in a locality (depending on the size of a housing society), the residents and non-resident population of which are committed to meet the conditions mentioned below:

- The residents form a 'Local Committee' (LC) with one representative from each *chawl* or building and select office bearers for interaction with the MCGM.
- The LC to meet once a week and maintain a register to note the problems of the society. The Ward Office to be then notified about the problem(s) and the residents' intention to form an ALM through a formal letter (Fig. 6.4) be presented to the MCGM.
- The residents to segregate waste, wet-waste (bio-degradable waste) and dry-waste (recyclable waste) and store the segregated waste separately.
- Dry-waste to be disposed off directly through rag-pickers and wet-waste to be converted through vermi-composting either individually or at community level. The excess waste to be handed over to the MCGM. The debris to be disposed off by the generator at the MCGM designated areas.

The ALM functions through the following process:

- An ALM Society is registered with the local Municipal Ward Office, which appoints a Nodal Officer to collaborate with the ALM and attend to citizens' complaints promptly by following up on them with the MCGM departments at ward level. This is the essential aspect of this partnership that ensures success of the ALM. In case the complaints are not addressed the Nodal officer passes them on to the Deputy Municipal Commissioner or the Additional Municipal Commissioner of the zone.
- The Nodal Officer addresses the first ALM meeting, during which the major and minor problems of the locality are identified and the problems registered with the ward office are attended to.
- A common fund is set-up at the rate of Rs. 1 per apartment per day, collected on a quarterly basis[10]. Contributions are received from the residents of the lane and are treated as contribution towards the 'Maintenance Fund'. All expenses are to be incurred from the Maintenance Fund.
- An ALM representative attends the monthly meeting with the Assistant Municipal Commissioner at the Ward Office where locality registers and/or unresolved problems are

screened by the latter. General meetings for the whole area are held on monthly or quarterly basis, whose minutes are recorded and then circulated in an ALM Newsletter. During this meeting, all the problems concerning MCGM with respect to the ALM area are discussed among the ALM Committee and other citizens of the neighbourhood.

Date ....

To,

The Assistant Municipal Commissioner,
(Full address of the ward)

Dear Sir,

Sub: Formation of an ALM in our area—ALM Name

We, the members of the above mentioned association have adopted 'Advanced Locality Management' concept in our area. The name of our ALM is ________________.

Our area comprises of a total of__________ families/flats and covers the following buildings/dwellings from _________ to ________. (give the exact location of your area).

Please inform us the name and telephone number of the Nodal Officer who is appointed for our area, the registration number of the garbage truck, the route and the timings for the route. Also let us know the timing and day of the Ward level monthly ALM meeting.

We are also pleased to inform you that we have initiated steps for segregation of garbage.

Hoping to hear from you soon.

Yours truly,

____

(signed) Convenor for
(Name of the ALM)

Fig. 6.4: Format for ALM Registration

Source: www.agnimumbai.org

Once the ALM scheme became successful and was accepted by the people, the MCGM appointed an Officer on Special Duty (OSD) for the purpose of educating people about the scheme. While the overall performance of the ALMs are coordinated by the OSD, their performance is reviewed by the Additional Municipal Commissioner. On ALM members' demand, Nodal Officers allocate a specific day and time on which the ALM can come to the ward office and, in the presence of all heads of department, can discuss issues like SWM. Various NGOs that are already engaged in the activities relating to local governance and groups of senior citizens involved in civic issues became part of the ALM process. Thereafter, the corporates joined the process.

The role of rag-pickers in the overall SWM in the city became important as a result of the ALM movement. Hence, the MCGM took a decision to support the activities of rag-pickers by providing them with sheds, vehicles for dry waste transportation, open spaces and also helped them in linking with the ALMs, co-operatives housing societies, corporates and ward office. The MCGM also extended its support to the NGOs like Stree Mukti Sangathana (SMS), Force, Akkar Mumbai, etc. engaged in training and organizing the rag-pickers for fairer wages, better health and insurance services. The private contractors are also involved in the process, to carry out the same functions that the rag-pickers do, in places where the latter are not engaged.

Scale

Starting with one locality in 1997, the ALM scheme expanded to include 1,000 such societies (Modi *et al* 2002). Then the number decreased to 375 societies in 2000 (Jain 2000) catering to 3.27 million population.

According to the recent estimates 578 ALMs are now registered in the whole city (Table 6.8) and 261 local vermi-composting units working, with the highest numbers in two western suburb wards, H and K (that has high income group residents). Most recently, the ALMs system is to be replaced with Local Area Citizens Groups (LACGs) [Karmayog 2006(b)]. An analysis of this will be presented later.

Table 6.8: Distribution of ALM Groups

Zone	*Wards*	*ALMs present*	*Vermi-composting Units*
1.	A-E	88	33
2.	F/North, F/South, B/North, B/South	72	38
3.	H/ West, H/East, K/West, K/East	181	86

(Contd.)

(Contd.)

Zone	Wards	ALMs present	Vermi-composting Units
4.	P/South, P/North, R/South, R/North, R/Central	67	42
5.	L, M/East, M/West	80	31
6.	N, S, T	90	31
	Total	578	261

Source: Bhagwat (2005).

A Case of Pestom Sagar ALM

Pestom Sagar locality, with a total population of about 20,000, is in the M (West) Ward of the MCGM. The locality had the usual problems of solid waste management; namely non-clearance of garbage from just 6 community dustbins in the locality resulting in littering, slum dwellers and hawkers too littering the streets and the municipal garden of the locality neglected and misused by the anti-social elements.

The locality consists of 6 roads. An ALM was first formed for roads 1 and 2. After its successful implementation, the occupants of roads 4, 5 and 6 joined the ALM. This ALM has a Residents' Committee, headed by a Chief Convener. Each street is under the charge of one committee member, who looks after the cleaning of that street. He/she is in charge of 6 trained sweepers deployed on a permanent basis with all the necessary equipments for road cleaning, collecting garbage from door-to-door, taking the wet garbage to the vermi-compost pit and dumping the dry garbage in the community bins. The residents have to segregate the waste. There is also a Joint Coordinator along with the Convener manage the entire activity. The household contribution for the work is Rs. 0.33 per day per family member, whereas that of the commercial areas is Rs. 1 per day.

The current membership of the scheme is a total of 2,295 people that includes people from slums, hospitals, clinics, shops and vegetable vendors. Members attend the monthly meeting of the ALM at the Ward Office. The vermi-composting pit of size 440 cu ft. is along the periphery of the local municipal garden Kangra Garg Garden that has been leased to the ALM for maintenance. The compost is used in this garden, turning it into one of the finest municipal gardens of the city. During one of the annual festivals of Lord Ganesh (that is held every year), this compost pit processed 100 MT of flowers and other bio-degradable material offerings to Ganesh from 500 *mandals* of the city. The garden had a security cabin for watching over the activities in the garden. Need for vigilance not being required after the ALM took over its maintenance, this cabin has been converted into ALM office. This has saved the municipal body the monthly salary of the security guard.

Slum Adoption Scheme

Genesis

The slum adoption scheme was first tried in Dharavi in October 1997. In the present form, it started with a 1998 study by Youth for Unity and Voluntary Action (YUVA) in 100 slum communities in Jogeshwari (East) which found that: (i) while the residents were aware of the problems related to inadequate practices of household disposal of waste and systems of collection and transportation of garbage in the community, there was very little community involvement in solving the problems; (ii) the slum dwellers lacked belongingness to the city because of heterogeneity of their communities; (iii) the piled up garbage was adversely affecting the health of the dwellers, and (iv) some incentives had to be offered to motivate and involve the slum dwellers in keeping their areas clean. As a result, the Slum Adoption Scheme (SAS) or *Dattak Vasti Yojana*, to be implemented through community-based organizations and resident participation was started by the MCGM. This programme is meant to financially support slum communities to form garbage committees that would then hire workers to clean their respective areas.

The MCGM provided necessary equipment for cleaning and then decided to pay the salaries of the slum cleaners. After the success of this pilot project, the MCGM came out with a 'Slum Adoption Policy' in 1999 (Jain 2000), followed by a circular in 2001, sent over to the 3.27 million slum dwelling population (Redkar 2005), announcing the appointment of an OSD to educate slum dwellers about SWM. Prem Nagar slum was presented as an example to other slums. It is important to note that SAS is only meant for the authorised slums.

The first SAS started in 1999 in Prem Nagar slum community, by formation of a CBO to whom the MCGM provided

The Process and System

The CBOs intending to participate in the SAS require recommendation from their ward councillor and a registration with the charity commissioner office. Thus, the CBO attains a formal institution status. The CBO, on registering, is appointed as an agency to keep the slum area clean on behalf of the MCGM. The CBO has to be representative of the slum dwellers.

Some of the salient features of the scheme are[11] :

- The scheme is to be implemented through established CBO with direct or indirect involvement of NGOs if sufficient experience is not available with such CBOs.
- Supportive grant is provided to CBO, which will be gradually reduced in a tapered form in a span of 2 years.
- The primary responsibilities of the CBO are to provide basic conservancy services, namely, daily sweeping the roads and collection of the garbage and bringing it to designated collection spot.
- The people are persuaded to deposit the garbage in containers placed at their doorstep or at a spot suitable to the community.
- The CBO to coordinate with the municipal authority for the collection and transportation of garbage.
- The CBO to undertake awareness programmes with the slum residents on health, education, water supply, family planning and social welfare activities.
- The CBO could expand its welfare activities for benefiting from other government and MCGM schemes.
- Periodic evaluation of the work of the CBO and timely release of subsidy to be the responsibility of the ward officer.

The waste is collected from door-to-door. A system of keeping plastic containers in the passages for short periods for doorstep collection of the waste has been also introduced in place of waste dustbin at household level, given that in many slums individual housing units are located cheek-by-jowl. The waste cleaners employed to do the task in the adopted slums remove waste from such common containers thus placed. But, because the slums have narrow pathways and alleys, it is difficult to take in large vehicles for waste collection. Hence, waste is collected through handcarts, tricycles, auto-rickshaws or other smaller modes, and then transferred to community bins. The community bins in turn are lifted by a dumper-placer / skip-loader vehicles or waste is dumped in the compactors, and carted away to the landfill.

Financing Mechanism

The initial source of finance for the scheme is the MCGM that provides some annual amount for the first two years, after which the amount is gradually reduced. Then, the CBO is expected to raise Rs. 10/ per household per month for the entire garbage management task and maintenance of toilet blocks. The scheme has been designed such that by the end of the third year, the CBO is self-sufficient in managing waste and sanitation at the primary level. High housing density in slums as compared to the housing societies where the ALMs work, explains how the SAS with one-third the contributions than that in the ALMs, could financially sustain the scheme.

Coverage

As of March 1, 2006, MCGM's official data shows that about 4.93 million slum dwellers (Table 6.9), (63 per cent of the estimated slum population in MCGM area) are covered under the SAS. The beneficiaries of this scheme have increased overtime. In 2001-02, the total beneficiaries of the programme were 2.57 million, which increased to 3.88 million in 2002-03, to 4.43 million in 2003-04 and to 4.53 million in 2004-05. The corresponding figures for slum pockets covered over this period are: 582, 845, 992 and 961 respectively. Rs. 45.88 millions were spent on this scheme by the MCGM in 2004-05[12]. In spite of such wide coverage, the effects are not visible on visiting Mumbai slums.

Table 6.9: Coverage of Slum Adoption Scheme, March 1, 2006

S. No.	*Ward Name*	*Population covered*	*Households covered*
1.	F North	369,581	84,099
2.	G North	196,000	NA
3.	H East	348,000	NA
4.	H West	59,000	NA
5.	K East	589,299	107,683
6.	K West	292,000	NA
7.	L	684,584	NA
8.	M East	495,572	98,743
9.	M West	256,224	NA
10.	N	342,422	NA
11.	P South	237,811	39,862
12.	P North	214,610	43,033
13.	R North	251,325	48,855
14.	R Central	154,520	36,126

(*Contd.*)

(Contd.)

S. No.	Ward Name	Population covered	Households covered
15.	R South	438,885	79,619
	Total	4,929,833	

Source: Based on http: //www.karmayog.com/slums/dvywarddetails.htm

Parisar Vikas Programme

Genesis

The Parisar Vikas Programme (PVP) (Neighbourhood Development Programme) has been initiated by the Stree Mukti Sanghathan (SMS), an NGO based in Chembur, Mumbai. SMS begun its activities in 1975, as part of the Women's Rights Movement, taking up awareness programmes on women's equality and challenging patriarchal culture. SMS has made significant contributions to the women's movement in Maharashtra through various activities, including developing a play about the life of women called '*Mulgi Zali Ahe*' (A Girl is Born). In a society where the birth of a girl is not welcomed, this is a significant step.

There is no estimate of total waste recycled through door-to-door collection by the informal sector waste collectors and the rag-pickers. What is known is that there are about 50,000 to 60,000 street and dump site rag-pickers in Mumbai city, of which 60 per cent are women, 20 per cent are men and 20 per cent are children (Modi *et al* 2002). Further, door-to-door waste collectors and recycling workers total 80,000 to 100,000 in the whole city (Modi *et al* 2002). Therefore, there is already a large section of the workforce in the recycling sector in Mumbai, part of whom could be trained to collect SWM.

Those employed in the collection of such recycling materials are uneducated and unskilled labourers (Modi *et al* 2002), mainly women and children. A survey of 2000 women rag-pickers in Mumbai by the SMS in 1998 found that, rag-picking was restricted to women (85 per cent of them being women and 5 per cent being children) and Scheduled Castes (SC), most of them being SC landless labour families coming from drought-prone areas of Maharashtra. The age groups of rag-picking women varied from 7 years to 70 years and 98 per cent of them were illiterate and therefore not finding any skilled job. The study also found that the rag-pickers suffered serious health hazards resulting from unhygienic work conditions; carried heavy loads and had no form of transportation; and were unaware of their rights as citizens and hence were a highly exploited lot.

The SMS decided to organise these rag-picker women to emancipate them from social discrimination and ostracism faced due to poor quality of their work and low incomes, besides their low caste status. The SMS started the Parisar Vikas Programme (PVP), aimed at making the underprivileged women aware of their rights. After the survey of 2000 women rag-pickers (mentioned earlier), SMS initiated a meaningful dialogue with these rag-pickers, and gave them a name '*Parisar Bhaginis*' (meaning Neighbourhood Sisters). Because the SMS had 25 years of working on women's emancipation issues, it could address comprehensively various aspects of women's exploitation through one programmatic intervention.

This comprehensive programme had components like literacy classes, health and social awareness campaigns, saving groups, and education for the children of *Parisar Bhaginis*. SMS also sought the help of many likeminded organisations for inputs especially in health and educational activities. As work expanded, it was realised that what was most important for *Parisar Bhaginis* was a steady income and safer or more hygienic work conditions. Given this concern, a new dimension was added to the SMS initiative in the institution of entrepreneurship training programs for women. In the process, SMS established partnership with MCGM and was able to bring about changes for the benefit of the *Parisar Bhagini* community. The duration of the project is from 2002 to the end of 2005.

The Activities

The main activities of the programme are:

- Organisation and training of the women rag-pickers.
- Improving the standard of living of women rag-pickers by understanding their problems.
- Developing new techniques for treatment of waste.
- Creating zero waste situations by appropriate waste recycling techniques.

The rag-pickers are given identity cards, with the endorsement from the MCGM and the SMS. They were trained in waste handling, waste collection, transportation of waste to pits and pit management. Such trained women were addressed as '*Trained Parisar Bhaginis*' (TPB). They are also trained in alternative skills such as gardening, vermi-composting, etc. Under the programme, any rag-picker or housemaid living in the nearby slums could join this programme. The training module is of 16 days, from 10 a.m. to 5 p.m., held only two days a week, Wednesday and Saturday. The women are made aware of the importance of health and personal hygiene as the nature of their work makes them prone to different infections. The SMS also held medical camps for them and got all those with health problems treated in local municipal hospitals.

The '*Parisar Bhaginis*' go from door-to-door in several localities in Mumbai and collect garbage already segregated into 'wet' and 'dry' waste. They then convert the former into compost to sell in the market and recycle the latter. The compost is used for maintaining the plant nurseries and gardens in housing societies. Specially designed bio-composting buckets costing Rs. 350 each have been sold to individual households along with a detailed pamphlet describing the process of waste segregation and its importance. Thus what is achieved is a situation of 'Zero Waste'.

At a later stage, Self-Help Groups (SHGs) or micro credit societies of these women were established. With the formation of waste cooperatives, they were also able to get the right price for the sale of the dry (recyclable) waste that they collected, better than if sold through interim dealers in the waste recycling market. This improved the daily earnings and also the working conditions of the rag pickers.

A monthly staff meeting of the SMS workers and *Parisar Bhaginis* is held to evaluate the work progress and to find solutions to problems being faced by them. A bi-annual report is sent to the funding organisation.

Financing Mechanism

The project is being funded by an UK based NGO named 'War on Want', which has given Rs. 20 lakhs for a period of 4 years. Some part of the funds comes from the Central Government's SJSRY.

The SMS charges a household Rs. 10 per month for the door-to-door waste collection and uses this amount collected for the training of *Parisar Bhaginis*. The Parisar Bhaginis earn through sale of recyclable wastes an amount of Rs. 10 per hour of waste collected.

Coverage and Achievements

The SMS has been able to secure solid waste collection, local treatment and recycling contracts from some major public and private residential colonies; of the Tata Power, Tata Consultancy Services (TCS), Reserve Bank of India (RBI), Navy, Bombay Electricity Supply and Transport (BEST) Undertaking, Pfizer, City and Industrial Development Corporation (CIDCO), MCGM, Bhabha Atomic Research Centre (BARC) etc. About 250 trained *Parisar Bhaginis* have brought 'Zero Garbage' status in these colonies spread over 13 wards in Mumbai. In 2004, their work spread to the suburban areas of Navi Mumbai, Kalyan and Dombivli.

Parisar Vikas is currently working with 2,000 women rag-pickers, organized in 200 groups of 10 *Parisar Bhaginis* each. It is implementing waste management schemes successfully at 40 places throughout the city. Parisar Vikas has constructed and operating two *Nisargruna* plants (biogas plants) using BARC technology for processing 5 MT of wet waste per day for MCGM. Today SMS is maintaining five such plants.

The overall achievements of the SMS and its PVP are:

- Construction of a training cum community centre on the MCGM land adjacent to the Deonar dumping ground.
- Starting of *Balwadis* (kindergartens schools) for pre-primary education of *Parisar Bhaginis'* children with the help of Pratham, an NGO working for universalization of Primary Education.
- 300 girls have been given special educational assistance with the help of local donors.
- Starting of a crèche for the children of *Parisar Bhaginis* in the community centre.
- Health camps for women and children with the help of Family Planning Association of India.
- Of the 200 groups of the *Parisar Bhaginis*, 150 are working as saving groups and a federation of these groups has been registered as an independent organisation called 'Parisar Bhagini Vikas Sangha (PBVS). The saving groups together had disbursed Rs. 3.2 million to its members upto 31st March 2004.
- The SMS has been recognized as a training institute, which has conducted vocational training of 50 women, trained all *Parisar Bhaginis* in bio-composting, vermi-composting and gardening, and trained 300 women in manure and gardening techniques through which 250 women have gained meaningful employment. The SMS has also developed 5 to 6

composting models to suit the space available in housing colony.

- Registration of six service cooperatives with 50 women in each to function as business enterprises.

Poverty Alleviation Component

The Parisar Vikas Programme (PVP) has been adopted as a formal poverty alleviation programme under the SJSRY by the MCGM. Before issuing the identity cards, a household survey of the rag-picking women was carried out and then those below the poverty line were included in the list of beneficiaries of SJSRY. Under the SJSRY, a SHG is formed, each one of them getting an initial capital grant of Rs. 10,000 to start a micro enterprise. Of the 200 groups, 63 groups with 678 women have received revolving fund of Rs. 678,000 disbursed under the 'Thrift and Credit Societies' component of SJSRY. Besides that, the MCGM has given a tempo to *Parisar Bhaginis* in 5 wards to collect dry waste. Under the infrastructure development component of SJSRY, shed construction for the storage of dry waste has been sanctioned in 7 wards. It was decided that these sheds would be operated on a cooperative basis under the aegis of PBVS and 5 such sheds are already in use.

A Nisargruna (Bio-gas) Plant at Shatabdi Government Hospital, Chembur

The PVP by the SMS is operating a biogas plant located at Shatabdi hospital, which is a government hospital located in the M/East ward at Chembur. The plant is using 1,500 kg daily of wet waste from nearby hotels. It is using solar energy. The biogas is used in the hospital reducing its energy requirements. The plant is functional since June 2003 and engages three 3 *Parisar Bhaginis* at present.

PVP's SWM at Tata Power Colony, Chembur

Tata Power Corporation's housing colony, located in Chembur and having 540 houses was facing a problem of 600 kgs of kitchen waste and 1 MT of garden waste (dry leaves) generated per day. The SMS was invited to the colony. It began the work by first spreading awareness and then educating the residents of the colony about the importance of waste segregation. Then, it set-up a SHG of 20 women, all former rag-pickers, and then identified a supervisor for monitoring

Photo 6.1: Waste Collection Bin at a Park Maintained by Pestom Sagar ALM, Mumbai

Photo 6.2: Compost Pit in Pestom Sagar ALM, Mumbai

Figure 6.3: Street side plantations and clean Drains, Joshi Lane, Mumbai

Photo 6.4: Mittal Apartment ALM, Juhu

Photo 6.5: Compost Units along Roadside for garden waste, Mumbai

Photo 6.6: Biogas Unit at BARC Mumbai

this SHG. The SHG was given the tasks of garbage collection, road and building cleaning, general upkeep of the colony premises and composting. The rag-pickers, who knew nothing except rag-picking were trained for the all these tasks and then were named *Parisar Bhaginis*.

The *Parisar Bhaginis* are collecting segregated waste from the households and picking up the garden waste mainly the leaves. The wet waste of the households and the leaves are taken to 13 pits made for vermi-composting. The dry recyclable waste is sold by them to earn extra income. The *Parisar Bhaginis* are trained to prepare layers of kitchen waste, dry leaves and cow dung and water it in such a way to check foul smell emission. The compost thus prepared is given back to the Tata Colony for gardening. For this work in the colony, Parisar Vikas earned Rs 88,000 per month in 2005 from which it paid salaries of the women, supervisor and met the cost of renting a truck for final garbage disposal.

The effort still lacks full participation by the residents, who do not fully segregate the waste. At times, *Parisar Bhaginis* have been injured by sharp objects in the kitchen waste, particularly glass pieces or sharp metals.

Decentralised Methods for Waste Disposal

The decentralised waste disposal methods are (i) vermi-composting and bio-methanation by the PVP discussed above, (ii) vermi-composting by the ALMs and (iii) composting by a private sector firm from the vegetable market waste (Table 6.10). Under the PVP, the BARC plant and the Shatabdi Hospital plant, wet waste, that from the canteen in the case of former and from the nearby hotels in case of latter, is used for producing biogas through bio-methanation process. The vermi-composting is done at a plant located in Dadar and the high quality compost generated is called 'Vermigold' under the PVP. All the functioning ALMs vermi-compost the wet waste and the recyclables are sold in the market by the door-to-door waste collectors. A private company, Excel industries is composting wet waste from the vegetable market dumped at Chincholi dumping site. Waste disposed in this way throughout the different waste management units under various organisations adds up to 370 MT per day, coming to just 5 per cent of the total waste generated in the city.

The MCGM has recently issued a work order for establishing a centralized treatment facility on a Built-Own-Operate-Transfer

(BOOT) basis. This treatment facility will be equipped with imported autoclave and shredder machines to deal with different categories of bio-medical waste (MCGM website, Jul. 2005). A Waste to Energy and Compost Plant for 3600 TPD has also been proposed, although no date for commissioning has been indicated (AIILSG 2003).

Table 6.10: Decentralised Waste Disposal, Mumbai

No	*Organisation*	*Method of Disposal*	*Quantity of waste disposed (MTPD)*
1.	M/S Excel Industries Ltd at Chincholi Dumping Ground*	Converting to organic manure	240.00
2.	200 active ALMs through out the city#	Vermi-composting (Individual or Community based)	50.00
3.	5 T plants at Dadar (market waste), Versova and Colaba#	Vermi-culture	15.00
4.	Stree Mukti Sangathan's (SMS) composting units+	Composting	21.00
5.	Waste collected by *Parisar Bhaginis* of SMS+	Recyclable dry waste	1.00
6.	Approx. 30—40 Municipal Gardens#	Composting	1.50
7.	Hotel Orchid#	Vermi-Composting	0.15
8.	Composting units under Force Foundation#	Vermi-Composting	20.00
9.	Churchgate Plaza#	Composting	0.60
10.	Units belonging to Force Foundation#	Composting/ Vermi-culture	20.00
	Total		369.25

Source: * MCGM, 2004.
Discussion with personals from Dept.
+ Data from SMS.

Achievements

The above cases are the some instances where the MCGM has involved other stakeholders such as the CBOs, and NGOs in addressing the problem of solid waste management. The other stakeholders have been involved in tasks that the MCGM was not performing itself, namely, door-to-door waste collection and sweeping inside the apartment complexes and slum colonies. And, thus at no additional cost, the MCGM has moved towards fulfilling the requirements of meeting the SWM Rules of 2000. Further, these rules cannot be implemented without the citizens' participation household level waste segregation is required. The MCGM is using both carrot and stick approach; it is facilitating the setting up of the ALMs and slum CBOs and also addressing all local level issues through these organisations on one hand and on the other imposing penalties if the individuals or organizations are violating these rules. The MCGM itself has come up with SWM Rules 2006, that makes it mandatory for the citizens as well as the MCGM to fulfil their responsibilities in the SWM system. Lastly, the entire SWM operation is coordinated by the MCGM and it is depending on the partners to contribute where it was not doing anything. After nearly five years of the implementation of these programmes, specific benefits in case of each of the three innovative methods can be visible. These are discussed below.

Achievements of ALM

Wherever ALMs have been formed and are functioning, the principle of three Rs, Reduce-Reuse-Recycle is being applied. The waste is segregated at the source, the recyclables are removed at the source to be taken away by the rag-pickers thereby providing them an income and then the wet waste is vermi-composted on the site. This leads to a 'Zero Garbage' situation, elimination of community dustbins and reduction in total waste generated at the city level. This scheme has considerably reduced the burden of primary collection, transportation and disposal of waste which in turn has reduced the MCGM's expense on the waste disposal process amounting to Rs. 1.5 per kg of waste (Jain 2000). This has been counter-balanced by the increased cost of door-to-door collection, which has been passed on to the residents. It is estimated that about 25 per cent of the ALMs are managing entire solid waste at the local level through vermi-composting and recycling of dry waste (Redkar 2005).

The other feature of the scheme is that, as a prerequisite to the ALM scheme, the housing societies are asked to register themselves with the MCGM. Weekly meetings and focus group discussions are held between the residents of an ALM society and the Ward Officer of MCGM, through which problems at the local level are identified and dealt with. Awareness programmes have also been conducted in the localities and in schools with the help of local NGOs and CBOs.

With the success of the ALMs in SWM, the Municipal Commissioner has delegated additional functions to them, such as beautifying the localities and maintaining gardens, parks and roads, in addition to tree plantation and even preventing encroachments on pavements and streets. The successful ALM movements have been given the responsibility of their immediate neighbourhoods, such as Juhu Beach development, along with improving their own areas (Kundu 2005). As well, ten corporate houses have come forward to take up waste management and beautification of some roads and these are registered as 'ALM Trust'. The ALM Trusts are encouraging vermi-composting. On the other hand, ALM movement has also provided a forum to the citizen's for participating in city governance. In 2007 municipal elections of Mumbai, one member of an ALM stood as election candidate.

People's habit of throwing garbage on the roads has reduced with increase in segregation of wastes at household level. Dustbins from the main roads have been removed in ALM areas, providing a cleaner look to the streets.

Achievement of SAS

Because of the SAS the slum CBOs have become more aware of the functioning of the ward system. They have become formal organizations with registration and audited accounts. They are now prepared to take on many more complicated and larger contracts of solid waste, mid-day meal and slum sanitation programmes. The scheme has also resulted in creating employment opportunities for the slum youth, women and rag-pickers. The CBOs headed by women are more than 35 per cent of the total CBOs (Redkar 2005) and this has enabled women to participate in development processes that would eventually lead to their empowerment in the society.

Some of the glaring examples of the success of this scheme are Meghwadi in Jogeshwari (East), Ekveera Nagar in Kanjurmarg (East) and Kachpada in Malad (West). A special drive has been taken to contact CBOs in different slum pockets and motivate them to come into the network of the scheme. Priority is being given to those slum pockets that are located along major drains, so that the problems of flooding could be reduced. About 63 per cent of the slums have been covered under this scheme.

Achievements of PVP

This programme has simultaneously achieved four goals: clean neighbourhoods by deployment of *Parisar Bhaginis*, poverty alleviation of the participating women, skill upgradation and capacity building of the *Parisar Bhaginis*, and contribution towards waste reduction at the city level and thereby reducing the waste reaching the landfill site. The NGO involved in the activity being a women's organization, it has introduced many welfare measures for these women, such as health programmes, children's education, opening of Public Provident Fund (PPF) accounts and life insurance for 500 women under *Janashree Vima Yojana* of the Life Insurance Corporation (LIC) of India. The Urban Development Department of Government of Maharashtra has noted the importance of this partnership and in its GR of May 2004 and it has recommended all other municipalities of the state to adopt this approach. It has empowered women through organising them and increased their bargaining power. Lastly, it has given women sustainable income and self-sufficiency.

Participatory Governance

Innovative initiatives such as ALM, SAS and PVP have provided a platform where the urban local body and the citizens work through consultation rather than confrontation. Such interactions help in developing people's faith in the city's governance. It brings in a sense of ownership of the city among the people and increases the possibilities of bringing transparency to the state structure.

Benefits to Rag-Pickers

Through these programmes, the MCGM has accepted the importance of the role of the rag-picker in the management of solid waste. They have been formally brought in the formal SWM system, with better wages than before and a social security cover, increasing their feeling of well being.

Improved Waste Management System

The involvement of MCGM in undertaking these experiments with citizens groups and NGOs resulted in innovations in methods of waste management. Nevertheless, there is still much ground to cover in achieving a fully functioning and effective waste management system.

Table 6.11: Stage-wise Partnership with MCGM for Solid Waste Management

Stage of Involvement	*Name of NGO/Govt/ Private body*	*Type/Extent of work*	*Nature of Contract*
Primary waste collection	ALM Society	Segregation of waste at household level	ALM Society is registered with MCGM
	CBO working with the adopted slums	Waste collected through hand carts, tricycles, auto-rickshaws or similar modes and transferred to community bins	A recommendation from their ward councillor and registration with the charity commissioner office
	SMS	Training of women rag-pickers Area allocation for waste collection	No signed contract as yet between MCGM and SMS, the latter annually reports to MCGM.
	Parisar Bhagini	Collection of segregated waste at household level	Each Parisar Bhagini is given an I-card, signed by MCGM and SMS
	AGNI	Assistance in formation of ALM	Voluntary
Collection point	MCGM	Collection of waste from community bins, loading it into transport vehicle	
	Rag-pickers	Waste segregation at collection points	No contract
Waste transportation	MCGM	Provides labour and vehicles. Transports directly to TS and/or dumping site	
	Private contractors	Provides labour and vehicles	Formal, renewable contract with MCGM
Waste treatment and disposal	MCGM	Dumping at landfill sites	
	ALM Society	Biodegradable waste disposed at independent or community vermi-culture units and recyclable waste handed over tq Parisar Bhaginis	ALM Society registered with the MCGM
	SMS	Vermi-culture units for composting, bio gas plants for waste to energy generation	Formal contract with the waste supplying units (hotels or canteens) and with the units using the biogas produced
	Private bodies	Application of waste treatment technologies	Contractual

Concerns[13]

Although the MCGM has passed its own rules for SWM, has brought in partners to abide by these rules and have created decentralized models that can take care of the long term sustainability issues, all these efforts are still quite fragmented and the city has still a long way to go. Just 63 per cent of the slums and 27 per cent of the non-slum areas are part of the new efforts and only 5 per cent of the total waste of the city is being managed in such decentralized manner. The deluge of July 26, 2005 has shown that the garbage littering that goes in the city's open storm water drains inundating large parts of the city during heavy monsoon, is still a problem to reckon. Even moving around the city one finds garbage piled up and dumped in large parts of the city, especially close to the slum areas. In fact, some slums are located on garbage dumps[14]. This study also shows that there are some issues of conflicts and sustainability that need to be addressed even in the new efforts.

Conflicts and Sustainability Issues of ALMs

The coverage of the ALMs is still low across the city as the many citizens are not yet convinced that they should be participating, through financial contribution and time, in what they view as the local government's responsibility. This argument comes across more from the property tax paying citizens than the slum dwellers. Many citizens were not even finding garbage littered around as a problem, as they were accustomed to it and themselves throwing it around indiscriminately. These problems were faced even when the first ALMs were being set-up.

Theoretically vermi-composting at the neighbourhood level is a good idea, in many of the ALMs did not want them in their locality for the fear of foul odours being emitted. Such units require a very good management at the local level. Many ALMs wanted that the MCGM also picks up wet waste. As this contentious issue has not been resolved, the number of functioning ALMs has reduced in the city.

Lastly, to note is the fact that the ALMs are largely middle and high income based groups, whose interest is in a 'Clean City', at the exclusion of the poor. Hence, although many of the initiatives of the ALMs are recommendable, some have become organised forums supporting the eviction of slum and pavement dwellers from their own localities.

To overcome these problems, the MCGM in partnership with the NGO Council (formed in 2005), has decided to set-up a city-wide system of decentralized SWM coupled with other civic engagement, based on a system of LACGs. A LACGs will take care of 5,000-10,000 people, have clearly defined service areas, manage their own affairs including fees for collection, and be formally recognized by the municipal authorities as a means of ensuring civic interaction, participation and a sustainable working process. Individual ALMs or groups of ALMs may become LACGs. To ease the transition, a manual has been prepared for citizens' use [Karmayog 2006(b)]. At the same time MCGM, on the urging of the NGO Council has adopted the Municipal Solid Waste (Prohibition of Littering and Regulation of Segregation, Storage, Delivery and Collection) Rules 2006, which have been framed keeping in mind the ease and effectiveness of their implementation (MCGM 2006).

Conflicts and Sustainability Issues in SAS

The SAS is not without its own concerns and problems. One of the most glaring drawbacks of this programme is that it is meant only for the authorized slums. Due to this, the non-authorized slums of the city and pavement communities do not have any means of effectively dealing with their waste and sanitation issues.

The CBOs formed in the slums have not been able to manage the entire slum area and are found to be serving only a small section of the areas allocated to them. Thus, the beneficiaries' total contribution does not add up to any great amount; it comes to just 8 per cent to 10 per cent of their total costs of keeping the slum clean. Hence, the CBOs face a great difficulty in the operation and management of the cleanliness drive. The CBOs therefore, have begun to take up other activities that would give them financial sustainability. One of the method resorted to is to sub-contract the cleaning job to smaller groups at low wages and in the process become agents in the system of service provision. This has been going on because there has been no supervision from the MCGM. It therefore appears that the MCGM has just passed off its responsibility of keeping slum areas clean to the communities.

In places where the SAS has worked, local politics has entered; the local councillors have gradually started to promote their own party youth groups for the latter to get financial benefits and the former to draw political mileage. In such situations, the control of the programme is with the councillor and not the CBO. Some instances of corruption have also been found. In some slums, the MCGM has provided the grants for the wage of workers at the rate of one labourer for 1,000 households. Instead of hiring the required number of workers to clean the slum, lesser numbers are hired. The balance amount is pocketed—the money shared by the local officials and a few slum dwellers in leadership positions. This has happened because the slum dwellers are not yet aware of this programme and those few who are aware have tended to make private gains.

In essence, SAS was designed to be a sustainable programme to implement SWM scheme in the slums, but, in reality, it has not succeeded much, unless, there has been an NGO working with the community for a long time, and which has been able to put up a functioning and sustainable CBO for the task.

Sustainability Issues in PVP

As of now, no conflicts have emerged in the case of PVP though the coverage of this programme is quite small as compared to the size of the city. In the long-run, the activity of the rag-picking has to be eliminated and these women (*Parisar Bhaginis*) need to be given the status of sanitation workers. Also, just a small proportion of the rag-pickers have been covered under the initiative. It is also not feasible that any one NGO can do this task. Some ways of eliminating rag-picking all together and creating a regular cadre of sanitation workers with better wages and social security is required to be created.

Overall Concerns

On the whole, the first major concern is that the efforts have been too fragmented. When a problem arises, a short-term solution is sought. The MCGM, like municipal corporations of Hyderabad and Bangalore, has not thought of any city level plan to deal with the solid waste problem. The biggest constraint in doing so is that the size of the city is too large. This calls for decentralised approach of MSW. This however, has not been done sincerely in Mumbai, given that the Wards committee size in some instances is as large as a million plus city.

The other major concern is that there is no single authority in Mumbai who has full responsibility for the city. The Municipal Commissioner acts on behalf of the state government, the Mayor is powerless, the Chief Minister of the state (of Maharashtra) intervenes into the affairs of Mumbai particularly taking decisions about land use (often guided by vested interest than the interest of the city[15]), a system of good governance is difficult to establish. Every monsoon deluge crisis brings some knee-jerk reactions and some directives from the state High Court. Otherwise, the problems sustain in the absence of keen interest from a responsible authority to keep the city clean for better health and hygiene of its citizens.

The major concern is the predominance of slums in Mumbai, where about 60 per cent of the population lives. It is called 'Slumbay' with good reason, if snidely (Mahadevia and Narayanan 2006). Slums are often located on low lying areas, marshy lands, or hill slopes where it is difficult to put sewer and storm water drains, resulting in water logging during monsoons. These are the areas not served properly by the city's sanitation services. The population diversity is so high that forming community organisations for participation is quite difficult. It is not that attempts are not being made. Many NGOs are working with the slum communities, but these efforts are far from adequate considering the sheer scale of the problem. In such large cities, the local government has to be quite proactive to address solid waste concerns along with the concerns of delivery of basic services. The lasting solution is certainly re-housing the slum population into regularised housing colonies; in other words having massive affordable public housing programmes for the city, and it appears that city is not moving any where close to achieving this goal.

The task for keeping the city of Mumbai clean is gargantuan. Decentralised institutional development in a systematic manner is required to be able to manage the city waste in a sustainable manner. The fragmented approach currently adopted, in spite of many remarkable and innovative developments, is not sufficient.

NOTES

1. Assisted by Amit Mistry, Student, School of Planning, CEPT, University.
2. We would like to acknowledge the help of the SWM Department of MCGM, Mr Ashish Devasthale and Mrs. Uma Chatterjee of All India Institute of Local Self Government (AIILSG), Mumbai, for providing us with relevant documents and reports and also access to their library. The discussions held with Mr. Bejoy Davis of YUVA, Mr. Sunil Banker

and Mrs. Jyoti Mhapsekar, President, Stree Mukti Sangathana, Mrs Seema Redkar, ALM Officer for Western Suburbs, Dr. Sangule of Pestom Sagar ALM, Mr. Viren Merchant of Joshi Lane ALM, have been very useful. We are also grateful to Mr S.S. Bhagwat, Ex-Executive Engineer, SWM Department for sharing his knowledge and experience regarding the process of formation of ALMs and also providing us with reading materials on the same.

3. http: //www.mcgm.gov.in/forms/Masterdisplay.aspx?slmno=Mzg%3d-wXIpbBtWrGI%3d
4. http: //www.mcgm.gov.in/forms/Masterdisplay.aspx?dno=MzQ%3d-nLurYdAdnAE%3d&slmno=MTA%3d-Hk%2buffz58TY%3d&tlmno=MQ%3d%3d-Yh62L2fcsbg%3d (accessed on March 14, 2007).
5. Ajit Kumar Jain's presentation at All India Institute of Local Self Government (AIILSG 2004).
6. http: //www.mcgm.gov.in/forms/Masterdisplay.aspx?dno=MzQ%3d-nLurYdAdnAE%3d&slmno=MTA%3d-Hk%2buffz58TY%3d&tlmno=MQ%3d%3d-Yh62L2fcsbg%3d (accessed on March 14, 2007).
7. http: //www.mcgm.gov.in/forms/Masterdisplay.aspx?dno=MzQ%3d-nLurYdAdnAE%3d&slmno=MTA%3d-Hk%2buffz58TY%3d&tlmno=MQ%3d%3d-Yh62L2fcsbg%3d (accessed on March 14, 2007).
8. http: //www.mcgm.gov.in/forms/Masterdisplay.aspx?dno=MzQ%3d-nLurYdAdnAE%3d&slmno=MTA%3d-Hk%2buffz58TY%3d&tlmno=MQ%3d%3d-Yh62L2fcsbg%3d (accessed on March 14, 2007).
9. MCGM website: http: //www.mcgm.gov.in/Departments/swmanage/wastegeneration.htm.
10. It is estimated by the residents of Joshi Lane that the cost of Integrated Solid Waste Management by the residents is Rs. 8 per capita per month or Rs. 96 per annum (Jain 2004).
11. http: //www.mcgm.gov.in/forms/Masterdisplay.aspx?slmno=Mzg%3D-wXIpbBtWrGI%3D.
12. Data in this paragraph from http: //www.karmayog.com/images/dvystatus0105.pdf. These data are from SWM department, MCGM.
13. This section is based on discussions with the city government officials, NGOs involved in SWM, and discussions on the site with the people themselves.
14. See study by Mahadevia *et al* (2006) for the description of one such slum in Ward M of MCGM.
15. For the discussion on this aspect, which is quite overwhelming in the physical development of Mumbai, see Mahadevia and Narayanan (2005).

ANNEXURE A

Penalty Structure under Municipal Solid Waste (Prohibition of Littering and Regulation of Segregation, Storage, Delivery & Collection) Rules 2006, of the MCGM

On and after the date of commencement of these Rules, there will be a familiarisation/warning period as stated below, after which, any contravention of these Rules shall be punishable with fines as per the Schedule of Fines (Schedule I) below for every instance of breach of these Rules and thereafter, on a daily basis, for repeat offences.

Provision

1. For the Rules related to the Prevention of Littering (Rule 4.1), Nuisance Creation (Rule 4.2), ensuring "*Saaf Aangan*" (Rule 4.3), and for the Specific Categories/Situations (Rule 7), the warning/familiarisation period is one month, during which time the fine charged will be half the fine specified in the Schedule of Fines.
2. For the Rules related to the Segregation, Storage, Delivery and Collection of Municipal Solid Waste (Rule 5), the warning/familiarisation period is two months, during which time no fine will be charged. This familiarisation period will not be applicable to those generators of waste to whom collection-at-source or point-to-point bell ringing waste collection services are provided or those areas identified and notified as 'Clean Mumbai Zones[15]'.
3. For breach of Rules in any notified "Clean Mumbai Zones", the Fines applicable will be twice the amount as stated in this Schedule of Fines.
4. For repeat offenders, the fines charged will be five times the Fine stated in this Schedule of Fines.
5. There will be an escalation in the fines every year by 5 per cent or Rs. 5, whichever is higher, or as deemed appropriate by the Municipal Commissioner.

Schedule—I (Schedule of Fines)

Sr. No.	Rule No.	Sub-division/Description of Rule	Amount of Fine applicable for breach of Rule
Rule No. 4: Littering, Creating Nuisance, and Saaf Aangan[15]			
1.	4.1	Littering	Rs. 100
2.	4.2	Spitting	Rs. 50
3.	Creating Nuisance	Bathing	Rs. 50
4.		Urinating	Rs. 50
5.		Defecating	Rs. 50
6.		Feeding groups of animals/birds in non-designated areas	Rs. 50
7.		Washing vehicles	Rs. 100
8.		Washing utensils/clothes/any other object	Rs. 50
9.	4.3	For not maintaining Saaf Aangan: for a) for owners/occupiers of single premises b) for others	 Rs. 100 Rs.1000
Rule No. 5: Segregation, storage, delivery and collection			
10.	5.1 and 5.2	For delivering waste that is not segregated and stored as specified in separate bins: a) individual b) bulk generator	 Rs.100 Rs. 500
11.	5.3	For not delivering bio-degradable waste in a segregated manner as specified	Rs.100
12.	5.4 and 5.5	For non-composting by bulk generators or in new constructions within 6 months of these Rules, and for others when applicable	Rs. 100/day
13.	5.6	For not delivering specified household hazardous waste in a segregated manner as specified	Rs. 500
14.	5.7	For not delivering biomedical waste in a segregated manner as specified	Rs. 500
15.	5.8	For not delivering Construction and Demolition waste in a segregated manner as specified	Rs. 1000
16.	5.9	For not delivering "dry" waste in a segregated manner as specified	Rs. 100
17.	5.10	For not delivering garden waste and tree trimmings as specified	Rs. 1000
18.	5.11	For depositing waste outside designated community waste storage bin or in any non-designated area	Rs. 100
19.	5.12	For disposal of waste by burning	Rs. 500

(Contd.)

(Contd.)

Sr. No.	*Rule No.*	*Sub-division/Description of Rule*	*Amount of Fine applicable for breach of Rule*
Rule No. 7: Specific Categories/Situations			
20.	7.2	For not delivering (non-household) fish, poultry and meat waste in a segregated manner as specified	Rs. 500
21.	7.3	a) For a vendor/hawker without a container/waste basket	Rs.100
		b) For a vendor/hawker who does not deliver waste in a segregated manner as specified	Rs. 100
22.	7.4	a) For not keeping a house gully clean	Rs. 1000
		b) For not delivering solid waste from a house gully in a segregated manner as specified	Rs. 500
23.	7.5	For littering by pet/owned animals	Rs. 50
24.	7.6	For not cleaning-up after public gathering/event within 24 hours	Forfeiture of the Cleanliness Deposit

7

NGO Initiated Solid Waste Management

Ranchi[1, 2, 3]

DARSHINI MAHADEVIA and APPEEJI PARASHER

City	Ranchi
Urban Local Body	Ranchi Municipal Corporation[4] (RMC)
Population—RMC, 2001	846,7332[5]
Population—Ranchi Urban Agglomeration (RUA), 2001	862,850
Sewerage and Storm water drains in RMC	Urban Development Department of state government
Total wards in RMC	37

Introduction

Ranchi is the capital of the newly created state of Jharkhand, carved out from Bihar state in 1999. Jharkhand is part of a mineral rich and tribal population dominated region called the Chhottanagpur Plateau. Because of its altitude of 654.5 m above the sea level, lush green hills, salubrious climate and scenic waterfalls, it was known as the 'summer capital' while Patna was the 'administrative capital' of Bihar. Jharkhand derives its name from the extensive forests which cover about 29.3 per cent of the total area of the state. Ranchi is also a well recognized tourist centre because of its climate and regional surroundings. The city contains a few lakes, one of which suffers complete eutrophication because of solid waste dumping and sewerage discharge. The average annual recorded rainfall is around 126 cm. About 30 per cent of city's population is composed of Scheduled Tribes (ST). Interestingly, the city has a quite high literacy rate of 85.3 per cent; 91.3 per cent for males and 78.3 per cent for females.

Ranchi is an interesting case study because the solid waste system is promoted largely by an NGO. The City government has been in supersession since 1991, and the municipality is run by a state-appointed commissioner. There is thus a completely different approach to management.

At present, most areas of the city have no sewerage system[6] or treatment facility other than the areas served by two industries, Heavy Engineering Corporation (HEC) and Metallurgical & Engineering Consultants (MECON). All the major and minor drains in the city are open and have deposits of silt and garbage. Most of the residential buildings, commercial and educational institutions, have on-site septic tanks and soak pits as temporary arrangements in the absence of a sewerage network. The overflow from septic/soak pits finds it way to the nearby storm water drains or lowlands, which ultimately discharge into the Subarnarekha river. In some cases, raw sewage is also being discharged directly to the drains or even open lands. The roadside drains have not been designed scientifically to take care of peak rainfall. Further, only 40 per cent of drains are '*pucca*'[7] and the drainage network has not been designed in an integrated manner. The random dumping of garbage (especially plastic) and silt leads to clogging of drains and subsequently flooding even during mild rains.

By the new century, the Ranchi Municipal Corporation was unable to cope with the accumulated solid waste because of financial and administrative constraints. Worse still, the Government of the state of Jharkhand (GoJ) was also unable to give grants to the RMC because of its own budgetary woes. If Ranchi city was to be cleaned, community action had necessarily become the only way, and so the Clean Jharkhand Project (CJP) was envisaged.

The CJP is a Community Based Solid Waste Management (CBSWM) initiative, started by an NGO named Nav Bharat Jagriti Kendra (NBJK) and funded by the bilateral funding agency, Indo-Canada Environment Facility (ICEF). When the former approached the latter for financial support for this initiative, it was only approved on the condition that the GoJ and the RMC were involved in the project as partners. The CJP works towards creating a better local physical environment through institutional building, mass awareness, capacity building and local resource recovery approaches[8].

The decision to undertake the Clean Jharkhand Project was taken at the state level. The Honourable Chief Minister launched the CJP on 16th November 2002; its envisaged duration is 5 years (Nov. 2002 to Oct. 2007). The Ranchi project is a demonstration project of a new system of SWM, and is intended to be used for advocacy, mass awareness and community action purposes in 14 other major municipalities of Jharkhand state[9]. Following this initiative, the GoJ hopes to adopt a policy document on SWM and move towards fulfilling its obligations under the MSW Rules 2000 in the entire state. When the project was launched, the NBJK along with the RMC organised a cleanliness drive employing 125 workers for 15 days. This was done to send a message across the community about the kind of change one could expect in the future.

Coverage of the Project

To decide on the coverage of the project the CJP, in collaboration with the RMC, conducted a survey in June 2004 to determine the quantity and composition of waste generated in Ranchi city from a sample of 412 households in 7 different wards representing three different income groups, high, middle and low. It was found that the total waste generation was 325 Metric Tonnes (MT) per day and per capita waste generation was 247 gm /day. From this it was calculated that the total waste generated from the household sector in RUA comes to 209 MTs per day (Table 7.1). The data also showed that wards 20 and 8 (the most populous) generated the largest amount of waste, 14 MTs and 10 MTs respectively per day. While the NBJK is currently working in ward number 20; ward number 8 is still managed by RMC.

The study also found that: (i) of the total waste generated, 24 per cent was inorganic and remaining 76 per cent organic; (ii) the proportion of recyclables in the inorganic waste was 19 per cent, comprising of 1 per cent rubber, 8 per cent paper, 1 per cent glass, 8 per cent plastic and 1 per cent metal, and remaining 62 per cent are other wastes; and (iii) of the total organic waste, 65 per cent was kitchen waste, 2 per cent cloth waste, 1 per cent wood and remaining 32 per cent is other waste. By income, the study found that high income groups generated the highest amount of waste at 38 per cent (NBJK 2005).

Table 7.1: Ward-wise Population and Total Waste Generated Per Day

Ward No.	*Households (2001)*	*Population (2001)*	*Total Waste Generated (in Kg/day) @ 0.247 Kg/day*
1.	5,089	30,153	7,448
2.	6,212	33,537	8,284
3.	2,061	12,656	3,126
4.	6,108	35,494	8,767
5.	1,842	12,700	3,137
6.	780	4,868	1,202
7.	3,891	24,638	6,086
8.	7,219	41,477	10,245
9.	6,384	35,551	8,781
10.	4,974	27,058	6,684
11.	3,557	27,058	6,166
12.	2,175	24,964	4,005
13.	2,350	16,215	4,045
14.	2,457	15,958	3,942
15.	3,387	23,624	5,836
16.	2,045	14,929	3,686
17.	2,925	19,821	4,896
18.	4,460	24,834	6,134
19.	5,882	34,363	8,488
20.	9,941	56,749	14,018
21.	5,162	29,826	7,367
22.	5,698	31,951	7,892
23.	3,462	18,770	4,636
24.	3,645	19,806	4,892
25.	3,970	23,266	5,747
26.	4,870	22,556	5,572
27.	3,756	22,077	5,454
28.	1,921	13,401	3,311
29.	3,558	19,519	4,821
30.	5,020	28,518	7,044
31.	3,364	18,434	4,553
32.	2,723	14,360	3,547
33.	2,051	12,157	3,003
34.	2,624	15,418	3,808
35.	2,989	15,727	3,885
36.	3,729	20,105	4,966
37.	2,601	15,173	3,748
Total RUA	144,882	846,733	209,216
Total NBJK		420,872	103,962

Source: Census 2001 also available from the website of RMC.

Of the 37 wards in the NBJK-RMC partnership, 16 are part of the CJP (✓) (Table 7.1)—two wards have been added recently to 14 wards selected earlier. The RMC has also contracted 10 wards to three[10] other NGOs, Creative International, Express Sewa Sanstha and Nirmal Bharat on a yearly basis for collection, transportation and disposal of waste. The remaining 11 wards are taken care of by the RMC itself. Thus, three different though parallel mechanisms of SWM exist in Ranchi. The details of ward population and waste generated in these wards are given in Table 7.2. The CJP of NBJK covers a population of 420,872, representing 49.70 per cent of the Ranchi Urban Agglomeration.

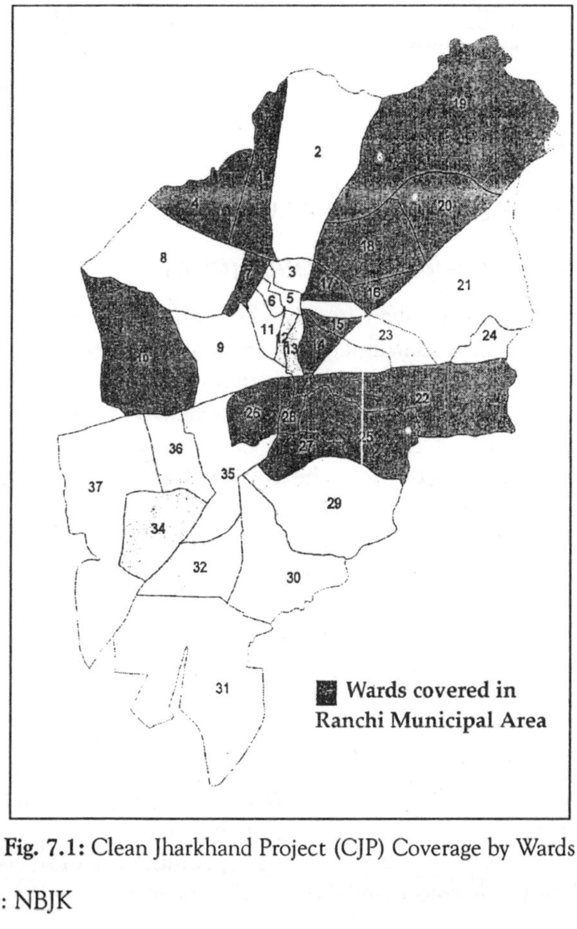

Fig. 7.1: Clean Jharkhand Project (CJP) Coverage by Wards

Courtesy: NBJK

Table 7.2: Chronology of Initiative Taken Under the CJP

S. No.	*Ward Number*	*Year and Month of Starting*
(a)	7	September 2003
(b)	1 and 4	December 2003
(c)	17 and 18	January 2004
(d)	14, 15, 16, 26 and 27	April 2004
(e)	10, 22 and 19	August 2005
(f)	3, 9 and 20	November 2005

Source: NBJK

The NBJK gradually expanded to cover 16 wards. It started with ward No. 7 in September 2003, and then added wards 1 and 4 in December 2003, and then the other wards as shown in Table 7.2. In addition to these activities, the CJP has also taken up vermi-composting at Chuttu Village and is soon to start a new landfill site at Zheeri village.

Institutional Arrangement and Mechanisms

The Ranchi Municipal Corporation (RMC)

The RMC was established in 1979 by merging the former Ranchi Municipality, Doranda Municipality and Ranchi Doranda joint water board[11]. It is the 2nd municipal corporation of the former Bihar state. The first election of the corporation was held in 1986. Since 1991, the elected body of the RMC has been superseded by the state government and elections to the corporation have not been held. The Municipal Commissioner, appointed by the state government, is in charge of the RMC. All the other functions of the RMC are performed by the state level departments (Fig. 7.2). SWM is under the health department and laying drainage and sewerage lines under the urban development department of the state government. The location of new landfill site has also been selected by the state urban development department and the district administration. Since the 74th CA[12] has not been implemented in either Bihar or in the new state of Jharkhand, most of the decisions regarding any infrastructure development other than routine work needs permission from the state government. This situation is likely to change soon as the state may be forced to hold RMC elections if it is to get funds under the National Urban Renewal Mission (JNNURM) of the central government.

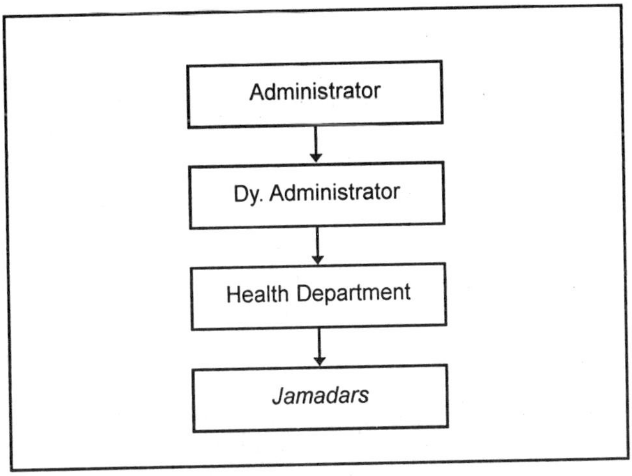

Fig. 7.2: Organization Structure of RMC

Presently, the health department of the municipal corporation, which is headed by the administrator, holds the responsibility of the SWM system. The middle level supervisory structure for sanitation does not exist. For example, the posts of Chief Sanitary Inspector (CSI), other sanitary inspectors, sanitary officers and environment engineer do not exist. Consequently the *Jamadars* (labour supervisory staff), who oversee the work of the sanitary workers, directly report to the health officer of the health department. This is therefore a highly centralised structure. It appears that the RMC does not have much of the necessary staff because of lack of funds. Among all the ULB cases presented in this book, the RMC is very particular: the ULB does not have any resources, is not able to carry out its normal functions, and is fully dependent on the state government. The situation continues despite Ranchi being declared the capital of the newly formed Jharkhand state.

As per the MoU of the NBJK and the other NGOs with the RMC, the municipality loads the garbage from designated collection points and transports it to the landfill sites. Oddly, the MoU also states that the NBJK supervisor must oversee the work of garbage lifting by the RMC staff (*jamadar* and two workers) from these designated points. Our field visits did reveal a strong co-ordination both between the RMC and the NBJK staff on this matter, and the NBJK's community coordinators are in direct contact with the health officer of the RMC.

The RMC provides office space to the NBJK so that regular contacts remain between them.

The Nav Bharat Jagriti Kendra (NBJK)

Nav Bharat Jagriti Kendra was formed in 1970 by four young engineers inspired by Gandhi's ideas. An important activity envisaged by the NBJK was to develop a Community Action Network. It has now made a network of 200 grassroots groups usually made up of 3 to 4 people and led by an experienced community worker, in Jharkhand and Bihar together. Each of the groups focuses on the most pressing issue (s) in their respective area. These groups work in rural areas through the formation of *Lok Samitis* (People's Committees) to motivate people to act for themselves. The *Samitis* debate village issues, help people to gain access to their rights, and work as 'vigilance committees' for monitoring the functioning of government programmes, including the implementation of legislation such as minimum wages at least on government construction projects. Over the years, NBJK has developed a network of village level volunteers who run its *Lok Samiti* network. Lastly, supporting the grassroots through its small groups support programme, NBJK also runs a wide range of activities on health, education, watershed development and appropriate technology throughout Bihar and Jharkhand. NBJK is now planning a district and state level support structure for the small groups programme, while promoting new groups and identifying their capacity building needs. The CJP is an important urban programme of NBJK.

For each of its programmes, the NBJK has a formal organisational structure. A project manager, in close coordination with the community coordinator and the supervisor, handles the project. In between are two other positions, Assistant Project Manager (APM) and Assistant Community Coordinator (ACC). For the SWM too, a similar structure exists within the NBJK. The ACC coordinates the work of four supervisors; one supervisor handles one ward and that means that one ACC handles 4 wards. Below the Supervisor are an Animator and *Safai Mitra* (The Cleaner—literally Cleaning Friend). The animator is responsible for collecting user charges from the community and deposit the amount with the supervisor. One *Safai Mitra* is appointed for one Pocket Development Committees (PDCs) of the ward. At present all *Safai Mitras* are males. The discussion on PDC is next in this section. Hence, the number of *Safai Mitras* in a ward (and hence under the supervisor) depends on the number of PDCs in that ward. At each level a written document is maintained for the purposes of transparency and accountability.

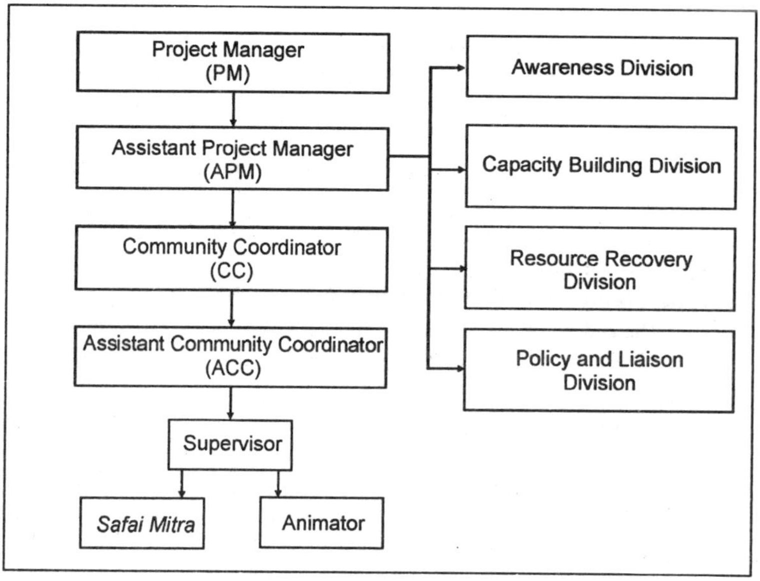

Fig. 7.3: Organization Structure of NBJK

The NBJK has four different divisions[13], each coordinated by the Assistant Project Manager (APP). The four divisions are: (i) Awareness division, (ii) Capacity building division, (iii) Resource recovery division, and (iv) Policy and liaison division. In addition, there are two more departments: (a) Training department—which provides training to the *Safai Mitras* on their work and work behaviour, namely on how to collect the garbage, how to go about door-to-door collection, how to talk to the house owners, and how to dress (b) Documentation department that regularly documents all the NBJK's work, including any workshop proceedings and newspaper reports on NBJK.

Besides ensuring financial contribution by the residents of the PDCs for the SWM service, the NBJK also acts as a mediator for the community to raise issues and concerns regarding the inadequate number of dustbins, poor drainage facility, and irregularities in garbage pick-up among other things. However, through orientation and training of the PDCs, communities are being slowly empowered to raise these issues on their own.

Pocket Development Committee (PDC) and Its Formation

The PDC is the key grassroots institution, through which the households participate in the SWM project. The PDC is comprised of selected community representatives who take decisions and act on behalf of the community. The group is set-up through the local mobilisation efforts of the NBJK. The PDC ensures that the participating households segregate their solid waste and the *Safai Mitra* collects the segregated waste from the households doorstep and then transports the collected waste to the nearest designated waste collection point. The PDC also ensures that the residents pay their dues for the SWM effort to the animator. The PDC may itself collects the dues from the households and then pay to the *Safai Mitra* and also for the equipment required, such as a cycle rickshaw for transporting the waste. Otherwise the CJP pays the *Safai Mitras* and also provides them with all the equipment. RMC helps with onward transportation of the waste collected. The NBJK, under the CJP undertakes capacity building of PDCs to develop and strengthen them as an effective local level institution of SWM.

As per the discussions with some of the PDCs, it was found that many of them have existed for as long as 30 years. These were the CBOs formed a long time ago for the purpose of organising local security arrangements, such as appointing security guards and paying them. These CBOs were first approached to become a PDC for the SWM purpose. Where such groups did not exist, the NGO had to adopted new approaches to form PDCs, such as motivating a key person of the community, like a senior citizen or a local leader, to organize a local group.

There is a set process to form a PDC. The steps followed are:

(i) Survey of the *Mohalla* (Area),
(ii) Identification of CBOs/ Club/ small NGO,
(iii) Informal meeting in colony by the CJP staff,
(iv) Identification of interested person in SWM,
(v) Meeting with interested people,
(vi) Door-to-door visit by CJP staff and interested residents,
(vii) Formal meeting with residents,

(viii) Deciding the office bearers of the PDC, namely the president, secretary, treasurer and the general members, and

(ix) Formation of a PDC.

In a single ward there may be 2 to 4 PDCs. The PDC size ranges from 18-20 households to 230 households. These PDCs are both homogenous and heterogeneous in nature. Some of these PDCs are self sustaining PDCs, i.e. the NGO provides them with *Safai Mitra* whose payment and costs of rickshaw for collection and transport of waste is directly borne by the members of the PDC while the others are dependent on the NBJK.

The PDCs provide training on SWM as well as institutional and financial management. The regular experience sharing meetings are organised in which PDC members discuss problems of their respective area and formulate solutions. At the initial stage, these meetings are facilitated by the NBJK through the CJP. The latter gradually withdraws and members start conducting the meetings independently.

From the field visit, it was observed that there was good communication between the various members in the system and also the community or PDC members. It was also found that the PDCs of Ward 7, the first one to come under the CJP, were working very efficiently. In all other wards part of the CJP, the number of PDCs have continuously expanded. Lastly, scattered PDCs were working in wards 11, 21, 23, 25, 29 and 30 on their own without even being part of the CJP.

Appointment of Safai Mitra

Applications are invited through a notice in the newspaper and the applicants are selected through a personal interview. The selected candidates are appointed for a probationary 3 month period. If any *Safai Mitra* is not working properly then he is permanently barred from working on the project. Each *Safai Mitra* is given a pair of uniforms which must be worn on the job. If they do not do so then their half-day's salary is cut. *Safai Mitras* in Ranchi are all men, unlike other cities covered in this book where they are mostly women.

Ward Development Committee and its Constitution

It was realised that the PDCs were working in an isolated and scattered form and the whole ward was not involved in this. In a ward there were 2 to 3 PDCs, which were administered differently. NBJK has started a process of federating all the PDCs of a ward into a Ward Development Committee (WDC). This was done through ward level meetings organised under the awareness campaign called '*Ranchi Nagar Nigam Apke Dwar*' (RMC at your doorstep)[14]. Members of ward committee are representatives of the PDCs and hence the WDC in essence became the federation of the PDCs. This has achieved two objectives, one is a singular representation of wards not only for discussing the problems related to SWM and reaching homogenous level of cleanliness across the ward but also for putting across issues related to other infrastructural facilities. The role of these WDCs would be formalised should the 74th CA be adopted in Jharkhand state and RMC elections held in the near future.

Major role of Ward Committee is to:

- Keep the RMC updated about the problems of the ward,
- Mobilize community support and co-operation in solving the problems of the ward,
- Liaison with other stakeholder on behalf of the community,
- Promote community level action for environment protection and SWM.

As per the information from NBJK, 11 ward committees have been constituted across 14 wards under the CJP. These WDCs hold regular monthly meetings to get acquainted with the problems of the community and plan out strategies to solve them.

Awareness

Awareness is the key to ensuring community participation in development projects. Towards this purpose, under the CJP, the NBJK undertook a mass awareness campaign among the citizens and the administrators to motivate them to perform their respective role. In the initial stage mass awareness through road shows, street plays in Ranchi city were carried out on a pilot basis, along with published and electronic media, and finally specific communities such as school children, teachers, housemaids and so on were targeted. Evidence of mass awareness campaigns can be found in the city.

Recently, NBJK has expanded mass awareness activities to other municipalities of the state through workshops and seminar. ICEF funding has been important in making this move ahead. Further, the NBJK holds press conferences every six months. It brings out a newsletter for circulation to all the PDCs and schools. NBJK also releases its quarterly and half yearly progress reports. It has also published articles and journals on all forms of waste. A documentary on CJP has also been made and released recently.

Capacity Building

One of the other main objectives of the project is to improve the present system through increasing the efficiency and capacity of the administration, improving the capability of the NGO workers and increasing the capacity of the community. Through this it is also anticipated that all the different actors will be motivated for effective participation in the SWM system being set-up. Capacity building is also required to orient different actors towards the concept of SWM, CJP objectives and methodologies and also to bring awareness regarding the MSW Rules, 2000.

In this regard, the CJP has arranged capacity building programmes for different actors. Different training programmes were held for the *Safai mitras*, the employees of RMC and other municipalities of the state, the NBJK staff and the PDCs (Table 7.3). The training imparted to the project team is with regards to: (i) Record keeping, (ii) GIS software, (iii) Budget planning, and (iv) Project team reflection. Workshops and seminars have also been held, largely with the participation of the RMC officials, to share the experience and then discuss different issues and concerns arising in the course of project implementaiton. CJP also arranged exposure visits for its workers to other cities in India including Madurai (Tamil Nadu), and Bhadreshwar and Kacharapara Municipality of West Bengal where innovative ways of MSW were emerging.

Table 7.3: Number of Capacity Building Sessions Organised under the CJP

S. No	*Activity*	*Number*
1.	Training of	
	(a) *Safai Mitras*	20
	(b) Municipality employees	17
	(c) Project team	15
	(d) PDCs	11
2.	Workshop/ seminar	15
3.	Exposure visits	12

Source: NBJK

Capacity building of the PDCs, is also accomplished through regular meetings where community members and representatives are present with the RMC officials and NGO staff. These meetings focus largely on issues related to SWM and other public services. The latter set of issues do not fall under the purview of CJP, though these issues also come up because of the presence of the RMC officials. Direct dialogue between the users and service providers develops transparency in the system is useful for the review of the work progress.

The System

Overall City

The Clean Jharkhand Project (CJP) initiative starts with a ward adoption procedure, which is an agreement signed with the RMC by the NBJK on behalf of the CJP. As per this agreement, the CJP agrees to deploy *Safai Mitras* for door-to-door collection of waste in the ward while the RMC provides labour for waste loading and transportation and the equipments and vehicle required for the purpose. All other labour required is to be paid for by the CJP. After the agreement, the NBJK proceeds to form a PDC, as explained earlier. Then it draws up a plan with the RMC for the lifting and transporting of the collected waste.

After the adoption of ward, massive cleanliness drive is undertaken to clean roads/streets/drains/other public places. During this drive long accumulated garbage on public places and blocked drains are cleared out. The RMC puts its staff and equipment to the service of the CJP. The requirement of extra labour and equipment is met by the CJP. The cleanliness drive is accompanied by awareness activities like distribution of pamphlets, awareness by cultural team and door-to-door contact to inform the people about the project and seek their participation. This is done to raise not just awareness but also build faith in the upcoming partnership so that the households participate through segregating their waste and making required financial contribution.

After the completion of first major cleanliness drive, the system of regular cleaning is put in place. *Safai Mitra* along with a trolley and other equipment is deployed for regular collection of waste from doorstep, sweeping of streets, roads and cleaning drain periodically. Each household has to pay an amount of Rs. 20 per month for the services of waste collection and cleanliness. One animator is deployed

Photo 7.1: PDC meetings in progress, Ranchi

Photo 7.2: Doorstep collection and special cleanliness during Diwali festival, Ranchi

Photo 7.3: A Clean Inner Street, Ranchi

in each ward to collect these monthly charges from the households.

The doorstep collection is carried out in the morning hours, between 7 and 11 am. During this time period the *Safai Mitra* sweeps the internal neighbourhoods, cleans open drains (this is done only once a month) and collects waste from each household, even from those who are within the PDC area but not members of the PDC. Many areas in Ranchi that are away from the main arteries are quite narrow. In the absence of dedicated doorstep collection this waste would have accumulated in small nooks and corners of the street. Thus, it is useful that the door-to-door waste collection is carried out. Further, the *Safai Mitra* goes around the lanes and by-lanes in a tractor for waste collection. In case of narrow roads and lanes, tricycles are used. Since most roads are cemented, the tricycles can traverse easily. Community dustbins are not located inside these smaller streets but somewhere in a corner or on a main road. In wards without doorstep collection, the waste finds its way in the open drains, which are sometimes cleaned by the workers of other NGOs on contract.

Within a ward under the CJP, the PDCs manage the door-to-door waste collection and road cleaning in the inner areas, whereas the RMC workers sweep and manage garbage on the main roads. Since, the RMC workers do not engage in door-to-door collection of waste, the households and establishments, particularly the informal sector workers, on the main road dump unsegregated waste wherever they want. Thus, within a CJP ward itself, there are two systems of waste management, working of PDCs in the inner areas and non-participation of households and establishments located on the main roads. RMC's service is free and PDC activities are self-financed, which accentuates this divide.

The RMC has put community waste collection bins on the inner roads as well as the main roads. The *Safai Mitras*, after collecting door-to-door waste in the inner areas, dump it in these collection bins. The households and establishments on the main road do the same on their own if they are conscientious enough to do so, after which the RMC workers, as per the direction of the NBJK's community coordinator and the supervisor, load the waste from the community dustbins into the tractors. At some collection points, metal dumpers have been placed. These are lifted and loaded on the hydraulic dumpers. The tractors and hydraulic dumpers transport the waste to the dumping sites. Due to a shortage of tractors, the supervisor has to plan the number of trips to be made from different points depending upon the quantity of garbage collected at each collection point. Normally 2 to 3 trips are made regularly from each of these wards.

The current dumping site is at a place known as Mukaddam that has now seen housing colonies developed around it. The 'people have come to the pollution' and have now started protesting against the dumpsite. A foul smell is emanating from it because of the lack of scientific treatment of the waste. The RMC, with the help from Ranchi Regional Development Authority (RRDA) and the urban development department of GoJ, has identified a new site for dumping named '*Zheeri*' located 10 kms away from the city. This site is undulating and a rocky terrain. The RMC believes that dumping of garbage and levelling it over the years would create flat land for some use later. The new site is to be maintained by the NBJK.

Infrastructure

The RMC has a staff of only 700 workers, which includes 500 main *Safai* Workers and 200 others including peons etc. There has been no new recruitment in the department since 1971[15]. Hence the system of waste collection, street cleaning, and waste lifting have been contracted out to the *Safai Mitras* under the CJP or the other three NGOs.

The RMC has a total of 44 tractors. Each ward is given one tractor in the CJP area. About 12 tractors are reserved for emergency purposes. Each tractor has 1 driver and 3 workers, all of whom are employed by the RMC. Each NBJK supervisor looks after the waste collection by one tractor, for which he gives a detailed route and scheduling of trips one day in advance. Hand trolleys are given to the *Safai Mitra* for door-to-door waste collection. Presently 139 hand trolleys are being used to collect garbage from 27,000 households and catering 125,000 people.

Waste Management in A Ward

Ward level SWM system is illustrated through the example of ward 18 located in the northern part of the city (See Fig. 7.1), where the CJP was started in May 2003. This ward has a population of 24,834 (in 2001) and generates 6.13 MTs per day of solid waste at the rate of 0.247 kg per person per day (Table 7.1). The ward has concentration of low and middle-income groups. Of the 4,460 households in the ward, only 1,600 (36 percent) are participating in the CJP through the formation of 16 PDCs of which one is called sustainable PDC[16]. The 250 to 300 shops located mainly on the main roads and a large number of hawkers and other informal sector workers in the ward are not participating in the CJP. Thus, the main streets become garbage dumpsites whereas the inner localities, which are part of the PDC, remain fairly clean. The RMC workers clean the main streets.

The ward has three slum pockets with an estimated population of 800 households, of which around 500 are part of the CJP paying from Rs. 5 to Rs. 30 per month per household. Their enthusiastic participation in the CJP is because they are happy about receiving some service in the area.

The ward also has around 10 schools, four small dispensaries and a hospital. Some schools do not pay any user charges. NBJK has installed compost dustbins near the schools for throwing tiffin waste. Awareness programmes have been held by the NBJK in the schools in the past.

The 16 *Safai Mitras* in the ward collect door-to-door garbage using a rickshaw trolley (costing around Rs. 5,000) and their work is monitored by one supervisor. The animator, collecting the household contribution for the work is paid 15 to 16 per cent of the total collection made from the ward. Even then, only 75 per cent of the participating households contribute their share of user charges. The total monthly collection from this ward is Rs. 25,000. The RMC has provided 1 *Jamadar*, 4 workers for road sweeping, and 1 driver with three workers for lifting garbage. Thus, in one ward, altogether 29 personnel are directly engaged in managing the solid waste.

There are four collection points in the ward, one on the main street and three at different places in the locality. From here, the collected garbage is lifted by the RMC staff under the supervision of the NBJK supervisor using 1 tractor of 3 MT capacity. The tractor makes 2 to 3 trips daily. NBJK pays a bonus to the RMC driver and staff (Rs. 25) for making additional trips if required.

Every morning the work starts at 6 am with the attendance of the *Safai Mitras* when the community co-ordinator, the supervisor, the RMC's *Jamadar*, and the animator are present. The supervisor keeps a watch on the work of the *Safai Mitras* and then move in the area to check out on the cleanliness and prepare a daily report consisting information on: (i) attendance of the *Safai Mitras*, (ii) no. of trips made by the RMC tractor, (iii) amount of collection done, (iv) vehicle repair and maintenance, (v) no. of complaints received and the reasons, (vi) no. of complaints addressed and (vii) no. of new members added in the PDC. From the daily reports, weekly, monthly and tri-monthly reports are prepared. An action plan for the following day is prepared based on the preceding day's performance and the complaints received. The NBJK has also prepared schematic maps of the PDC and the wards as the detailed ward are not available with RMC.

Composting and Recycling

Given that the recycling of the city waste is the most economical option given the Indian situation, the CJP also has a resource recovery plan. It has proposed the following for the purpose:

- To promote centralised and decentralised composting,
- Segregation of compostable, recyclable, re-usable material thereby reducing the quantum was for final dumping, and
- Identification and promotion of recycling options.

The Resource recovery division of the NBJK has set-up a centralised composting plant 14 kms away from Ranchi at village Chuttu in Kanke block of Ranchi district, since April 2004. This plant is on 5 acres of land acquired on a lease for Rs 50,000 per year for 5 years from the government. Local people had protested against setting up this site. Yet, after local consultations and appropriate routing of the garbage carrying tractors, the plant has started functioning. The NBJK has also committed to take up some community development activities in the local community.

The compost yard receives a specified amount of vegetable market waste free of cost as per the agreement of the CJP with the RMC. The waste is transported free of cost by the tractors of the RMC. The compost yard has a capacity of 10,000 MTs of waste annually but only half of its capacity and 2 acres of land is being utilized at the moment. Waste is recycled using two methods, one is vermi-composting and the other is through spraying of inoculums such as effective microbes solution and cow dung slurry on the waste. Normally compost requires 70 days for its formation, however, the process is delayed if waste received is not in segregated form, adding another 8 to 10 days for separation.

The price of sieved manure is Rs. 3 per kg while non-sieved is Rs. 2 per kg for local farmers. The manure is sold both in retail and wholesale for which the rates are different. The buyers are the farmers (who generally buy in bulk), and horticulturists and individual garden owners.

Besides the opposition from the local people, there was another glitch in the beginning; the compost plant was receiving irregular quantity of garbage and the reason was found to be irregularity of trips by the tractor drivers. This emanated from their indifference towards their job. The NBJK organised a meeting with the tractor drivers, who are RMC employees, to firstly build awareness among them about the entire system by sharing all the information about the plant with them and then explaining to them through video films the all encompassing benefits of the composting activity. The RMC staff also had apprehensions of working for the NGO and had fears of loosing their jobs because of the CJP.

The NGO responded positively. It did the following: (i) As a part of the CJP initiative, the tractors, which were defunct, or were having problems, were repaired on war-footing, which enabled the trips to be made even during the night time. CJP also announced some incentive and welfare scheme for RMC staff, such as: (i) Group Insurance for RMC staff; the cost of premium to be shared by the RMC, the staff themselves and the CJP in ratio of 50: 30: 20; (ii) Free health camps for the staff and (iii) Cash Voucher of Rs. 25 to the tractor drivers per waste carrying trip to Chuttu processing site, encashable at the NBJK office.

Financing Mechanism and Institution

ICEF, the main funding agency has given a grant of Rs. 6,570 million for a 5 year period to the CJP. The funds are sanctioned annually after an annual budget, prepared by the SWM experts of the CJP, is presented to the ICEF. Twenty-five per cent of the annual budget is allowed for administration purposes, which includes salary, office expenses, operation and maintenance, the remaining seventy-five per cent is allocated for the project expenses such as awareness building programmes and other project expenses.

The other source of funds for the project is the contribution of the participating households used towards meeting the salaries of the *Safai Mitras* and the animator. Each household is expected to contribute Rs. 20 per month. In reality, however, the per household collection ranges from Rs. 5 to Rs. 35 per month. The *Safai Mitras'* salary is Rs. 1,820 per month and is paid by the CJP. If the household contribution is not adequate to pay the *Safai Mitras'* salaries then main funding is used for the purpose. Salary is cut if the *Safai Mitra* is found to be irregular while he is rewarded for good performance.

Currently, the total collection from the households comes to Rs. 300,000 per month, from a meagre amount of Rs. 25,000 per month earlier. Still, this is just 21 per cent of the expected income from household collection, although 60 per cent of the beneficiaries' are contributing (Table 7.4). This means that either the number of residents contributing is less than 60 per cent stated by the NGO, or their contribution is less than Rs. 20 per household mandated. Further, the amount collected is not yet enough to meet the salary costs of the 160 *Safai Mitras* at the rate of Rs. 1,820 per month (the total coming to Rs. 291,200 per month) and other additional costs such as purchase of a trolley (costing Rs. 5,000) for door-to-door collection and its maintenance (average Rs. 100 per month). The NBJK is striving hard to make households contribute and be self-sufficient in meeting the costs of the entire PDC functioning, as the wards 4, 7, 10, 17, 18 and 26 have done. Otherwise, the innovative effort would come to an end in 2007 when the project funding would end.

Table 7.4: User Charge Contribution under CJP

Particular	*Units*	*Number*
1. No. of HHs catered under CJP (in 16 wards)	no.	71,374
2. Estimated collection @ Rs 20/ HH/ month	Rupees	14,27,480
3. Present collection (monthly) (in 2005)	Rupees	3,00,000
4. Gap	Rupees	11,27,480
5. Collected as proportion of estimated	%	21

Other source of finance for the CJP is the RMC's own contribution in terms of providing salaries of the workers lifting garbage from the collection point and transportation of the waste to the dumping site or composting site using its own tractors and drivers. The RMC also pays the contractual amount to the three other NGOs for tasks that the *Safai Mitras* are performing in the CJP wards. The corporation makes a payment of Rs. 150,000 per month for 10 wards, amounting to Rs. 1.5 million per year, to the NGOs (functioning as contractors) for cleaning non-CJP wards. No finance data of the RMC is available except that it generated total revenues of Rs. 51.30 million (in 2003-04) from property tax, rents from municipal markets, professional tax and others and of which Rs. 39.7 million (77%) came from the property taxes[17].

Evidently, the total internal resources of the RMC are even not adequate to meet salary and wage expenses of the staff despite having just 700 sanitary staff in place of necessary 2000 as per national norms. The all India[18] average property tax is Rs. 150 per capita per year, and therefore the income from property tax alone should have been Rs. 150 million instead of about Rs. 40 million in the RMC. The RMC officials agreed that just 40 per cent of the estimated property tax is being collected from the city today. The RMC is considering expanding its revenue base from Rs. 200 million to Rs. 250 million in the very near future by improving its revenue tax collection and introducing other sources of revenue generation in the city. The RMC wants to expand its boundaries and cross the one million population mark to access state level funds.

Partnership/Alliances

One of the important aspects of the CJP is the involvement of different stakeholders in the initiative. Although the project started as a result of initiation by the NGO and the ICEF, various stakeholders have been brought in through establishing collaboration and cooperation among them. The NGO, the NBJK has established a liaisoning division to built alliance with various stakeholders. The various partnerships that have emerged in this initiative are discussed below.

The ICEF and the NGO (NBJK)

The initiative started as a result of ICEF funding the NBJK. The latter's role was to establish a sustainable community based SWM system in Ranchi, to act as a pilot case for all other municipalities in the state.

The NGO NBJK and the RMC

The ICEF funds were made available only on the condition that the RMC was involved in the project. In any case, as per the MSW Rules, 2000, it is the ULB that is responsible for the SWM and hence, it cannot pass on this responsibility to an NGO. Thus, the CJP required that the NGO sign an agreement with the RMC before the project was launched. The NGO took major responsibilities in this partnership, building awareness among the community groups for SWM, motivating them to participate through the formation of PDCs and then financially contributing to support door-to-door collection of waste and transfer to the designated collection point through *Safai Mitras*, appointing *Safai Mitras*, monitoring and evaluating their work through supervisors, monitoring and scheduling the waste pick-up from the collection point by the RMC workers, keeping the RMC workers motivated to participate in the CJP, convincing the village residents around the vermi-composting site and then managing the entire vermi-composting activity. The RMC only participates in this project through providing waste transport vehicles and equipments and workers for the lifting and transporting of collected waste. In a way, the ULB has no role in the logistics of the entire project. Probably such situations would arise where the ULBs are defunct. Civil society organisations have to bear the responsibility of activating the public body for the latter to take up their own responsibilities.

Other NGOs and the RMC

The RMC has contracted out the non-CJP wards to three other NGOs for the tasks that the NBJK is coordinating in the CJP wards. This partnership does not fall under the CJP project.

The NBJK and the Community

The NBJK has been instrumental in setting up PDCs in project wards, wherein the PDCs are community groups participating in the project. It is expected that these PDCs would take over and be able to sustain the new SWM initiative, once the project is over. The PDCs could then also participate in the Ward Committees when the 74th CA is implemented in the state. PDCs help in mobilising community abilities and finances in keeping neighbourhoods clean and in the future will also manage all local level basic services. This partnership is built on trust and the respect that the NBJK has in the city. To be able to coordinate ward level functions, PDCs have been federated into WDCs, which could then become Ward Committees in the future. The NGO, NBJK has played a pioneering role in motivating the community and building the latter's capabilities in operating the PDCs in a sustainable manner. It is however not known how long this system will last once the NGO withdraws on the completion of the project.

Achievements

Innovation in Practice

Innovation in the Project is with regards to its management strategies. Firstly, the project has created community involvement and ownership in the project of SWM. Herein lies the potential for sustainability of the project once the funding organisation and the NBJK withdraws. The other innovation is the partnership approach, that is build on the development of capacities of all the partners, including that of the *Safai Mitras*, supervisors, the trolley drivers, the RMC officials, the community itself, and even the residents of the village where the vermi-composting plant has been set-up. Thirdly, the NBJK, under the CJP, has taken an all-inclusive approach. There has also been a great dependence on awareness building at the city level and in specific communities, particularly in the context when the RMC has become a defunct organisation by its suppression for a long period of time. Lastly, the project has addressed the issue of bringing in higher order of professionalism than before through the development of Management Information System (MIS), regular reporting of the activities and addressing grievance redressal system. In contrast, in the wards where other NGOs are working, the grievances have to be reported to the RMC, and the problems reported take a long time to be addressed, if they do get addressed.

Beneficiary Participation

About 60 per cent of the population in the wards under the CJP are participating through the PDCs and are financially contributing to the project. The NGO has brought beneficiary participation from nil to about 60 per cent in the CJP wards, which is quite remarkable in a situation where the citizens had not even experienced any functioning of the local municipal body. Some of the PDCs have also become financially sustainable. Since the beneficiaries are paying the salaries of the *Safai Mitras*, the former are demanding accountability of the system.

Two specific groups that have shown enthusiastic participation are the women and the slum dwellers. The NGO is utilising this opportunity to mobilize women, who mainly deal with the household waste.

Local Institutional Development

The PDCs formed under the CJP could act at neighbourhood level institutions of urban governance. In some places, these have been federated to become WDCs, which could then become the basis for the formation of the Ward Committees when the 74th CA is implemented. It is not far when the 74th CA would be implemented in the state, as it is mandatory for accessing 12th Finance Commission grants and also participating in the Jawaharlal Nehru National Urban Renewal Mission (JNNURM).

Up-Scaling

The CJP is being implemented in Ranchi city as a pilot case, to send across a message of importance of urban SWM throughout the state. The project intends to replicate the efforts and learnings of Ranchi in other ULBs in the state. In Ranchi, the NGO has expanded its working to 16 wards out of total 37 wards within a span of 3 years. Thus, some upscalling of the CJP itself has happened.

Table 7.5: Replication in Ranchi and in other ULBs

Particular	*Ranchi*	*Hazaribagh*	*Jhumri Tilaiya*
Wards	15*	4	2
PDCs	149	8	5
Households	29,427	1,460	896

Source: NBJK (2005).

* Recently a new ward has been added. The number of PDCs has also increased to 157 in Ranchi.

In terms of replication in other ULBs, the NGO has extended similar initiative to Hazaribagh and Jhumri Tilaiya municipalities of the state (Table 7.5). In Hazaribagh municipality the CJP initiative started in December 2003 (NBJK 2004), with an awareness campaign and cleanliness drive. Doorstep collection with a similar system like Ranchi has been set up in 4 wards through the formation of 8 PDCs. Similar alliance has been established with Jhumri Tilaiya, Griridh and Medininagar municipalities by involving the special programme officer of the municipality and the District Collector. Similar replication is intended in all other municipalities of Jharkhand. The project has not only been helpful in entering in partnership with these municipalities but intends to develop networks with other NGOs too in other cities. The NGO has also started vermi-composting in these municipalities with the assistance of the authorities.

Concerns

Although this effort has built something from nothing and the project has upscaled to cover 16 of the 37 wards of the city, there are many concerns with regards to the sustainability of the SWM initiative after the CJP's funding comes to an end and also with regards to SWM in the entire city itself. But, even at the present, with the ICEF funding and the NBJK active, the success is limited.

Limited Success

Even after upscaling of the CJP, the wards covered under the project are less than half. Within the CJP wards, households located on the main roads and the commercial establishments including the informal sector, are not participating in the PDCs and on the contrary littering the roads. Thus, while the inner neighbourhoods look clean, the main streets continue to remain filthy.

The households participating in the PDC are not contributing the mandated amount and hence the total financial collection from the households is just 21 per cent of the expected amount. Except a few PDCs, in all others, the ICEF project has to meet the basic costs of running the project at the moment. Where the PDCs are able to collect substantial funds, these are just enough to meet the salaries of the *Safai Mitras*. The salaries of supervisor, coordinator and other administrative staff are still made from the project funds. So from where would the SWM initiative meet the funds once the ICEF funding is over?

In some wards, sustainable[19] communities (or the PDCs) have come up. But, these are mostly of the higher middle class society. This is generally because in spite of small groups they have higher paying capacity which covers the salary of the *Safai Mitra* and the cost of the rickshaw trolley.

Limited RMC Participation

Currently, the RMC participates in the CJP project is only for lifting and transporting the garbage. The ICEF funds have totally subsidized the ULB for the CJP. Because the RMC has not recruited any new staff since 1971, and has also been unable to increase its own revenues after the formation of Jharkhand state, there is a big question concerning how the RMC be able to cope, financially as well human resource wise once the CJP is over. The CJP is far too dependent on NBJK.

No Source Segregation

At the moment, the households participating in the PDC are not segregating the waste at source. Thus, reduction in the quantity of waste going to the landfill sites has also not been attempted in the city.

Waste Recycling

The efforts at waste recycling are still weak. The NBJK has started vermi-composting unit for the bulk waste from the vegetable market. Also, since the household waste is not segregated, it is has to be disposed of on a landfill site. The RMC is thinking of utilizing plastic wastes for road surfacing purposes. But, this is just in the planning stage at the moment. For a small city like Ranchi, it should be possible do develop a good waste recycling system, either in a decentralised way, as Hyderabad Exnoras have or a centralised way as Suryapet has done.

Overall Lack of Sanitation Infrastructure

The drains in the city are open. There is no sewerage system. Open drains are exposed to defecation by households not having latrines. There is a need to construct basic infrastructure for the city, without which the SWM would not be very successful. The RMC and the state government's urban development department have prepared plans for the same, but lack finances at the moment for implementing these projects.

No System of Bulk Garbage Producers

At the moment, there is no system of waste collection from the bulk producers, except vegetable markets from where waste is being regularly lifted for NBJK's vermi-composting plant. The hoteliers are selling their garbage to the pig-rearing communities in the city and the rest of the garbage is thrown either on the road or in the municipal dustbin during night hours. Scientific treatment of bio-medical waste has not been thought off. Some institutional framework is necessary to address the problem of waste from the bulk producers.

Different SWM Systems

Currently, in Ranchi three different SWM mechanisms are working simultaneously: the CJP project, which is based on external funding and NGO initiative, the outsourcing of the road sweeping and door-to-door waste collecting system to three other NGOs than the NBJK, and the RMC's own waste cleaning activities.

The level of achievements of each system varies, with the first delivering the cleanest neighbourhoods. At the same time, some households are paying user charges for the SWM in their neighbourhoods while others are not. This has created a situation wherein some households in the CJP wards have a free-ride over other paying households. This study has not gone into the details of the second and third system and hence refrains from commenting on them.

Non-Implementation of the MSW Rules 2000

Although the MSW Rules 2000 states that the ULB has to ensure segregation of waste at the source and all households to be covered under the door-to-door waste collection, the RMC has not been able to implement these rules. Further, the principle of 'polluters pay' is being differentially applied, with some households are paying for the garbage they are generating while some are not.

Non-Implementation of 74th CA

The city of Ranchi has been divided into wards for ease of management; however the implementation of the 74th CA for the ULBs to act as third tier of government and for the decentralised governance in a city, has not yet been implemented. The biggest lacuna today in the innovative SWM is that the overall structure of governance in the city is not yet democratic.

Overall Assessment

We selected this case because much was written in the literature about this model. On researching in detail we found that this innovative SWM model is also unusual. The ULB itself is not carrying out its functions. It does not have trust in its own staff. It has contracted out the waste collection to the community through the NGO, NBJK. The NGO, on behalf of the community, appoints waste collectors who carry out door-to-door waste collection and for which the community pays. The NGO monitors the staff of the ULB for waste pick-up from the collection point and transport to the dumping sites. And the NGO is funded by a bilateral organisation, whose funding would discontinue from 2007 calendar year because of the Government of India's decision. In all these, there is no spirit of the 74th CA, wherein the ULBs would perform all their functions through decentralised governance. This model is in fact a case of failure of urban governance. It should not be adopted as a policy for the urban SWM in the state of Jharkhand. NGO can be a partner assisting the ULB in performing its functions but cannot take over the ULB function. It is a short-term solution in a situation where the ULB is dysfunctional. A long-term solution is to get the ULB functioning, which would then take on various partners on board.

While assessing this case, the entire context of the city has to be kept in mind. The city government lies superseded. There has been no elected body for a long time. The city government does not have any finances to carry out its mandatory functions. All its functions are being carried out by the state government departments. The municipal employees are not working. To make matters worse, they have many grievances. The situation is of all-out loss, where nothing seems to be happening. And the municipal body does not even have an ability to perform. In a situation of total darkness, the CJP came as a ray of hope in Ranchi city, demonstrating that the city had the capability to implement the MSW Rules. In such an atmosphere of mistrust, between the ULB and the people, between the ULB employees and the ULB, between the ULB employees and the NGO and between the people themselves (some wanting to pay for the services and some not), it was possible to keep some parts of the city clean and also think about a sustainable solution to city waste is an achievement and on which the city has to build on by reviving the Solid Waste Mangement.

NOTES

1. We have documented here the Clean Jharkhand Project (CJP) pilot project in the city of Ranchi. Two field visits were made to the city, one

during monsoon and other before the Diwali festival. In the hindsight we realised that both the time points were important because, most urban centres in India get badly affected during rains (waterlogging due to solid waste) and a large amount of waste is generated during Diwali festivities.

2. Information in this case study is largely based on discussions with various groups of people viz., the NGO (NBJK), the municipal corporation officials and community members.
3. We would like to thank Mr. Rajesh Kumar Das, Project Manager and Branch Head, Nav Bharat Jagriti Kendra (NBJK), for his generous discussion on his experience of community based issues; Mr. Awadhesh Prasad, Assistant Project Manager (APM) for giving an initial insight on the project, Mr. Vishwa Ranjan, Policy and Liason Officer (NBJK), Mr. Suresh Yadav, Community Coordinator (NBJK), for assisting in field visits across various wards in Ranchi and sharing their experience on ground realities while working with communities, Mr. Santosh Kumar, Supervisor, ward no. 18 of Ranch for giving a detailed information of his ward; Mr. Shashi Ranjan (IAS), Administrator, Ranchi Municipal Corporation (RMC) for providing time for discussions and with the necessary information, and Mrs. Rekha Rani, Health officer RMC for providing details related to finances of RMC. We would also like to express our sincere thanks to communities and other supervisory staff of the NBJK working under Clean Jharkhand Project, animators and *Safai Mitras* for providing valuable insights on issues in SWM in Ranchi. Finally, Mr. Amit Mistry, Student, School of Planning, CEPT University, assisted us in collecting the information from Ranchi.
4. It is the only Municipal Corporation in the state of Jharkhand.
5. As per data downloaded from census of India website (provisional figures).
6. The city has a topographical advantage for taking care of its storm water run off. The city slopes away in all four directions more or less from centre. The storm water run off from the town thus finds it way to various drains and tributaries of river Subernarekha though roadside drains. But, wherever the waterlogging takes place is because of unplanned expansion of the city. (*Source*: Minutes of the meeting held under the Chairmanship of Deputy Commissioner, Ranchi to approve the Project Proposal Report of integrated sewerage system submitted by M/S Span Travers Morgan. See Urban Development Department letter No. 119 dated 22.01.2004 and RMC letter No. 1992 dated 08.09.2004.)
7. *Op cit*. note 6.
8. Clean Jharkhand Project—Half Yearly Report (1st April 04-30th Sept 04) prepared by NBJK, Ranchi & submitted to ICEF, New Delhi.

9. A task force was also constituted to prepare a draft policy on SWM in Jharkhand State. The members of the taskforce are from: (a) NBJK, (b) Member, Supreme Court Committee on SWM, (c) ICEF, (d) Urban Development Department, (e) RMC, (f) Ranchi Regional Development Authority (RRDA), and h) Indira Gandhi Institute of Development and Research (IGIDR).
10. Creative International, Express Seva Sanstha, Nirmal Bharat.
11. RMC website.
12. However, the State has implemented the 73rd amendment of the constitution for establishing Panchayati Raj System.
13. The Resource recovery, awareness and capacity building division are discussed later in this chapter. Policy and Liaison is discussed under partnership/ alliance head.
14. This mega campaign was organised from 21st July to 13th August 2004 to facilitate interface between RMC and community, whose central focus was organisation of ward level meetings. RMC's Administrator with Dy. Administrator and Health Officer were present in the campaign.
15. As as told by the RMC officials.
16. Where the PDC is able to purchase the rickshaw cycle on their own and are able to manage the waste on their own in a sustainable manner.
17. http: //ranchimunicipalcorporation.org/revenueSources.php.
18. Article in *Hindustan Times* dated 25th August 2004.
19. Presently there are four sustainable PDC in the city of Ranchi viz., Kozi Corner Society, Residency Estate Apartments, Bank Colony and Viber Porish Colony.

8

A Litter Free Small Town
Suryapet

DARSHINI MAHADEVIA[1] and MANISH PATHAK

City	Suryapet
Urban Local Body	Suryapet Municipality
Population, 2001	93,239
Population Growth Rate, 1991-2001	4.40% per annum
Floating Population	30,000 per day[2]
Area	35 sq km
Total wards in Municipality	28

Introduction

Suryapet is a litter-free town. With a population of 93,239 in the 2001 census, Suryapet is historically known as the 'Gate Way of Telangana' and it is the headquarters of the Nalgonda district of the Telangana region of Andhra Pradesh (AP). The town has become known as a 'Dustbin Free and Zero Garbage Town', recognised as such by the Supreme Court in its deliberations on SWM in the context of Public Interest Litigation on the subject[3] and received a 'Good Practice' award from UN-Habitat in 2004.

The main economic base of the town is trade and commerce, because of its location on the main Hyderabad—Vijayawada corridor (NH 9), and its role as district headquarters, agricultural marketing through government set up yard, and education and health services to the district. There is only one industrial estate located within the municipal limits, which has units manufacturing PVC pipes, granite and polished stone, chemicals and plastic goods. The estate is administered by Andhra Pradesh Industrial Development Corporation (APIDC) Ltd. The town has one engineering college, one post-graduate college, four degree colleges, 23 junior colleges, 75 high schools and 142 primary schools, along with Government hospitals and several private nursing homes with modern medical facilities.

In terms of infrastructure, 80 per cent of drains in the town are open, 30 per cent of the 240 km length roads are concrete, another 30 per cent are tarred and the remaining 40 per cent are *kachcha*. The municipality has 7 circles with 4 wards each making total of 28 wards in the town. The municipality also has the *e-seva* (e-service) centre for collection of various bills and taxes. Suryapet's finances are not in deficit and the town is able to collect on average 85 per cent of its targeted taxes on an average[4].

Before the launch of the SWM programme in 2000, the town had no proper system of managing the waste, dustbins were not regularly cleaned, streets were dirty and the drains (of which 80 per cent are open) blocked. The programme is called 'Zero Waste Management' (ZWM), which was launched in the December 2002 by the newly appointed commissioner at the time Mr. S. A. Khadar Saheb. Incidentally, he has been transferred in July 2006 to another town in the state.

Solid Waste Management Status

Suryapet generates about 43 MT of waste every day from households, vegetable markets, fish markets, hotels, lodges, *kalyana* mandapams (community halls for marriages or religious functions), restaurants, hospitals, schools, business establishments, and all other economic activities (Table 8.1).

Out of the total 43 MT waste generated, 21 MT is separated at collection and at the disposal point. 2 MT of the remaining 22 MT of waste is collected by the pig rearing community from the hotels and restaurants for feeding their animals, as this waste is nourishing. The municipality takes the 20 MT left over to their treatment unit both for composting and sale of recyclable waste. On an average, 70 per cent of the city waste generated is reused or recycled and the remaining 30 per cent is dumped in low-lying lands.

Table 8.1: Daily Waste Generation (MT)

	Waste Sources	*In MT*
1.	Average household waste generated per day @ 0.200 kg/capita	18.00
2.	Shops and Establishments	1.50
3.	Vegetable Market	2.50
	Total (A)	*22.00*
4.	Hospital & slaughter house waste	5.30
5.	Debris	7.50
6.	Drainage silt	9.00
	Total (B)	*21.80*
	Total A+B	*43.80*

Note: Waste generated varies 20 per cent on an average from the figures in the table.

Source: Municipal Council, Suryapet (2003).

Since the start of the project in Suryapet, within two years, the target of 100 per cent garbage collection was achieved (Table 8.2). In 2002, the first year, just 71 per cent of the generated garbage was collected: by the next year, 100 per cent was reached and the city made income of some Rs. 26,000 from recycling. Since then, the income of the city from SWM has increased. Thus, rather than garbage clearing requiring expenditures by the urban local body, Suryapet made some income. To the gratitude of the residents the entire town is now served by the system.

Table 8.2: Basic Figures of ZWM, Suryapet

No.	*Particulars*	*2002*	*2003*
1.	Garbage generated	42 MT	43 MT
2.	Garbage lifted	30 MT	43 MT
3.	Annual income from garbage	Nil	Rs. 26,000
4.	Households covered	19,000	21,000
5.	Area	24 sq km	34 sq km
6.	Public satisfaction	25%	100%

Source: Khadar Saheb (2004).

The SWM System

A door-to-door waste collection system has been introduced under the ZWM system. The waste is collected such that the need for roadside waste collection bins is eliminated. Hence, approximately 360 roadside dustbins have been removed. When the waste is segregated at the source, the recyclables are available for sale—a revenue source from which the municipality earned Rs. 26,000 in the year 2003. In the second phase of the project, two-bin system has been introduced at the household level, wherein all the houses have been given two types of bins, one green and other red by the municipality, at no cost. This is for better source segregation of waste. If the households wish, the municipality has also offered to give individual households a vermi-composting bed to recycle bio-degradable waste at home. A comprehensive system therefore has been worked out.

Sweeping and Cleaning

Suryapet municipality does not have enough conservancy workers. Thus, part of the sweeping and cleaning job in the ward has been contracted out to DWCUA groups, in much the same way as in Hyderabad. The DWCUA groups are registered under registrar of societies as the womens' savings groups. They sweep the roads, clean the drains, collect the segregated waste from the households on their tractors and transport the waste to the treatment unit. The silt taken out from the drains is collected by a separate tractor and is dumped in low lying land. Presently, four circles (16 wards) have been given to the DWCUA groups, one DWCUA handling one circle[5]. That is, 4 DWCUA groups are working in the city.

Each DWCUA group has 13 sweepers, 14 drain cleaners, 1 tractor with a driver and 4 persons on a tractor to lift the waste, adding up to 32 individuals per group. Each group gets Rs. 44,000 per month as reimbursement, which is calculated taking into consideration all the activities involved in the task[6].

The DWCUA group is given a subsidy of Rs. 125,000 and sufficient funds are made available through a bank loan to buy a tractor. The subsidy is given as an incentive. A one-year contract on non-judicial bond paper is signed by the municipality and the group, and is revised every year. The same group gets the work every time. A bank sanctions loan for a tractor; instalments of it are deducted from the monthly bill paid by the municipality to the DWCUA. In essence, the municipality gives a bank loan guarantee for a DWCUA group. If the group fails to carry out the job, municipality repays the loan and recovers the tractor.

Photo 8.1: A Slum Neighbourhood, Suryapet

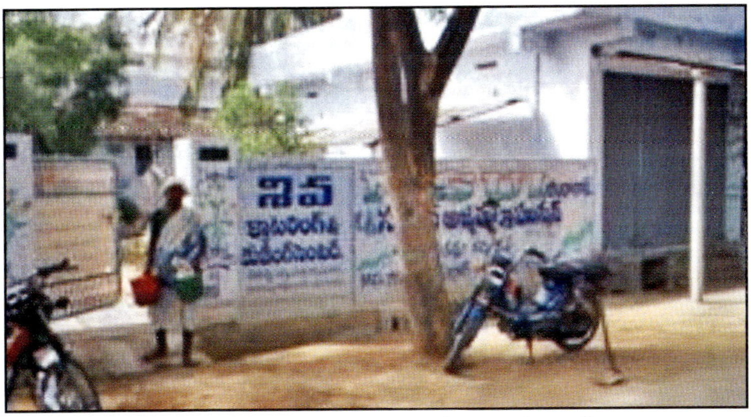

Photo 8.2: Household Waste Collection, Suryapet

Photo 8.3: A Clean Street, Suryapet

Photo 8.4: Tractor with Two Partitions, Suryapet

Photo 8.5: Red & Green Dustbins Distributed, Suryapet

The remaining three circles (12 wards) are maintained by the 228 municipality workers, called the Public Health (PH) workers. The commercial streets are swept by a group of 18 PH workers. The main roads are swept in the night. The working capacity of a PH worker is first tested before allotting him/her the work.

Collection

Every day, the municipality workers or the DWCUA groups move around in a tractor throughout the allocated ward to collect segregated waste from the households. At collection points, it announces its arrival by a whistle. The same vehicle covers all public institutions and commercial complexes in the ward. This system covers the whole town and there is no differentiation between the slums and non-slum areas. Currently, 75 per cent source segregation is being achieved daily.

For the hospitals and nursing homes meetings are conducted periodically with all the doctors registered with Indian Medical Association (IMA). They are instructed to give the segregated waste. The bio-medical waste is collected from the hospitals separately and buried into the earth. Presently, hospitals are not charged, but a proposal has been put before the municipal council by the commissioner to levy them user-charges. Until that happens the hospital offers free medical check-ups to the municipality workers in lieu of payments.

The dairy and cattle shed waste is collected every day separately and is used for vermi-composting process, 50 per cent of the Vermi-compost mixture being the cow dung. Trade and institutional waste is collected between 10 am to 12 noon, keeping in mind that these establishments start the day at around 10 am. A separate system has been set-up to collect waste from the vegetable vendors every day. The slaughterhouse waste is collected in a tractor in the evening. Some of this is used for the composting and the rest is buried in the earth. Hotels and restaurant waste is collected by the pig farming community in the town. This system has made the municipality's work easier.

The debris collection is looked after by the Town Planning (TP) department. Minimum of Rs. 500 is collected at the time of sanctioning the plan for building a house, which is used for collection and transportation of debris. TP department monitors the disposal of debris.

For the sustainability of the system, the commissioner was in the process of setting up informal ward committees at the time of our study. These committees are expected to monitor the system of road sweeping, drain cleaning and waste collection and lifting in each of the wards, along with monitoring other municipal works. The major *nallah* passing through the town is being developed so that the sanitation can be improved. To avoid littering on streets and footpaths by the vendors and pedestrians, small pole dustbins have been installed (sponsored by a private company, Suwen Pharmaceuticals).

Transportation

The transportation system has been designed so that the waste is picked up from the wards and is directly taken to the recycling unit. Motorized three-wheelers are used for collecting waste from households located on narrow lanes. On larger roads, the tractor is used. Both the tricycle

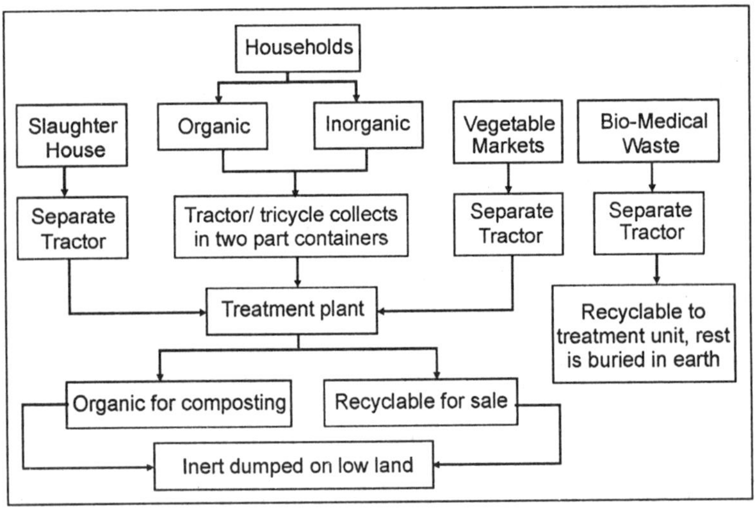

Fig. 8.1: SWM System of Suryapet

and the tractor have two compartments, one of which is used for wet waste and the other for dry waste. The wet waste is transported to the vermi-composting unit while the dry waste is taken to a recycling unit. The condition of the tractors owned by the municipality is not good as they are 25 years old and have been declared misfit for use by the Road Transport Organisation (RTO) that registers vehicles. The municipality has requested the State government for funds to purchase the tractors and other items. The waste thus collected is taken to the treatment unit, where the wet waste is composted and dry waste is recycled. From there, the waste that cannot be recycled at all, is taken to the dumping site. In all 9 tractors are deployed, 2 belonging to the water supply division and 3 that have been declared unfit for use, and 4 hired. In addition, 2 tricycles are deployed.

Treatment Unit

The municipal treatment facility is located at Ambedkar Nagar Road, at a distance of 3.5 km from the city centre. The area of the unit is 2000 sq yards. It has both dry waste storage unit, from where the dry waste is sold to the recycling unit. The organic waste is composted here.

Vermi-composting

The wet waste collected daily is used for making compost, using different methods such as vermi-composting and windrow system[7]. Compost produced is sold at Rs. 2.50 per kg. The organic waste is directly put in the beds and moisture is maintained by spraying water mixed with cow dung in suitable proportion. Such beds are maintained till the bed is decomposed to a maximum of 50 per cent, after which the manure is removed and the worms are left in the beds. The worms then multiply and then are reused in other beds. The daily waste collected is brought to the plant. Wet garbage is put into the beds (pits windrow system) and Vermi-compost pits. Therefore the daily waste is used on the same day and there is no need for storage of the waste.

Recycling

Households themselves segregate valuable recyclable waste and either sell it to, or exchange it for onions with the waste collector. But, if they throw it away in the garbage, it is stored as dry waste and taken to the treatment unit. Here, the papers, bottles and plastic bags are segregated from the dry waste. These recyclable materials are sold to the recycling industries at the rates finalised by the municipality. There are no rag pickers in the town at any level of the SWM system.

Landfill Site

The inert waste which is of no use is dumped on the land around the treatment unit, which is low lying. The quantity of such waste is around

30 per cent of the 20 MT waste brought to the recycling unit. Suryapet Municipality is able to manage source segregation, collection, and transportation of solid waste but the landfill aspect is not being managed properly. For a proper landfill site, 10 acres of government land has been identified on Kasarabad road, 4 km away from town, the acquisition of which is yet to be secured. The site will have a concrete base as per the MSW 2000 rules. Leachate treatment will be done on it and the leachate water will be used on the windrows to make compost out of waste. Other infrastructure, such as a weighing bridge, roads, and a compound wall among other things, are being planned at the landfill site.

Work Norms

The Public Health (PH) workers are employed 8 hours a day. Their attendance is taken at 5 am and again at 2 pm. The morning session ends at 10 am and the evening session ends at 5 pm (Table 8.3). The manpower and the machinery available have been allocated keeping in mind the size of the wards.

Table 8.3: Work Schedule

Time	*Task*
5: 00 am	attendance
5.15-10 am	door to door collection of segregated waste
8-8.30 am	tea break
10-12 pm	collection of segregated waste from commercial establishments collection of hospital waste
2-5 pm	lifting the drainage silt
Evening	collection of slaughter house
Arrangement is changed as per requirement as the vehicles are less	

The Initiation Process

The main components of the initiative are, stakeholder consultation and communication strategy, entrepreneurship development programmes, training courses, learning seminars, awareness campaigns, and documentation and dissemination of the experiences. The commissioner involved all the stakeholders including the chairperson and councillors of Suryapet municipality, staff, unions, specific user groups and other agencies and promoted the concept and practices of converting the waste to wealth.

The programme started with meetings of the Municipal Commissioner with various associations such as that of residents', commercial establishments', and so on. A programme of personal hygiene was taken up in the schools for two months. On realizing that the municipal staff members were discourteous and irregular in discharging their duties, group meetings and meditation and *yoga* programmes were organized for staff at different levels. The commissioner also provided counselling to the public health workers on their problems. Special efforts were made to provide salaries, tools/ equipment and other benefits as rewards for their hard work.

Training imparted to PH workers consisted of:

- Maintenance of discipline, courteous behaviour, keeping good health, discarding of bad habits, (i.e. smoking, alcohol drinking, playing cards, illegal sexual contacts).
- Courteous behaviour during the work hours and also with the households.
- Free health check up camps at public health centres periodically.
- Meditation, medication and *yoga* (MMY) programmes for 10 days, so as to set the right the behavioural attitudes.

Awareness programmes were designed and conducted amongst the residents, school children, college going youth, shop owners, fish and vegetable market associations, hawkers and pavement dwellers. Suryapet municipality also distributed pamphlets among traders to make them aware of the segregation of waste at their door itself. Also, pamphlets were distributed in the households telling them about the two-bin system.

In order to spread awareness among the residents, street meetings, street plays, distribution of pamphlets, stickers pasted on the doors of each house with slogans favouring primary segregation, house-to-house visits by groups of five or six women volunteers were undertaken. This was followed by meetings with the residents. The meetings not only focused on the benefits of source segregation and on-site composting, but were consciously designed to impart other important messages such as the right age of marriage, compulsory education, importance of adult education, reasons for the spread of communicable diseases, implementation of family planning, antenatal doses for pregnant women, problems related to child labour, and other important social messages.

Needless to mention that the messages were location-specific, that depended on the physical condition and living standards of the locality. The main message that was delivered across the town was that if each and every person maintains cleanliness, communicable diseases will reduce and therefore monthly expenditure on health will decrease, leading to savings and poverty alleviation.

Institutional Mechanism

The entire effort is headed by the Municipal Commissioner. Below the Commissioner, are Town Planning Officer, Sanitary Inspector and Executive Officers. Under the Sanitary inspector are 7 health assistants, one each for one circle. Below the health assistants are 4 DWCUA groups or the 228 municipal workers (Fig. 8.2).

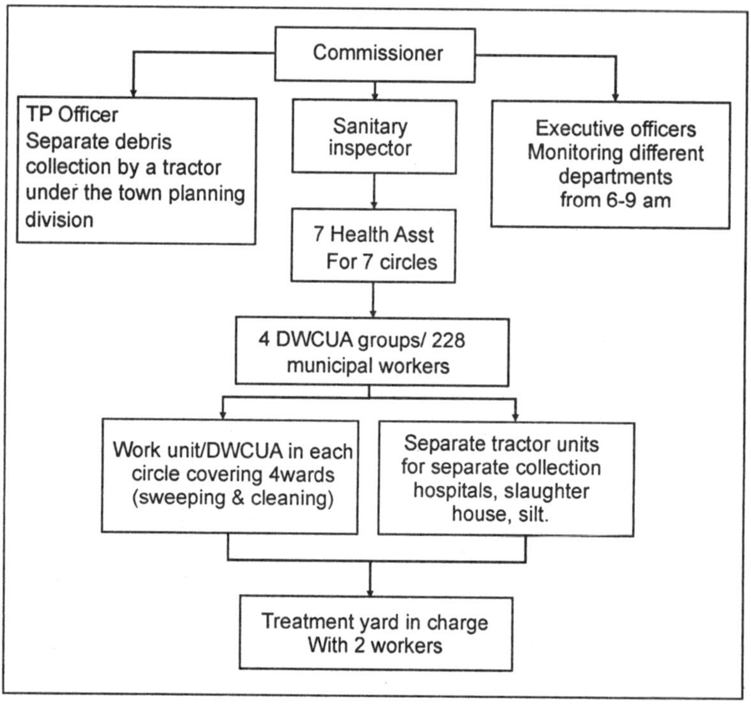

Fig. 8.2: Institutional structure of Suryapet Municipality

All the policy decisions are taken in the council meetings through resolutions. The day to day working and management decisions are taken by the commissioner. Decision of issuing work to private contractor is taken through the council resolutions. These decisions are sent to the District Magistrate for approval. The district collector is also involved in the decision.

Sanitary inspectors (3 in no.), the circle health assistant, assistant engineer, T P officer, revenue officer and working inspectors monitor the entire process. They are required to put in 2 additional work hours in the morning, from 6 am to 8 am, for this purpose. All the health assistants in charge of the particular circle have to report to the sanitary inspector and the inspector has to report to the commissioner. The health assistant takes the attendance and reports to the sanitary inspector.

The job of the sanitary inspector is to monitor whether there are adequate workers deployed in a given ward. The sanitary inspector also takes account of the vehicles, drivers, tools, and tackles. All the field level executive officers including the health assistants of the sanitary section have been provided cell phones for easy communication.

The municipality has devised an innovative way of imposing a penalty on 'defaulters', or those who litter the street. Firstly, they are imposed a fine of Rs. 100 after passing this decision in the municipal council. Then the penalised person's name is announced on the mic in the market area, to embarrass the defaulting person in front of the public, expecting that this would discourage the person from doing the same again. Such penalties would also discourage others also from littering the streets.

A people's organisation called *Lok Satta* is active. It also has its representatives in the legislative assembly. Its main job is to create awareness among the citizens related to civic commitments, their rights and rules. They have made ward committees with 4-5 people each in Suryapet, though they remain informal. These committees look after the day to day working of the municipality and guide other people in terms of their rights and duties. Suryapet Commissioner has supported these committees but no formal recognition has been given. Also, there is an inspection committee, which comprises a Member of Legislative Assembly (MLA), Municipal Chairman, prominent citizens, bank officials, advocates, and others.

Actors and their Roles

Municipality

The municipality is the main agency in charge of the entire SWM system. It provides its workers all the facilities they need so that they have no complaints and can work efficiently. It has also ensured that the workers have regular health check-ups at a local hospital, whose waste is collected by the Municipality free of cost. It monitors the DWCUA group and pays them on time. Also, it checks the condition of women working in the group. The municipality makes sure that the tractor for garbage pick-up reaches the streets daily on schedule, because this is the most important link in system. Lastly, it spreads awareness among the citizens regularly, especially among the children. It keeps in touch with the various associations and groups in the town to keep them motivated and maintain the efficient working of the system. The municipality workforce maintains the system daily and keeps a check on each other and also on general population with regard to waste littering.

Contractors (DWCUA)

Contractors (DWCUA) deal with all the official work, paying wages, monitoring the work, instructing the supervisor and facilitating the workers with the required infrastructure. It is the contractor's responsibility to get the work done on time and as per the contract otherwise a penalty is deducted from his bill.

Citizens

Citizens segregate the waste at their doorstep in the bins provided by the municipality and give it to the tractor team daily. They do not throw any garbage on the streets or drains. The commercial establishments maintain their street clean. The footpath vendors and pedestrian throw the garbage in the dustbins hanging on the poles. Senior citizens involve themselves in collecting suggestions and monitoring the work. Also, a people's organisation, *Lok Satta* has made ward committees which monitors the working of the municipality and informs the municipality if any irregularity is found.

Partnerships

Municipality—DWCUA

DWCUA groups sign the contract of maintaining the sanitation of the particular circle as per the decided conditions and rates. The rates are calculated considering all the items and functions involved; mainly three functions are considered—sweeping, drain cleaning and collection on tractor. If any dispute arises the DWCUA group consults the commissioner and resolves the problem. Since these groups are under the SJSRY scheme, the municipality doesn't generally act strictly but in severe cases penalties are imposed on them.

Municipality—Citizens

Citizens make sure that they bring the segregated waste to the tractor on their doorstep. This habit has been developed among the citizens after an effective awareness campaign by the municipality. Citizens keep the streets and drains clean and avoid littering. In turn they get the efficient sanitation services free of cost from the municipality. Also, citizens monitor the workers and report any irregularities to the commissioner.

Municipality—Recycling units

The recycling units established in Suryapet buy the waste from the municipality sale unit on the prescribed rates. This has generated recycling entrepreneurship and employment in the town. This partnership has eroded the practice of rag-picking.

Pig Farmers—Hotels

The pig farming community collects the organic waste from the hotels and restaurants and feed their pigs. Hotel owners keep this waste segregated from recyclable waste. The recyclable waste is given to the municipal tractor. The pigs do not roam around the town for food.

Funding

Municipal Budget

The total estimated expenditure of the municipality for the year 2004-05 was Rs 155.37 million. But, the revised estimate came down to Rs. 88.06 million in the same year (Table 8.4). Once again, the budgeted expenditure for the year 2005-06 is higher than the revised estimates for the previous year, at Rs. 112.02 million. In per capita terms, the revised expenditure in 2004-05 was Rs. 855, which has increased to Rs. 1,057 in 2005-06's proposed budget (Table 8.4). The per capita expenditure figures appear to be higher than average per capita expenditure figures for towns with population 50,000 to 200,000 in Andhra Pradesh, a category to which Suryapet belongs. In 2000-01, average per capita expenditure in this size class of towns in AP was Rs. 496 (Mahadevia and Mukherjee 2003), which at 2004-05 prices would be about Rs. 625. Therefore, Suryapet has, on an average, higher per capita expenditures than the towns/cities of the same size class in the state.

Table 8.4: Income Details, 2004-05 & 2005-06

Item	*Rs. million*
Estimated Income 2004-05	
Initial Balance 2003-04 (Revised Estimates)	10.38
Revised Income 2004-05	81.94
Total	*92.32*
Revised Expenditure	88.06
Final Balance (Surplus)	4.26
Estimated Income 2005-06	
Initial balance 2004-05	4.26
Estimated Income 2005-06	113.62
Total	*117.89*
Estimated Expenditure 2005-06	112.02
Estimated Balance	5.86
5 % Compulsory Saving	5.68
Final Estimated Balance (Surplus)	0.18

Source: Municipal Budget, 2005-06.

The SWM expenditure comes under the head health in the municipal budget. In 2004-05 (revised expenditure), Rs. 4.86 million (5.51 percent of the total expenditure) were spent, under the head health in 2004-05 (Table 8.5). In 2005-06, the proposed expenditure on health has been increased to Rs. 6.04 million or 5.39 per cent of the total expenditure. Thus, expenditure on health head has remained nearly the same in the last two years. Thus, the municipality is spending quite little on the SWM.

Further, as mentioned earlier in the text, and what we can see from Table 8.4, this municipality tends to have a surplus budget.

Rs. 10.38 million were carried forward from the previous financial year (2003-04) into the next financial year (2004-05). In 2004-05, there was surplus budget of Rs. 4.26 million. Since the town has higher per capita expenditure levels than the towns of similar size in AP, the financial situation of the town appears good. But, a surplus budget can be also be interpreted as a municipality's inability to make expenditures, which is not a good situation.

Table 8.5: Estimated Expenditure 2005-06 *(Rs. lakhs)*

	Budgeted expenditure 2004-05	*Revised expenditure 2004-05*	*Proposed expenditure 2005-06*
Salaries	16.10	14.96	18.30
Operation and Maintenance	3.20	2.72	2.04
Health*	6.26	4.86	6.04
Water supply	11.71	5.45	7.87
Lighting	1.50	0.29	1.60
City Planning Section	5.13	3.04	1.13
Others	3.88	4.90	15.81
Development activities from Taxes	107.60	51.85	59.24
Total	155.37	88.01	112.02

* This is expenditure on Solid Waste Management.

Source: Municipal Budget, 2005-06.

Suryapet has been able to collect 85 per cent of the targeted taxes. But, the town has a high dependency on the grants and transfers from the state government in general. In 2004-05, of the total Rs. 81.94 millions collected, 56 per cent (Rs. 46.20 million) came as plan and non-plan grants from the AP State government. This is mostly the case with all the small towns (up to population 50,000) and medium sized towns and cities (between population 50,000 and 200,000) in the state (Mahadevia and Mukherjee 2003). In the proposed budget of 2005-06, 47.13 per cent is expected to be grant component.

After the successful implementation of SWM in Suryapet, the Chief Minister of AP had announced an incentive grant of Rs. 2.5 million for the city, which however has not yet been released at time of writing. It needs to be noted that there has not been any change in extent of expenditures incurred on health sub-head of the municipal budget after this new initiative[8].

Table 8.6: Details of Expenditure on Health and Sanitation, 2005-06

(in Rs. Lakhs)

	Budgeted expenditure 2004-05	*Revised expenditure 2004-05*	*Proposed Budget 2005-06*
Private Sanitation Contract	30.00	24.00	30.00
Lifting work/Sanitary labour charges	6.00	6.20	6.50
Private contractors + Proclaims rents	5.00	2.80	1.50
Contracted Vehicles	8.00	8.95	9.00
Maintenance	1.00	0.90	2.00
Chemicals for Sanitation	3.00	1.50	2.00
Tools	2.00	0.80	1.50
Dresses	2.20	—	2.80
Accessories for workers	1.00	0.87	1.00
NGOs	1.00	0.83	1.27
Unidentified dead bodies	0.25	0.14	0.20
Pigs	2.00	0.71	1.00
Stray dogs	0.15	—	0.15
Vermi composting	0.50	0.25	0.50
Contingencies	0.50	0.60	0.75
Others	—	—	0.25
Total	62.60	48.55	60.42

* 1 lakh = 100,000

Source: Municipal Budget of 2005-06.

Table 8.7: Fund Requirement for Upgrading the System

Items	*Cost (in Rs. Lakhs)*
2 tractors, 1 hydraulic	20.0
Pulverizer	8.5
2 tractors trolleys 25 hp	6.0
2 tillers	3.0
200 wheel barrows	4.8
Total	*42.3*
Open windrow system	80.00
Landfill site (10 acres has been acquired, further 10 acre land is required)	296.12
Total	418.42

Source: From Khader Sahab (2004).

The highest expenditure within the health sub-head, at about 50 per cent of the budget is on a private sanitation contract, which is for the payments to the DWCUA groups (Table 8.6). It is likely that the salaries of the 228 sanitation workers of the municipality are not accounted in health sub-head but in salaries sub-head in Table 8.5. Just 16 to 17 per cent of the total expenditure in the municipality is on the salaries (Table 8.5). In 2001-02, Rs 14.36 million (about 30 per cent of the total budget) of Suryapet was spent on salaries[9]. Thus, by, giving out SWM activities to the private contractors, the municipality seems to have lowered its salary bill.

Fund Requirement

Fund Requirement for the entire SWM system for Suryapet Municipality, which includes a proper landfill site, mechanized lifting of garbage bins and a new windrow-composting pit comes to Rs. 41.84 million. If the town gets Rs. 2.5 million as an incentive grant, it would not be enough for the purpose.

Income

Initially the Zero Waste Management (ZWM) project was planned to generate income of Rs. 15,000 per day from the garbage collected from the town through source segregation. That did not happen. In the first year, the total income that accrued to the municipality came to just Rs. 26,000. At the end of the second year (2003), the income from the ZWM reached Rs. 1,000 per day. In 2005, income from recyclables and compost has reached Rs 2,000 per day.

Costing

The Operating cost are as follows:

- Cost of collection per MT per day Rs 260.
- Transportation per MT per day Rs 180.
- A municipal sweeper gets maximum salary up to Rs. 6,000 based on the seniority.

To equip the residents with two separate dustbins and vermi-compost sheds, the municipality has raised funds sources, like Lion's Club, big industrial houses and government resources.

Achievements

Suryapet Municipality has achieved its target of Zero Waste Management (ZWM), as it calls its programme. This has reduced the waste going to the dumping site and minimised the requirement of a landfill site. However, a landfill site requirement cannot altogether be eliminated as there is some waste that is non-recyclable and which needs to be disposed off in a scientific manner. But, the Suryapet experience shows that the landfill site requirement can be minimised if waste is segregated at household level, all of it is collected and then appropriately recycled. Considering, the future requirements, the municipality requires only 10 acres of land for dumping non-recyclable wastes.

This case study also shows that ZW is possible without any extraordinary efforts, expenditure, or labour, if there is a will to do so in the local government. Suryapet municipality has been spending between 5-6 per cent of its annual budget on SWM, which is quite low when compared to the cleanliness the city has achieved. Further, the city has begun to earn from the waste, although at the moment the income is not so high, at just Rs. 26,000 per year. This is just 0.43 per cent of the total expenditure proposed on health in 2005-06 budget, but it can certainly meet half the proposed expenditure on vermi-composting proposed in this year's budget. The municipality expects to improve its income from the waste in due course of time.

The new system is environmentally friendly and economically. This system is based on segregating the household waste, which provides for optimal recovery of wastes at the source. Recycling leads to reduction in resource depletion. Waste minimisation leads to conservation of valuable land, which otherwise would have got used up for land filling. This system has also created a situation of low environmental pollution at the waste dumpsite through reduction in soil, water, and air pollution. Lastly, it has created employment for women and enterprise opportunities in the 'green' industry.

The new system has other indirect benefits as well. At this moment, there are no rag-pickers in the city. A dehumanising employment category has been done away with. Every development process should in fact aim at reducing such jobs and creating new employment opportunities in better paying and better quality jobs.

The new type of employment for the poor is sought through an urban employment programme called SJSRY, under which SHG groups of women are formed for livelihood purposes. Formation of DWCUA groups is also for women's empowerment. Engagement of DWCUA in this ZWM system provides a sustainable and decent employment opportunity for the women of low-income households.

These women not only get comparatively decent wages but also a work uniform and by that, most importantly, some dignity. With time, if better technology is applied to this work and women are provided social security coverage, there would be even more dignified work for these women.

The ZWM system also fulfils the commitment of meeting the requirement of the SWM rules. Interestingly, the Indian Supreme Court has pronounced in its Record of Proceedings in W.P. (Civil) No. 888/1998, dt: 24-08-2004, that everyone should go and learn from Suryapet as it provides an excellent example of a smoothly running SWM system.

The municipality is doing this programme free of cost and efficiently, which has created its positive image among the citizens. This helps them in doing other functions without any protests from the citizens. Suryapet, a small town with one lakh population, has therefore become an internationally known town.

Lastly, the most important achievement of Suryapet is that the municipality has taken the initiative to implement ZWM and has brought in the support of the people and has incorporated functions for other actors. It has also formalised some of the partnerships like that between the hotel/ restaurants and the pig-rearing community. Like other cities of the AP state, it has dovetailed urban employment programme with the waste management programme. All these have been synchronised because the ULB is in charge of the entire effort and has not outsourced its responsibilities as many ULBs have begun doing with the encouragement of the urban policy makers. One quarter of the 5169 urban centres in India fall in the range of 50,000 to 200,000 population, wherein, it is very easy for an ULB to put in place a sustainable SWM system at practically no additional costs.

Concerns

In spite of so many achievements, some issues remain. The first and most important one is that the entire ZWM has emerged as an initiative and interest of one person, the then municipal commissioner of the city. He has personally participated in the entire system; he has gone around the city carrying out awareness campaigns, setting up the system, monitoring the system, and resolving the conflicts as and when they arise. He has been instrumental in motivating the municipal workers to perform their duties with these new ideas, without any opposition from the labour union. A lingering concern is whether the present system will function as well as it has under a different commissioner.

The second concern is with regards to the assignment of duties to the municipal workers. There are some duties that most workers want to do and there are some duties that most workers do not want to do. At present, the Commissioner personally solves these conflicts. A systematic rotation of duties is required to minimize such conflicts. Still, some way of allocating work to the municipal workers needs to be created.

The DWCUA groups carry out the waste management tasks in 4 out of 7 circles of the city. Their contract gets renewed every year, which makes them in essence permanent workers. Due to this, a monopoly in particular areas is created. Also, they are under the jurisdiction of state government (SJSRY) and the municipality cannot take strict action against them in case of irregularities. With regards to the employment conditions, the DWCUA workers have to be given social security, which is not there in the present system.

Although, the system is working well for now, the citizenry's participation in the system is crucial, something that is not formalised at the moment in Suryapet. The current commissioner is very committed to the ZW system. It may happen that in the future, one of the new commissioners appointed does not have as much commitment. (As mentioned earlier, the commissioner who had set up the system has been transferred and we do not know the inclination of the new commissioner who has taken over the administration of the city). The ward committees formed by the *Lok Satta* organisation, but they are informal in nature at present. It is necessary that the Ward committees are formalised as per the provision of 74th CA so that the system of citizens' participation in waste segregation and monitoring of the system remains sustainable.

NOTES

1. We are thankful to Khader Saheb, the then Municipal Commissioner of Suryapet. He has been transferred to another municipality in July 2006. It needs to be seen whether the same initiative continues with the new Municipal Commissioner taking charge of the city.
2. On account of it being a district headquarters.
3. The Hon'ble Supreme Court of India in its record of proceedings in

Writ Petition (Civil) No. 888/1988 filed by Mrs. Almitra H. Patel and another, have noted that "Lack of funds is no excuse for inaction. Smaller towns in every State should go and learn from Suryapet in Andhra Pradesh and Namakkal in Tamil Nadu, which have both become Dustbin Free Zero Garbage Towns complying with MSW Rules since 2003 with no financial input from State of Centre, just good Management and a sense of commitment" dated 4th. October, 2004.

4. As per discussion with Mr. Jagannath, Sanitary Inspector, SPM, Suryapet Municipality.
5. As per discussion with the commissioner.
6. As per the notification issued by the municipality.
7. Windrow system is where there is a pit created (in Suryapet it is of 20 m × 2.5 m × 1.5 m (depth). There is a concrete slab at the bottom with a lining and a slop of 2 per cent. The waste is dumped in these pits on which there is leachate spraying.
8. As per discussions with the Commissioner.
9. Based on data from Directorate of Municipal Administration, Government of Andhra Pradesh.

9

Managing Waste in a Peri-urban Locality
Vejalpur, Ahmedabad[1]

SHRAWAN KUMAR ACHARYA and APPEEJI PARASHER

ULB	Vejalpur Nagarpalika
Population, 2001	113,263
Population Growth Rate 1991-2001	5.76% p.a.
Area	12.5 sq km
Water supply, sewerage lines, storm water drain—Capital works	AUDA
Water supply, sewerage lines, storm water drain—Operation and Maintenance	Vejalpur Nagarpalika
Wards	12

Introduction

Note: Vejalpur become part of the Ahmedabad Municipal Corporation in May 2006. At the time this research was undertaken, it was a suburban municipality, well known for its early adoption of *e-governance*.

Vejalpur is one of the Nagarpalikas (municipalities) located outside Ahmedabad Municipal Corporation (AMC) limits but within Ahmedabad Urban Agglomeration (AUA), and comes under the jurisdiction of the Ahmedabad Urban Development Authority (AUDA), the planning agency[2]. It is located half a kilometer west of the important Sarkhej-Gandhinagar Highway, which formed the western-most boundary of the AMC. As a suburb, Vejalpur's growth rate, at 5.76 per cent p.a., is much higher than that of the AUDA (1.95 per cent p.a.) and the AMC (2.04 per cent p.a.). Such high growth exerts tremendous pressure on its environment and infrastructure.

Vejalpur Nagarpalika was formed in April 1994. It is considered to be India's first e-governed Nagarpalika having a fully computerized administration system. As a result administrative work is supposedly more effective and efficient. In addition to the main Civic Centre at the Nagarpalika office, there are two decentralized civic offices, namely Su Ganga and Akshardhara, where the citizens can pay their bills, obtain information and register complaints. This case study has been selected because of its known success in e-governance and waste management practice.

The major objective of providing access to e-governance through decentralized civic centres is to make information available and thus empower the citizens. The obligations and rights enumerated in the 'citizen's charter' have been implemented and made user friendly. The services offered in the city civic centres include; online payments of Nagarpalika dues, registration and issuance of birth and death certificates, and information about infrastructure projects and any other municipal matter.

In addition to these services, the Nagarpalika also maintains databases, including those dealing with solid waste management and sanitation. One of the objectives of the E initiative was to facilitate on-line monitoring and efficient delivery of civic services, including an on-line complaint redressal system for civic amenities. All the 12 wards of the Nagarpalika are connected through internet to receive complaints on civic amenities which are directly received by the Chief Officer and then directed to the concerned department(s) for speedy action. However, the redressal system is yet to realize its full potential as the user base is still limited.

Waste management has been one of the priority concerns of the Nagarpalika, especially for the current elected representatives and including the President of the Nagarpalika and the Chairman of the committee looking after waste and sanitation. Of the many Nagarplaikas in AUDA, Vejalpur is one of the few that has implemented the waste management initiative effectively.

The Vejalpur waste management scheme covers 9.53 sq. kms and 113,000 residents in 11 of the 12 wards. A total of 40,000 households in 1700 housing societies, 9 slum colonies housing about 11300 people, 110 hotels and small eateries, 115 medical utilities and 872 shops are covered by the project, as are 80 percent of the Vejalpur roads. The sweeper to population ratio in Vejalpur is lower (1: 6) than the AUDA average (1: 7)[3]. Total solid waste generated is estimated to be around 60-70 MT/day, of which wet waste (biodegradable) is 52-60 MT and dry waste (recyclable) is 8 to 10 MT. Per capita waste, household generation, is around 0.400 kgs/ per day. Waste generation is highest in wards 1 and 7 where the growth rate and concentration of population is highest.

Partnerships and Institutional Framework

As a part of the waste management initiative the Nagarpalika has succeeded in forging partnerships with two civil society institutions, namely the Self-Employed Women's Association (SEWA)[4] and the Vishnu Khadi Co-operative Society. SEWA approached Vejalpur Nagarpalika about providing assistance in SWM, and negotiated an agreement with the municipality for door-to-door collection from residential and commercial areas. This initiative was started in February 2004. Maintenance and cleanliness around community waste containers has been contracted out to Vishnu Khadi Co-operative. Waste from vegetable markets, construction debris, bio-medical waste from hospitals and clinics, and wastes from hotels and restaurant are being managed by the Nagarpalika. Street sweeping is carried out by the departmental sanitation staff.

The Nagarpalika has a formal agreement with both its partners. The partnership with SEWA is forged through annual renewable contract. Under this agreement SEWA through the '*Arogya Bhaginis*' (Health supporter/ sister) provides door-to-door collection service in 11 wards.

SEWA has formed a co-operative of the women rag-pickers and sanitary workers called '*Shri Karya Siddhi Kargal Kam Mahila Sewa Sahkari Mandli*'. It is registered under the co-operative act and was established in 1998. The co-operative regulates the waste management initiative of SEWA. Initially its activities were confined to the AMC. Vejalpur is one of the first Nagarpalika to provide SEWA with an opportunity to extend its services outside Ahmedabad. Unlike AMC, in Vejalpur SEWA operates in a greater number of wards, which gives it the scope to employ more marginal and destitute women and provides it with much needed economies of scale and operational efficiency. Geographical contiguity also helps its operation.

Photo 9.1: Clean Streets, Vejalpur

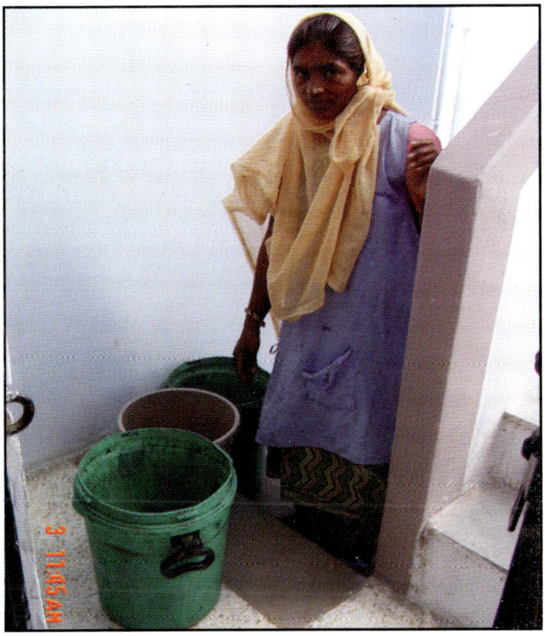

Photo 9.2: Arogya Bhaginis collecting segregated waste, Vejalpur

Photo 9.3: Dumping Site, Vejalpur

SEWA[5] provides training to women workers. The role of the women, the *arogya bhaginis*, is to collect segregated waste from households and shops and to deposit it in secondary storage containers. Work is monitored by women supervisors appointed by SEWA. Besides monitoring the work of the waste collectors the supervisors also act as an intermediary between the households, SEWA and the Nagarpalika regarding complaints, other civic issues and information dissemination. They can play this role effectively because they are better qualified and more aware than the waste collectors. Safety and security of the workers is also their concern. One supervisor is appointed for every 15 *arogya bhaginis* to monitor daily attendance and the quality of the work. If any *arogya bhagini* is absent, the supervisor is responsible for making alternative arrangements. Every month a feedback form is given to residents to evaluate the work process. Apart from the feedback form, SEWA also distributes leaflets to improve citizen awareness. The workers and supervisors are systematically rotated in different wards to encourage greater responsibility, and accountability.

The Ngarpalika has seven committees looking after different services, one of which is sanitation. It has six elected members. Management of waste is principally handled by the sanitary inspector who is assisted by one supervisor, one clerk and 4 *mukadams* (labour contractors) to supervise the work of NGOs, contractors and their own workers. The Sanitary Inspector and Chairman of the Nagarpalika monitor the entire operation through regular field visits and the e-based complaint redressal system. The Chief Officer gives required suggestions and orders to the Sanitary Inspector. The President of the Nagarpalika is the key person in the decision-making process and as indicated before, the credit for instituting the SWM system goes to the present incumbent. However, it should be noted that the effectiveness of the e-complaint system is doubtful. In fact discussion with the households indicated that complaint through the supervisors and sometimes even the *arogya bhagini* is more effective. It appears that the percentage of households and neighbourhoods registering complaints through the e-based system is very marginal.

Manpower and Infrastructure

The manpower from the Nagarpalika involved in the process of waste collection and disposal at different levels is limited to 60 workers of which 26 are permanent and 34 temporary. The Vishnu Khadi Co-operative, which is the private partner, employs 30 workers and SEWA 344 *arogya bhaginis* along with 20 supervisors. Each *arogya bhagini* caters to 100-125 households daily from 8 am to 11 am. Each supervisor has a circle of 1,500 to 2,000 households and they supervise 100-125 households every day on a rotational basis.

Presently there are 25,000 women rag-pickers associated with SEWA in the Ahmedabad region. Of these only 1,500, i.e. 6 per cent, are formally employed and of these, around 344 (23 per cent) are employed in SWM in Vejalpur municipality and the remaining work at other offices or societies in the AMC. Each of the *arogya bhagini* has to become a member of SEWA by paying an annual fee of Rs 5. All the workers have to be above 14 years of age. The majority of the women belong to Valmiki and Dalit communities.

The Khadi Gram Udyogh employs men for taking care of waste around community dustbins. Their employment is however not permanent and depends upon the contractual period. Many of the workers were formerly employed by the Nagarpalika as daily wage workers and were reemployed by the private contractor. Many such workers consider working with the Nagarpalika a better option because the salaries are higher and working conditions more humane. The private contractors employ less people to do the same job: as a result the work load per worker is high, while the pay is less.

The person taking care of waste around the dustbins works from morning 8 a.m. to noon and reports to the Nagarpalika office (by 2 p.m). The workers are also entrusted to see that the residents put their litter in the proper place and also trace and monitor the lifting of waste by the tractors in different locations. One worker takes care of 7 to 8 community dustbins, but in the case of Nagarpalika employees this is 3,—less than half the work. The workers reported earning around Rs. 1,200 to Rs. 1,500 per month if working for the contractor and somewhat higher, Rs. 1,800, if with the Nagarpalika. Further, there is always a hope of permanent job while working with Nagarpalika. It was interesting to note that in some families the husband was working for the Khadi Gram Udyogh and the wife with SEWA which helped their household income.

In order to facilitate the work of their staff, NGOs and the contractor, the municipality has made a capital investment of Rs 16.07 million for procuring 6 tractors, 3 container lifting systems, 180 containers, 325 handcarts, 1 road sweeper machine, 1 JCB loading machine, 1 loader and 5 tankers. Despite such investment, infrastructure is still inadequate, especially during emergencies.

System of Waste Management

The *arogya bhaginis* are involved at the primary stage of door-to-door waste collection. They collect segregated waste from every household. As a part of their training at SEWA, they are taught methods of collecting waste in a segregated form, and how to behave in dealing with the householders in a pleasant, courteous and confident way.

The waste is collected on a daily basis throughout the year for three hours from 8 a.m. to 11 a.m. The workers also help during emergencies such as a disaster or epidemic, as in the case of dengue fever in Ahmedabad and its suburbs in 2005. However, no additional payment is made by the Nagarpalika for such support.

It is supposedly mandatory for residents to provide segregated garbage to the workers, otherwise the garbage is not collected. However, field visits showed that many households are not complying with this requirement. In many of the cases it is unsorted, which forces the *arogya bhagini* to segregate the waste. Initially, the Nagarpalika had provided High Density Polyethylene (HDPE) plastic bags to the residents free of cost for storing dry waste but despite this the households have failed to comply. The major incentive for the workers to separate the wet and dry waste is because they can sell the dry waste and earn extra income.

The Nagarpalika workers collect vegetable market waste, construction debris, bio-medical waste from hospitals and clinics, and hotel and restaurant waste. Construction and bio-medical wastes are not allowed to be mixed with the household garbage. Construction debris is lifted for a fee of Rs. 300 per truck by the Nagarpalika. The bio-medical waste is separately collected and incinerated with the help of a private partner. Main roads in the municipal jurisdiction are cleaned with the help of a mechanical road sweeper.

The collected garbage from households and shops is taken to a community dustbin (located within a radius of 1 km) by push carts fitted with separate containers for wet and dry waste. The maximum walking distance an *arogya bhagini* walks to dump waste is 50 to 100 meters from the neighbourhood. The garbage is dumped manually in the community dustbin as is waste from street and road sweeping.

The worker employed with the Vishnu Khadi co-operative maintains the cleanliness near the dustbin.

The *arogya bhaginis* sell the dry recyclable waste to an itinerant buyer, separating it if need be. By this they earn additional income of Rs. 600 to Rs. 700 per month, over and above their regular monthly income of Rs. 1,500.

The waste collected in community dustbins is lifted, with the help of a hydraulic dumper-placer, and transported by tractor trolleys to the dump site. This work is done by the Nagarpalika. The amount of garbage transported by different modes is recorded on a daily basis by the Nagarpalika officials and uploaded to the website for reference.

The Nagarpalika does not have any engineered landfill site under its jurisdiction and this is a major lacuna in the waste management initiative. Often the waste is dumped in open plots and water bodies.

Financing Waste Management

As noted, the Nagarpalika has made a capital investment of Rs. 16.07 million for the procurement of equipments and vehicles. The annual operating expenditure is Rs. 9.12 million. The average cost of one municipal worker is Rs. 4000 per month and that of SEWA is Rs. 1,250 p.m. Therefore, by employing the *arogya bhaginis* of SEWA and other workers the Nagarpalika saves around 66 per cent annually on labour costs. Financing waste management services is through the property tax. The property tax is low and recovery is only 66 per cent as per the municipal officials. In the absence of other sources of revenue, procuring more finances is a major problem. At present the Nagarpalika spends around 4 per cent of its total budget in sanitation and health, which is incidentally higher than other Nagaprpalikas in AUDA which spend around 2 to 3 per cent on these services. The sanitation tax only contributes around 2 per cent of the total expenditure in Vejalpur.

The payment to SEWA is at the rate of Rs.11 per household per month (Rs.10 to each *arogya bhagini* and Re.1 as a service charge to pay the supervisors). Since all *arogya bhaginis* are the members of SEWA union, 5 per cent of their monthly earnings is deducted as contribution to SEWA co-operative to meet other welfare needs of members. The members are entitled to take small loans to meet their needs as a part of the micro finance initiative started through such contributions. Loan disbursement amongst the *arogya bhaginis* in

Vejalpur is still small, both in number and amount, as compared to other areas. This is mainly because it is a new association. The Nagarpalika does not charge the residents for waste services: they are financed from the property tax.

Table 9.1: Operating Expenditure

Operating Expenditure	*Rs in Lakhs*
Salary (*Arogya Bhagini* + Sweepers)	4.50
Salary (*Safai Kamdar* + Daily wage workers)	1.25
Maintenance works for containers	0.60
O & M cost for infrastructure	1.25
Total	7.60

Source: Vejalpur Nagarpalika

Table 9.2: Annual Budget of Vejalpur Nagarpalika

(*in Rs.*)

Particular	*2002-03 (up to Dec) Actual*	*2002-03 (up to Dec) Budget Est.*	*2003-04 Budget Estimates*
Sanitation Work			
Salaries to Regular Sweepers	1,176,277.0	1,600,000	1,569,000
Wages to Daily Sweepers	649,781.0	890,000	833,000
Daily wages to tractor drivers	91,129.0	125,000	108,000
Tractor/ auto rickshaw maintenance expenditure	46,048.5	200,000	200,000
Tractor/ auto rickshaw diesel expenditure	168,529.9	200,000	250,000
Regular Tractor driver salary	55,755.0	75,000	75,000
Sanitary Inspector's salary	71,370.0	96,000	98,000
Other equipment maintenance		25,000	25,000
Total Sanitation Work	2,258,890.0	3,211,000	3,158,000
Health			
Cleanliness Personnel Salary	1,800.0	50,000	50,000
Health Personnel Salary		50,000	50,000
Swachha Nagar Abhiyan	97,500.0	150,000	150,000
Cleanliness During Monsoon		100,000	100,000
Total Health	99,300.0	350,000	350,000
Total (Sanitation work + Health)	2,358,190.4	3,561,000	3,508,000
Total Budget Expenditure			77,645,815

Source: Vejalpur Nagarpalika

Achievements

The whole practice of waste management was designed as per the MSW Rules rules of 2000. Despite bottlenecks, the Nagarpalika has succeeded in forging partnership with other civil society actors in addressing SWM issues. As indicated, Vejalpur is one of the few Nagarpalikas within the AUDA jurisdiction to implement the act. The Nagarpalika has learnt and adapted the practice initiated by the AMC, which should be appreciated and recognized. Following Vejalpur, local bodies from other cities are adopting similar strategies for waste management. *Arogya bhaginis* from SEWA are now being sent to other parts of Gujarat to train sanitation workers.

As indicated above, the residents of Vejalpur do not have to pay any extra charges for the services provided at their doorstep. At present they are paying Rs. 75 annually as sanitation charge which is very nominal and can be raised. This however also weakens the system and questions long term sustainability of the initiative. However, despite low property tax rates and limited resources, the fact that the Nagarpalika authorities have implemented the SWM directives reflects their concern for the health and hygiene of the citizens.

The Vejalpur Nagarpalika through e-governance has created a Waste Management Information System to encourage public participation and also good governance. Information related to the amount of garbage collection and disposal (ward wise) is available on the website which helps authorities to regularly monitor garbage collection and disposal. However, how effectively it is being used by the public still needs to be ascertained, but there is no doubt that the attempt needs to appreciated and supported.

Through this initiative the low caste *dalit* women, involved in rag-picking and sanitary work, have been organized and economically empowered. Most of the 344 odd *arogya* bgahinis in Vejalpur have witnessed rise in their household income. The initiative has instilled dignity in their work which is a significant achievement for them. This initiative also covers three slum areas, Thakurvas, Bharwardwas in ward number 9 and Juhapura, which are normally deprived of such facilities. This initiative has attempted not to discriminate against minority (Muslim) locality like Juhapura, a communally sensitive area.

Encouraged by the support of the Nagarpalika officials, SEWA is planning to start vermi-composting at a neighbourhood level. For this purpose, it has sent its workers to Mumbai for training on vermi-composting from Stree Mukti Sangathana (SMS), an NGO. However, the Nagarpalika is unable to provide land due to scarcity within its jurisdiction.

SEWA has also contacted one of the recycling companies to establish a neighbourhood level crushing plant in Vejalpur for plastic waste collected by *arogya bhaginis*. The crushed material will then be send to the main plant for manufacturing new products. The role of the middle man, who at present buys the recyclable waste and supplies it to the waste recycling units will not be necessary. With the absence of the middleman, the *arogya bhaginis* will get a better price for their materials and thus benefit substantially.

Sustainability Concerns

The main feature of the waste management system, as observed by the Nagarpalika officials, is compliance by nearly 90 per cent of households in segregating waste. Waste sorting at the household level is mandatory for the residents as the *arogya bhaginis* are instructed to collect segregated waste. This was encouraged by the Nagarpalika by subsidizing the households through the provision of plastic buckets, and by not imposing additional payments. But field observation certainly does not support the claim of 90 per cent achievement in source segregation. Compliance is quite low, although it was high in the initial period. Today most households do not sort the waste: they put it all together in the bucket and it becomes the responsibility of the worker to separate it. It is a different matter that the waste collector segregates it for her own economic reasons. But the fact remains that the households have failed in fulfilling their responsibility to participate in the programme effectively. This household behaviour is observed irrespective of educational level and income status. Clearly, renewed public awareness campaigns are necessary, perhaps with the aid of neighbourhood and resident welfare associations, service clubs and schools. Perhaps irresponsible behaviour requires the introduction of sanctions or fines.

The residents of Vejalpur do not have to pay any extra amount for the services. The Nagarpalika claims that the increased efficiency of property tax collection through e-governance has enabled the Nagarpalika to pay for the cost of managing the waste and sanitary services. In the first year of the operation of the computerized system, the municipality reported that its tax collection rates improved by 400 per cent.

The workers from SEWA involved in this programme are paid by the Nagarpalika through the NGO. As a result many neighbourhoods and households have done away with their former private arrangements. Householders find it much easier to deal with the workers of SEWA and are also saving some amount of money which otherwise they would have to pay to their workers. (This could be said to indirectly subsidize households, many of whom certainly have a much higher income than to need such a subsidy). In fact, for the long term sustainability of the programme, the Nagarpalika should take the success and acceptance by the neighbourhoods of the programme as an opportunity to raise the contribution from the household. Many neighbourhoods and households will be willing to pay more for improved services.

The issue of financing the services is a critical issue in sustaining the programme. More so in the suburban Nagarpalikas of the AUDA area where residents pay less than half the taxes of what AMC residents pay. The property and water taxes, which 34 ULBs under AUDA collect, fall short of expenditure in supplying services like water, solid waste management and street lighting. For instance, for a residential property of 100 sq. yards, on average an ULB under AUDA collects Rs. 450 in property tax annually, while the AMC collects Rs. 1,200. Besides which, the AMC also has other sources like *octroi* (Rs. 5,000 million) which is not levied in the municipalities[6]. It is estimated that Nagarpalikas in the periphery spend only 2 per cent to 3 per cent of their annual budget for SWM, of which around 80 per cent is spent on wages and salaries. Therefore, any augmentation in services like waste management, as initiated by the Vejalpur Nagarpalika, certainly calls for higher charges in order to be sustainable. Though the Nagarpalika may be saving up to 60 per cent of waste management labour costs by employing NGOs and other actors, the requirement for additional funds is still there, especially in face of the crying need for a sanitary landfill and composting facilities.

A critical assessment of the SWM initiative in Vejalpur shows that the system is effective at the primary collection (despite problems of segregation!) and disposal at the secondary storage at the neighbourhood level. But the problem starts from the point of community bin storage. Though the waste is transported and disposed from the community bins there is complete mismatch between the amount of waste disposed at the bin and transportation to the final disposal site. This is especially true in the case of community bins located catering to high density and slum areas. The bins are not being emptied sufficiently often. As a result, the waste starts spilling out, or is spilled out by dogs and cattle. Improper location of bins is also a problem. In many cases there seems to be mismatch between the number of bins and waste generation in the locality resulting in nuisance spots which numbered as high as 39 in the entire Nagarpalika. Given the limited number of workers under the private contractors, they are over stretched. Therefore, it is very common to see a clean site in the morning and a very filthy site in the evening. The problem becomes acute during the monsoon. This may not be true for all locations but the problem exist in many places and therefore may require a site specific approach.

Perhaps the major problem of waste management in Vejalpur is the absence of a proper landfill site within the Nagarpalik. Though it is part of the same urban agglomeration as AMC (parent City) it is not allowed to dump the waste in the AMC land fill site!! This issue, common to many peri-urban municipalities, underlines the need for cooperation between the local bodies, whether large or small. It also raises questions regarding the planning process and the relevance of AUDA Development Plan. The requirement of disposal sites for all ULBs has been highlighted in the MSW Rules, which clearly have not been implemented. Recently AUDA has identified a site near Sarkhej Village as a landfill site, but meanwhile waste disposal into the natural water bodies and open plots continues with severe environmental risk. One such site is just next to the Nagarpalika office and defeats all the noble work of the Nagarpalika. Filling up of the water body also appears to be the result of real estate greed.

Dumping of considerable amounts of biodegradable waste near water bodies is a major problem. Waste is also being dumped on open ground in some localities. This poses serious environmental problems. In 2005 dengue fever was reported in Vejalpur because of waste dumped in water bodies. Further, many of the filled former water bodies in Vejalpur have also witnessed considerable construction activity. These are unsafe sites because of the threat of seismic activity as was experienced in the 2001 Gujarat earthquake. Many buildings constructed in such areas collapsed both in Vejalpur and surrounding areas. There is thus a need to select a proper site and also adopt appropriate technologies for the construction of an engineered landfill.

The problem of effective waste management is also observed in transitional and overlapping jurisdictional areas like the boundary zone between the AMC and the AUDA, rail tracks and flyovers. In all such areas the waste management is below the standard as compared to other areas in Vejalpur. These zones are characterised by the not-in-my-back-yard syndrome. The road near Rajiv Nagar slum is a classic case where the residents live in the AMC area but dump their waste in Vejalpur Nagarpalika. Should the AMC compensate Vejalpur for the negative externality generated by its residents? How the problems of such transitional areas should be addressed is a major challenge in the provision and management of civic services.

A further problem is that of how to service "*ghettoised*" areas, where resident trust in local government is almost non-existent. The jurisdiction of Vejalpur Nagarpalika also includes one such community known as Juhapura, an area worst affected during the communal riots in 2002. Juhapura is dominated by the minority community (Muslims), however, there are a considerable number Hindus as well. In a large part of the area the status of infrastructure is extremely poor, especially in the inner localities, with regards to the condition of roads, drainage and solid waste management. Inadequate administration and poor services have resulted in haphazard development and encroachment. There is complete absence of any drains, either open or covered in the locality (most of the houses have soak pits), so waterlogging is a major problem during the monsoon. Waterlogging and garbage result in dengue and malaria which has caused illness and some deaths in the locality. This entire problem is further compounded by the administrative structure of the area. Administratively, Juhapura falls under four different jurisdictions, namely, Vejalpur Nagarpalika, Sarkhej Nagarpalika, Makatampur Panchayat and Jhaliwadi Panchayat. It is only in the area under Vejalpur Nagarpalika that SWM has been implemented. The Nagarpalika officials and elected representatives claim that they do not discriminate between the Hindu and Muslim areas in providing services. They also argue that their SWM initiative was also a strategy to bridge the communal divide and integrate the minority areas. However, interviews with a few of the Muslim residents reveals that the SWM services introduced in the area have not been very good. Initially they were better, but have deteriorated over time. Recently the *aroghya bhaginis* have not been regular in collecting waste from the locality. Instead of going to houses to collect the waste the households have to deliver the waste to them. This has been reported from row housing colonies. It appears that SEWA supervisors have not been able to monitor the work properly and address people's grievances: this is a community ruled by fear. Effort to integrate minority areas is commendable; however it needs to be implemented properly.

There are additional issues which need to be taken into account; problems related to dogs and stray cattle need serious consideration. In most of the cases the spilling out of waste from the secondary storage bins in Vejalpur is caused by dogs and cattle, and has not been addressed at all. The AUDA area including Vejalpur is home to the *Rabari* community who are pastoralists and cattle rearing is their principle occupation. The increasing number of cattle is a public health nuisance, and despite the High Court pronouncement to take action it has not been successful. Controlling dogs and cattle would go a long way in addressing littering.

Further issues that need attention are related to the inadequate provision of other infrastructural facilities especially drainage and sewerage systems. The absence of these, along with frequent road digging and construction activities, explain much of the state of the environment in the Nagarpalika. The point needs to be emphasised that waste management is not only the collection of solid waste from the household and its appropriate disposal; it encompasses much wider issue of human waste and needs proper introspection and an integrated approach.

The problem related to scaling up and sustainability of SWM in Vejalpur is indicative of much larger issue related to the governance and administration of the urban peripheries. The whole issue needs to be seen in the context of multiple institutions existing in the area with unclear and sometimes overlapping powers and responsibilities. There is lack of proper delineation of administrative responsibilities in the area. The ULBs collect tax but have no role in infrastructure development, but as a local body they are answerable to the public than bodies like AUDA. Even the tax which they collect is much lower than what should be collected to manage basic services like solid waste. With the government grant and taxes Nagarpalikas like Vejalpur are barely able to meet their administrative expenses[7]. Most ULBs face shortage of staff and funds and depend not just on the Collectorate and AUDA but also AMC. This is also true of health and SWM. Scaling up is also constrained by the scale (small) of the ULBs and fragmented administrative structure in the peripheries

As per the Act [U/S 23 (1) (2)] AUDA has to perform the co-ordination function. However, the Nagarpalikas do not get any assistance from AUDA. But on the other hand U/S 17 (3), 39 (4) and 49 (3) of the Act, they have to follow the proposals of development plan and Town Planning Schemes and AUDA's bye laws and there is no separate department of SWM in AUDA (Patel 2004). Therefore, there is a need to rethink the governance of these peripheries either by amalgamation with the AMC or by creating Separate Corporation in the AUDA area. It may also be possible to privatize or create parastatal organizations to provide specific service, like the SWM, in both the AMC and the AUDA. However, from public welfare function bringing all the ULBs under a centralised authority appears to be more reasonable and also a viable (enlarged tax base) option. It is commendable that despite the constraints Vejalpur Nagarpalika has gone ahead and sustained the solid waste management efforts for last few years which needs to be recognized and supported.

NOTES

1. We would like to acknowledge the help of Mr. Kishore Chauhan, President, Vejalpur Nagarpalika and other officials from the Vejalpur Nagarpalika for providing us with the relevant information which have helped us in our study. We would also like to thank Mrs. Yamini Patel of SEWA, Coordinator, Vejalpur Project. Discussions held with Mr. Prashant Vora, a builder from the same locality have proved very helpful.
2. In early 2006, Vejalpur comes under
3. Gujarat Municipal Finance Board, Status Report, 2003.
4. SEWA is a trade union registered in 1972 with a workforce of 530,000 women. SEWA works for the holistic development of women employed in the informal sector. For more details on the organisation see website www.sewa.org.
5. For more details about the organisation log on to www.sewa.org
6. Octroi was abolished from the municipalities some years back, but it continues in the Municipal Corporations; *Times of India*, September 19th 2005.
7. Ramesh Joshi, Chief Officer Vejalpur Nagarpalika quoted in the, *Times of India*, September 20th 2005.

Bibliography

Agarwal, B. (2001). 'Participatory Exclusions, Community Forestry and Gender: and Analysis for South Asia and a Conceptual Framework'. *World Development, 29*(10).

Agrawal, V. S. (2005). 'Sustainable Waste Management: Case Study of Nagpur, India' [Abstract]. *Asian Development Bank* Retrieved June 18, 2007, from *http: //www.adb.org/Documents/Events/2005/Sanitation-Wastewater-Management/paper-agrawal.pdf*

Anand, P. B. (1995). 'Decentralizaton and Municipalities in India: A Mistaken Case of Murphy's Law?' *Urban India, 15*(1), 1-26.

Anand, P. B. (1999). 'Waste Management in Madras Revisited'. *Environment and Urbanization, 11*(2).

Anand, P. B. (2000). 'Co-operation and Urban Environment an Exploration'. *Journal of Development Studies.*

Ananthaswamy, H.C. (2005). *'Municipal Solid Waste Management in Banagalore Mahanagarpalika—Innovative Practices'*. Paper presented at the National Workshop on Urban Governance and Environmental Management: Action Oriented Policy Studies on Waste Management in Jabalpur and Raipur, Bhopal.

Ansari, J. H. (2001). 'Public-private Partnership Arrangements in Solid Waste Management'. *Quarterly Review of Regulatory Development, TERI Newsletter* (9), 4-7.

Anshutz, J. (1996). 'Community—Based Solid Waste Management and Water Supply Projects: Problems and Solutions Compare' [Working Paper], *Community Participation in Waste Management*: Urban Waste Expertise Programme (UWEP).

Appaswamy, P. P. (2005). *Multistakeholder Cooperation in SWM: The Case of Chennai City.* Paper presented at the Seminar on New Forms of Urban

Governance in Indian Mega-Cities, Jawaharlal Nehru University, New Delhi.

Asnani, P. U. (2001). *Modernization of SWM Practices in Jabalpur City*: United States Agency for International Development.

Asnani, P. U. (2005). *'Status of SWM in India'*. Paper presented at the National Workshop on Urban Governance and Environmental Management: Action Oriented Policy Studies on Waste Management in Jabalpur and Raipur, Bhopal.

Asnani, P. U. (2006). 'Solid Waste Management'. In A. Rastogi (Ed.), *India Infrastructure Report 2006*. New Delhi: Oxford University Press.

AusAID. (2002). 'Bangalore Water and Sanitation Master Plan Project'. *Australian Government AusAID* Retrieved June 11, 2007, from http: // www.ausaid.gov.au/projects/bangalore.cfm

Aziz, A., N. Sivanna and Nageshwara M. Rao, (2002). *Sustainability in Practice: Exploring Innovations in Domestic SWM in India [Research Project]*. Bangalore: ISEC.

Baindur, V. (2005a). 'Breathing New Life into Ward Committees'. *India Together* Retrieved June 1, 2007, from http: //www.indiatogether.org/ 2005/jul/gov-wardcomm.htm

Baindur, V. (2005b). 'For the People, by Diktat'. *India Together* Retrieved September 5, 2005, from http: //www.indiatogether.org/2005/sep/gov-nurm.htm

'Bangalore Still a Favourite for IT Companies'. (2005, September 21). *The Hindu*.

Bannerjee, T. and S. Chakravorty (1994). 'Transfer of Planning Technology and Local Political Economy: A Retrospective Analysis of Calcutta's Planning'. *Journal of the American Planning Association*, 60(1).

BARC. (2007). 'Kitchen Waste (Biodegradable) Based Bio-Gas Plant [Nisargruna]'. *Bhabha Atomic Research Centre* Retrieved June 11, 2007, from http: //www.barc.ernet.in/webpages/technologies/home.html

Barse, S. (2006). 'Pune's Draft Development Plan under a Cloud'. *India Together* Retrieved April 25, 2007, from http: //www.indiatogether.org/ 2006/apr/gov-punecdp.htm

Bartone, C., J. Bernstein, J. Leitmann and Jochen Eigen. (1994). *Toward Environmental Strategies for Cities, Policy Considerations for Urban Environmental Management in Developing Countries*. Washington, DC: World Bank, UNCHS, Urban Management Programme.

Baud, I. (2000). *Collective Action, Enablement and Partnerships: Issues in Urban Development*.

Baud, I. (2004). 'Reuse, Recovery and Recycling of Urban Organic Solid Waste; Modalities, Commodity Chains and Sustainable Development'. In J. P. and C. F. Isa Baud (Ed.), *Solid Waste Management and Recycling:*

Actors, Partnerships and Policies in Hyderabad, India and Nairobi, Kenya (pp. 115-132). The Netherlands: Kluwer.

Baud, I., Stelios Grafakos, Michaela Hordijk and Johan Post, (2001). 'Quality of Life and Alliances in Solid Waste Management Contributions to Urban Sustainable Development'. *Cities, 18*(1), 3-12.

Baud, I. and Johan Post (2003). *Between Market and Partnerships: Urban SWM and Contributions to Sustainable Development*. The Netherlands: University of Amsterdam.

Baud, I. and R. Dhanalakshmi (2005). '*Governance in Urban Environmental Management: Comparing Accountability and Performance in Multi-Stakeholder Arrangements in South India*'. Paper presented at the New Forms of Urban Governance in Indian Mega Cities, Jawaharlal Nehru University, New Delhi.

Baud, I. and H. Schenk (Ed.). (1994). *Modes, Assessments and Linkages in Bangalore*. New Delhi: Manohar Publishers and Distributors.

BDA. (2005). 'Draft Master Plan 2015'. *Bangalore Development Authority* Retrieved May 14, 2007, from http: //www.bdabangalore.org/brochure.pdf

Beall, J. (1997). 'Policy Arena: Social Capital in Waste-a Solid Investment?' *Journal of International Development*, 9(7), 951-961.

Beall, J. (2001). 'Valuing Social Resources or Capitalizing on Them? Limits to Pro-poor Urban Governance in Nine Cities of the South'. *Journal of International Development*, 6(4), 357-375.

Berthier, H. C. (2002). *Garbage, Work and Society*. Mexico: UNAM Institute for Social Research.

Beukering, P. V., Madhushree Sekher, Reyer Gerlagh and Vijay Kumar. (1999). 'Analysing Urban Solid Waste in Developing Countries: A Perspective on Bangalore, India' [Working Paper], *CREED*. London: International Institute for Environment and Development.

Bhattachariya, S. and C. Shylajan (2004). *Economic Instruments for Managing Municipal Solid Waste in India* [Occasional Paper], Centre for Development and Environmental Policy. Kolkata: Indian Institute of Management.

Bisoyi, L. K. (2005). 'Solid Waste Management in Puri Municipality'. *Orissa Review*.

BMP. (2005a). *A City that Beckons...* Bangalore: BMP Public Publication.

BMP. (2005b). *Budget 2005-2006*, Bangalore Mahanagara Palike. Bangalore.

Bose, A. (1996). 'Private Markets for Waste: The Case of Calcutta'. In R. A. and J. Toye (Ed.), *Challenging the Orthodoxies*. New York: St. Martin's Press.

Boulanger, P.M. and Thierry Bréchet (2005). 'Models for Policy-making in Sustainable Development: The State of the Art and Perspectives for Research'. *Ecological Economics, 55*(3), 337-350.

Broekema, J. (2004). 'Trial and Error in Privatisation; The Case of Hyderabad's Solid Waste Management'. In J. P. and C. F. Isa Baud (Ed.), *Solid Waste Management and Recycling: Actors, Partnerships and Policies in Hyderabad India and Nairobi Kenya* (pp. 93-114). The Netherlands: Kluwer.

Case, E., Dinesh Mehta and Cecilia Kinuthia-Njenga (Ed.). (2001). *Participation to Partnership, Lessons from UMP City Consultations*. Nairobi: UNCHS (Habitat).

Chakravorty, S. and G. Gupta (1996). 'Let a Hundred Projects Bloom—Structural Reform and Urban Development in Calcutta'. *The Third World Planning Review, 18*(4).

Chamaraj, K. (2006). 'Urban Renewal: Mission Not Accomplish-able'. *India Together* Retrieved September 28, 2006, from http: //www.indiatogether.org/2006/jul/gov-jnnurm.htm

Chambers, R. (1988). *Sustainable Rural Livelihoods: A Strategy for People, Environment and Development* [Discussion Paper]. Sussex: University of Sussex, Institute of Development Studies.

Chatoopadhyay, S.S. (2005). 'Capturing Kolkata'. *Frontline, 22*(14).

Chatterjee, P. (1996). *'Mega City Management in the Asian and Pacific Region'*. Paper presented at the Seminar on Mega-Cities Management in Asia and the Pacific, Istanbul.

CLGF. (2006). *Local Government in India. Commonwealth Local Government Forum* Retrieved August 26, 2006, from http: //www.clgf.org.uk/index.profiles.htm

CMAG. (2002). 'Best Practice Catalogue—2002': City Managers Association Gujurat.

CMAM. (2005). 'Compendium of Best Practices in Municipal Waste Management 2004-2005 Volume I', *City Managers' Association of Maharashtra*. Mumbai.

Coad, A. (Ed.). (1997). *Lessons from India in Solid Waste Management*. Mumbai: WEDC and the All India Institute of Local Self Government.

Cointreau, S.J. (1984). 'Solid Waste Collection Practice and Planning in Developing Countries'. In J. R. Holmes (Ed.), *Managing Solid Wastes in Developing Countries*. New York: John Wiley and Sons.

Cointreau-Levine, S. (1994). 'Private Sector Participation in Municipal Solid Waste Services in Developing Countries', *The Urban Management Programme (UMP), Volume 1, The Formal Sector*. Washington, DC: World Bank.

CPCB. (2000). *Status of Municipal Solid Waste Generation, Collection, Treatment and Disposal in Class 1 Cities* [Report]. New Delhi: Central Pollution Control Board.

CPCB. (2005). 'Management of Municipal Solid Wastes'. *Central Pollution Control Board Highlights* Retrieved November 10, 2005, from http: // www.cpcb.nic.in/highlight2004/highlight2004ch3.htm

CPHEEO. (2000). *Manual on Municipal Solid Waste Management*. New Delhi: Ministry for Urban Development and Poverty Alleviation, Central Public Health and Environmental Engineering Organization.

CSE. (2003). *State of the Environment Report—Waste Management*. New Delhi: Centre for Science and Environment.

Dahiya, B. (2003). 'Hard Struggle and Soft Gains: Environmental Management, Civil Society and Governance in Pammal, South India'. *Environment and Urbanization, 15*(1).

Das, D. (2004, February 8). 'Do your Veggies come from Dhapa?' *Calcutta Times*.

Davis, B. (n.d.). 'Solid Waste Management and Debris Recycling'. In *Understanding our Civic Issues: SWM in Mumbai*. Mumbai: The Bombay Community Public Trust.

Desai, M. (2001). 'Party Formation, Political Power, and the Capacity for Reform: Comparing Left Parties in Kerala and West Bengal, India'. *Social Forces, 80*(1).

Devi, K. and V. Satyanarayana (2001). 'Financial Resources and Private Sector Participation in Solid Waste Management in India', *Technology Advisory Group on Solid Waste Management*. New Delhi: Government of India, Ministry of Urban Development.

Dillinger, W. (1993). *Decentralization and its Implications for Urban Service Delivery*. Washington, DC: World Bank, Urban Management Programme.

Divay, G. and Jeanne M. Wolfe (2002). *Metropolitan Governance Background Study: What do we need to know? A rapid foray into operational concerns*. Washington, DC: World Bank.

Drescher, S. (2002). Decentralized Composting-Fact Sheet from 5 Cities in Southern India. *(EAWAG-SANDEC) Aquatic Research Institute-Water and Sanitation in Developing Countries* Retrieved March 17, 2007, from http: //www.sandec.ch/SolidWaste/Documents/01-Composting/city_matters-markets_for_compost.pdf

Dutt, S. (2000). 'Megacities of Joy: A Case Study of Calcutta's Environmental Problems in the Age of Globalization'. *Australian Journal of International Affairs, 54*(3).

FICCI. (2007). 'FICCI Survey on Scope of Privatization of Municipal Solid Waste in India' [Press Release]. *Federation of Indian Chambers of Commerce and Industry* Retrieved April 20, 2007, from http: // www.ficci.com/press/243/5.doc

Friedmann, J. (1998). 'The Common Good: Assessing the Performance of Cities'. In H. C. Dandekar (Ed.), *City Space* and *Globalization, An International Perspective* (pp. 15-22). Ann Arbour: University of Michigan.

Furedy, C. (1998). 'Sustainable Consumption and Municipal Solid Waste Reduction in Developing Countries of Asia', *Promoting Sustainable Consumption in Asian Cities*. Nairobi: UNCHS (Habitat).

Gandhi, H. V. (2005). '*Waste Management Business Perspective*'. Paper presented at the National Workshop on Urban Governance and Environmental Management: Action Oriented Policy Studies on Waste Management in Jabalpur and Raipur, Bhopal.

GDRC. (2007). 'Urban Waste Management—Documents and Info Sheets'. *The Global Development Research Center* Retrieved June 8, 2007, from http: //gdrc.org/uem/waste/z-doc.html

Gerlagh Reyer, P. v. B., Madhu Verma, P.P. Yadav and Preety Pandey. (1999). *Integrated Modeling of Solid Waste in India* [Working Paper], *CREED*. London: International Institute for Environment and Development.

Ghosh, A. (2000). 'Urban Environment Management: Local Government and Community Action', *Urban Studies Series* (pp. 105). New Delhi: Institute of Social Sciences.

Ghosh, A. (2005). 'Public-Private or Private-Public? Promised Partnership of the Bangalore Agenda Task Force'. *Economic and Political Weekly*, 40(47), 4914-4922.

Ghosh, A. and M. Mitra (2003). *Functioning of Ward Committees in West Bengal: A Case Study*. Kolkata: Institute of Social Sciences.

Ghosh, T.K., T. Chakrabati, G. Tripathi (Ed.). (2005). *Biotechnology in Environmental Management Volume I*. New Delhi: A.P.H Publishing Corporation.

Gidman, P., I. Blore, J. Lorentzen and P. Schuttenbelt. (1995). *Public-Private Collaboration in Urban Infrastructure Services* [Working Paper], UMP Working Paper Series. Nairobi: UNDP/UN-Habitat/World Bank.

Godard, F. (1996). *City Partnerships for Urban Innovation* [Discussion Paper], Discussion Paper Series. Istanbul.

Goetz, A. and G. (2001). *Bringing Citizen Voice and Client Focus into Service Delivery* [Working Paper], IDS. Brighton: Institute of Development Studies, Sussex.

GOI. (1995). *Report of the High Power Committee on Urban Solid Waste Management in India*, Bajaj Committee. New Delhi: Government of India, Planning Commission.

GOI. (1996). *Report of the Expert Committee on Commercialization of Infrastructure*. New Delhi: Government of India.

GOI. (1998 and 1999). *Solid Waste Management in Class 1 Cities in India*, Interim and Final Reports, Asim Burman Committee. New Delhi: Government of India.

GOI. (2000). *Municipal Solid Waste (Management and Handling) Rules 2000*. New Delhi: Government of India, Ministry of Environment and Forests.

GOI. (2001a). *Good Governance Campaign*. New Delhi: Government of India, Ministry of Urban Development and Poverty Alleviation.

GOI. (2001b). *Recommendations for National Action Plan for Good Urban Governance*. New Delhi: Government of India, Ministry of Urban Development.

GOI. (2002a). *Agenda 21: An Assessment*, Paper prepared for the World Summit on Sustainable Development, Johannesburg. *Government of India, Ministry of Environment and Forests* Retrieved August 29, 2006, from http: //www.envfor.nic .in/divisions/ic/wssd/doc2/ch15.html

GOI. (2002b). *'Empowering People for Sustainable Development'*. Paper presented at the World Summit on Sustainable Development, Johannesburg.

GOI. (2002c). 'National Human Development Report 2001'. New Delhi: Government of India, Planning Commission.

GOI. (2004a). *Implementation of Part IX-A of The Constitution*, Third Report: Standing Committee On Urban Development (2004-2005) (Fourteenth Lok Sabha), presented to Lok Sabha on 7.12.2004. Government of India, Ministry of Urban Development Retrieved May 14, 2007, from 164.100.24.208/ls/CommitteeR/urban/3rep.pdf

GOI. (2004b). *Report of the Twelfth Finance Commission*. Government of India Retrieved April 15, 2007, from http: //finance.rajasthan.gov.in/doc/12fcreng.pdf

GOI. (2004c). *India National Environment Policy 2004* [Draft]. Government of India Retrieved June 18, 2007, from www.geocities.com/shankarb.geo/nep_2004.doc

GOI. (2005a). *JNNURM (Jawaharal Nehru National Reform Mission)*. Retrieved April 15, 2007, from http: //urbanindia.nic.in/moud/programme/ud/jnnurm.htm

GOI. (2005b). *Slum Population Series-1, Census of India, 2001*. New Delhi: Government of India, Office of the Registrar General & Census Commissioner.

GOI. (2006). *National Environment Policy 2006*. New Delhi: Government of India, Ministry of Environment and Forests.

GOI. (2007). *Eighteenth Report: Standing Committee on Urban Development (2006-2007)* (Fourteenth Lok Sabha). Government of India, Ministry of Urban Development, from http: //164.100.24.208/ls/CommitteeR/urban/18threport.pdf

Grinspun, A. and Mercedes Gonzales de la Rocha (Ed.). (2001). *Choices for the Poor: Lessons from national poverty strategies*. New York: United Nations Development Program.

Gupta, S.K. (2004a). 'Managing Waste in Educational Institutes: A Case Study of Two Institutes in Delhi'. *Toxics Link* Retrieved June 8, 2007, from http: //www.toxicslink.org/art-view.php?id=9

Gupta, S.K. (2004b). 'Rethinking Waste Management'. *India Together* Retrieved May 14, 2007, from http: //www.indiatogether.org/2004/apr/env-rethink.htm

Gupta, S.K. (n.d.). *Municipal Solid Waste. A Complete Synopsis*, Centre for Environmental Science and Engineering. Mumbai: Indian Institute of Technology Bombay.

Hanrahan, D., Sanjay Srivastava and A. Sita Ramakrishna. (2006). *Improving Management of Municipal Solid Waste in India: Overview and Challenges*. New Delhi: Environment and Social Development Unit, South Asia Region, World Bank.

Hasan, S. and M. Khan (1999). 'Community-Based Environmental Management in a Megacity: Considering Calcutta'. *Cities, 16*(2).

Heitzman, J. (2004). *Network City*. New York: Oxford University Press.

Hordijk, M. (2000). *Of Dreams and Deeds, the Role of Local Initiatives in Community-based Urban Environmental Management—A Case Study from Lima, Peru*. Amsterdam: Thela Thesis.

HUDA. (2003). *Hyderabad 2020, Draft Master Plan for Hyderabad Metropolitan Area*. Hyderabad: Hyderabad Urban Development Authority.

Hunt, C. (1996). 'Child Waste Pickers in India: The Occupation and its Health Risks'. *Environment* and *Urbanization, 8*(2), 111-119.

Huysman, M. (1994). 'Waste Picking as a Survival Strategy for Women in Indian Cities'. *Environment* and *Urbanization,* 6(2), 157-174.

ICLEI. (1996). *The Local Agenda 21 Planning Guide: An Introduction to Sustainable Development Planning*. Ottawa: International Research and Development Centre, International Council for Local Environmental Initiatives.

IPE. (2004). Management of Solid Waste in Indian Cities [Draft Report]. Infrastructure Professional Enterprise Private Limited Retrieved May 14, 2007, from http: //fincomindia.nic.in/12reports.htm

Jain, A. K. (2000). 'Management of Solid wastes in Mumbai'. *Bio Energy News,* 4(1).

Jain, A. K. (2004). 'Solid Waste Management in Mumbai', *Bombay Chamber Bulletin*. Mumbai: All India Institute of Local Self Government.

Jain, V. (2004). 'No Wasteful Business This!' *India Together* Retrieved June 8, 2007, from http: //www.indiatogether.org/2004/mar/env-recycling.htm

Jalal, J. (2005a). *Administrative Reforms: Towards Sustainable Practices*. New Delhi: Sage Publications.

Jalal, J. (2005b). 'Good Practices in Public Sector Reform: A Few Examples from Two Indian Cities'. In A. Singh (Ed.), *Administrative Reforms: Towards Sustainable Practices*. New Delhi: Sage Publications.

Jana, A.K. (2004). *Administering District Plans in India: Issues, Constraints and Choices*. New Delhi: Concept.

Janaagraha. (2003a). *Participatory Planning: A Citizen's Handbook*. Bangalore: Ramanathan Foundation.

Janaagraha. (2003b). *Shaping Vibrant Cities-Neighborhood Vision Campaign*. Bangalore: Ramanathan Foundation.

Janaagraha. (2003c). *Ward Vision Document for Ward 54 Version 1*. Bangalore: Janaagraha Publications.

Janaagraha. (2004). *Ward 54-Srinivasa Nagar Community Pamphlet*.

Jayanta, B. (n.d.). *Strengthening Municipal Governance Through GIS: A Decentralized West Bengal Model*. GIS Documentation Team Retrieved May 16, 2007, from http: //www.gisdevelopment.net/application/urban/overview/urbano0031.htm

Jayaraman, S. (n.d.). *Role and Functions of Urban Local Bodies under 74th Constitution Amendment Act, and State Finance Commission Etc.* from http: //www.tn.nic.in/tnudp/images/42.PDF [website closed]

Jha, G. (1996). 'Contemporary Issues in Municipal Institutional Capacity'. *Urban India, 16*(2), 43-59.

Jha, G. and U. Raghupathi (1994). 'Facets of Effective Urban Management in the Context of New Economic Policy'. *Urban India, 14*(1).

Joardar, S. D. (2000). 'Urban Solid Waste Management in India: Issues Related to Institutional Arrangements'. *Public Works Management and Policy, 4*(4), 319-330.

Joshi, R. (2003). 'Succeeding in Municipal Accounting Reforms in the Gujurat, Tamil Nadu and Karnataka Experiments'. In S. Morris (Ed.), *India Infrastructure Report* (pp. 202-212). New Delhi: 3i Network, Oxford University Press.

Jouve, B. and Christian Lefèvre (Ed.). (1999). *Villes, Metropoles: Les nouveaux territoires du politique*. Paris: Anthropos.

Kanade, A., Deepanand, G.C., Prachi Patnaik, and Santosh Dwivedi. (2006). 'Status Report of ALMs in Mumbai'. *Karmayog* Retrieved April 24, 2007, from http: //www.karmayog.com/cleanliness/almreport.htm

Karande, S., Dipak Gandhi, Madhuri Kulkarni, Renu Bharadwaj, Sae Pol, Jyotsna Thakare and Anuradha De. (2005). 'Concurrent Outbreak of Leptospirosis and Dengue in Mumbai, India, 2002'. *Journal of Tropical Paediatrics, 51*(3), 174-181.

Karmarkar, D. (2005). 'Jagruit Project' [Reference Material], *Orientation Workshop on SWM*. Pune: SIUD/YASHADA.

Karmayog. (2006a). 'Letter to the NGO Council Members dt. April 24, 2006'. *Karmayog* Retrieved May 8, 2007, from http: //www.karmayog.com/ngocouncil/disletterngo council.htm

Karmayog. (2006b). *Local Area Citizens Group Manual 2006—Mumbai.*

Karnataka. (2001). *Census of India 2001—Bangalore Through the Ages*. Retrieved May 10, 2004, from http: //www.bangaloreit.com/html/govtinformation/Censuspaper2/ Paper-2%20Chapter-2.pdf

Karnataka. (2003). 'Waste Management Status of Karnataka'. *State of the Environment Report* Retrieved March 17, 2007, from http: //parisara.kar.nic.in/PDF/Synthesis. pdf

Karnataka. (2005). 'IT & ITES Scenario of Karnataka 2004-2005'. *Department of Information Technology and Biotechnology 2005* Retrieved November 21, 2005, from http: //www.bangaloreit.com/html/register/contactitsec.htm

KCDC. (2004). *Twenty-Eighth Annual Report 2003-2004*. Bangalore: Karnataka Compost Development Corporation Limited.

KEIP. (2007). *Kolkata Environmental Improvement Project*. Retrieved May 16, 2007, from http: //www.keiponline.com

Klink, J. (2002). *Recent Perspectives on Metropolitan Organization, functions and Governance*. Washington, DC: InterAmerican Development Bank.

Klundert, A. and D. A. Justine Anschiitz (2000). *The Sustainability of Alliances between Stakeholders in Waste Management* [Working Paper]. Gouda, The Netherlands: Urban Waste Expertise Programme.

KMC. (2003). 'Winds of Change—Operations and Initiatives 2003', *Capacity Building Programme (CBP)*. Kolkata: Kolkata Municipal Corporation and DFID, UK.

KMC. (2004). 'Budget Statement 2004-2005'. In S. Mukherjee (Ed.), *Kolkata Municipal Corporation*. Kolkata: Central Municipal Office.

KMC. (2005a). 'Budget Statement 2005-2006'. In S. Mukherjee (Ed.), *Kolkata Municipal Corporation*. Kolkata: Central Municipal Office.

KMC. (2005b). *Corporatising the Mindset*. Kolkata: Kolkata Municipal Corporation.

KMC. (2005c). *Dhapa Disposal Groung Hourly Vehicle Entry Report for October 10th 2005*, SWM Department—Unit II. Kolkata: Kolkata Municipal Corporation.

KMC. (2005d). *Kolkata Environmental Improvement Project—Master Plan on Solid Waste Management*. Kolkata: Kolkata Municipal Corporation.

KMC. (2007). *Kolkata Municipal Corporation*. Retrieved May 16, 2007, from http: //www.kolkatamycity.com/index.asp

KMDA. (2001). *Vision 2025—Perspective Plan of* K.M.A. Kolkata: Kolkata Metropolitan Development Authority.

Korten, D. (1987). 'Third Generation NGO Strategies: A Key to People-Centered Development'. *World Development, 15*, 147-150.

KPCB. (2005). *Annual Report 2003-2004*. Bangalore: Karnataka State Pollution Control Board.

Krishna, G. (2001). 'Electricity, But At What Cost?' *Toxics Link* Retrieved June 8, 2007, from http: //www.toxicslink.org/art-view.php?id=1

Krishna, G. (2005). 'Indian Municipal Garbage: How to Manage it?' *Toxics Link* Retrieved June 8, 2007, from http: //www.toxicslink.org/art-view.php?id=87

Krishna, G. (2006a). 'Burning Biomass is not Green'. *India Together* Retrieved June 8, 2007, from http: //www.indiatogether.org/2006/jul/env-timarpur.htm

Krishna, G. (2006b). 'Waste-to-Energy or Waste-to-Pollution?' *Infochange India* Retrieved June 20, 2007, from http: //www.infochangeindia.org/agenda5_14.jsp#top

KUIDFC. (2004). *Status of Urban Infrastructure in Karnataka* [Report]. Bangalore: Karnataka Urban Infrastructure Development and Finance Corporation.

Kumar, S. (2005). 'Municipal Solid Waste Management in India: Present Practices and Future Challenges'. *Hands-on Workshop on Sanitation and Wastewater Management, ADB Headquarters, Manila* Retrieved May 14, 2007, from http: //www.adb.org/Documents/Events/2005/Sanitation-Wastewater-Management/paper-kumar.pdf

Kumar, S. and S. A. Gaikwad (2004). 'Municipal Solid Waste Management in Indian Urban Centres: An Approach for Betterment'. In K. R. Gupta (Ed.), *Urban Development in the New Millenium*. New Delhi: Atlantic Publishers and Distributors.

Kundu, A. (2001). *Urban Development, Infrastructure Financing and Emerging System of Government in India: A Perspective*, Discussion Paper No. 48. *UNESCO, Management of Social Transformations (MOST)* Retrieved August 26, 2006, from http: //www.unesco.org/most/kundu.htm

Kundu, D. (2005). *'Elite Capture in Decentralised Governance in Metro Cities: Lessons from Delhi Experience'*. Paper presented at the Workshop on Inclusive Mega-Cities in India—The Learnings from Mumbai, Mumbai.

Kundu, D. and S.A. Gaikwad (2005). *'Governance in Delhi—A Study of the Decentralized System with Specific Reference to Ward Committees/Civil Society Organizations'*. Paper presented at the Seminar on New Forms of Urban Governance in Indian Mega-Cities, Jawaharlal Nehru University, New Delhi.

Kuttaiah, P. (2003). *Hospital Waste Management* [Term Paper], Centre for Environmental Education and National Law School: India University.

Lardinois, I. and R. Marchand (2000). 'Technical and Financial Evaluation of Composting Programmes in the Philippines, India and Nepal' [Internet Conference]. *Material Flow Analysis of Integrated Bio-Systems* Retrieved May 14, 2007, from http: //www.ias.unu.edu/proceedings/icibs/ic-mfa/lardinois/index.html

Lowndes, V. and Chris Skelcher (1998). 'The Dynamics of Multi-organizational Partnerships: An Analysis of Changing Modes of Governance'. *Public Administration, 76*, 313-333.

Mahadevia, D. (2001a). 'India's Highly Indebted Cities'. *Habitat Debate, 7*(4), 25-26.

Mahadevia, D. (2001b). 'Sustainable Development in India: An Inclusive Perspective'. *Development in Practice, 11*(2&3), 242-259.

Mahadevia, D. (2003). *Globalisation, Urban Reforms and Metropolitan Reforms*. New Delhi: Manak.

Mahadevia, D. (2005a). 'Globalising at Any Cost'. *Infochange Urban India, February* Retrieved August 26, 2006, from http: //infochangeindia.org/urban_india_11.jsp

Mahadevia, D. (2005b). *Metropolitan Employment in India* [Working Papers No. 4]. Ahmedabad: Centre for Development Alternatives.

Mahadevia, D. (2006). 'NURM and the Poor in Globalising Mega Cities'. *Economic and Political Weekly, 41*(31), 3399-3403.

Mahadevia, D. and Anurima Mukherjee (2003). *Municipal Finances and Development in Small and Medium Towns: Comparing Gujurat, Punjab and Andhra Pradesh* [Working Paper]. New Delhi: Institute of Human Development.

Mahadevia, D. and Harini Narayan (2005). Shanghaing Mumbai—Politics of Evictions and Resistance in Slum Settlements [Working Paper No. 7]. Ahmedabad: Centre for Development Alternatives.

Mahadevia, D. and Harini Narayan (2006). *'Slumbay to World Class City—Dynamics of Mumbai'*. Paper presented at the International Conference on 'Tackling Exclusion: Shelter, Basic Services and Citizens' Rights in Globalising Mega-Cities of Asia', Centre for Development Alternatives, Ahmedabad.

Mahadevia, D. and Sandeep Sarkar (2003). *Poverty, Levels of Living and Employment Structure in Small and Medium Size Towns* [Report], Central Statistical Organisation, *Government of India*. New Delhi: Institute of Human Development.

MASS. (1999). *MASS Education as Players for Solid Waste Management & Poverty Alleviation*. Kolkata: MASS Education Publications.

Mathur, M. P., Rajesh Chandra, Satpal Singh and Basudha Chattopadhyay. (2007). 'Norms and Standards of Municipal Basic Services in India'. *National Institute of Urban Affairs, New Delhi* Retrieved June 20, 2007, from http: //www.niua.org/Publications/working_papers/N&S%20 working%20paper.pdf

Mathur, O. P. (2003). *Financing Municipal Services: Reaching Out to Capital Markets* (pp. 1-8). New Delhi: National Institute of Public Finance and Policy.

MCGM. (2003). *Budget, 2003-2004*. Mumbai: Municipal Corporation of Greater Mumbai.

MCGM. (2004). *Budget, 2004-2005*. Mumbai: Municipal Corporation of Greater Mumbai.

MCGM. (2006). *Municipal Solid Waste (Prohibition of Littering and Regulation of Segregation, Storage, Delivery and Collection) Rules 2006*. Municipal Corporation of Greater Mumbai Retrieved June 11, 2007, from http: //www.mcgm.gov.in

MCGM. (2007). Solid Waste Management Department. Municipal Corporation of Greater Mumbai Retrieved May 16, 2007, from http: //www.mcgm.gov.in/forms/departmentmain.aspx?dno=MzQ%3d-nLurYdAdnAE%3d

MCH. (2005a). *Budget 2005-2006*, Municipal Corporation of Hyderabad. Hyderabad.

MCH. (2005b). *Solid Waste Management Action Plan, 2005*, Municipal Corporation of Hyderabad. Hyderabad.

MCH. (2005c). *Hyderabad City Development Plan 2005*, Municipal Corporation of Hyderabad. Hyderabad.

Mehta, M. (1999). *A Review of Public-Private Partnerships in the Water and Environmental Sanitation Sector in India* [Report]. New Delhi: Department for International Development.

Mehta, P. B. (2007). 'A Bleak Urban Future'. *Center for the Advanced Study of India, University of Pennsylvania* Retrieved April 11, 2007, from http: //casi.ssc.upenn.edu/india/index.html

Mendes, R. M., Ahmed Memon Mushtaq and Yedla Sudhakar. (2005). 'Issues and Challenges for Urban Environmental Management in Asia'. In *Urban Environmental Management Challenges in Asia*. Kitakyushu: Institute for Global Environmental Strategies.

Mhapsekar, J. (n.d.). *Environmental Entrepreneurship Program for Rag Pickers of Mumbai*. Mumbai: Sree Mukti Sangathan.

Mitlin, D. (1999). *Civil Society and Urban Poverty—Theme Paper 5* [Working Paper], Urban Governance, Partnership and Poverty Programme. Birmingham: University of Birmingham.

Mitra, S. (2005). *TDR as a Tool for Slum Redevelopment—Geography and Experience in Context of Mumbai [unpublished Master's thesis]*. Centre for Environmental Planning and Technology, Ahmedabad.

MNES. (2004a). *National Master Plan for Development of Waste-to-Energy in India* [Report]. Ministry of Non-Conventional Energy Sources Retrieved June 11, 2007, from http: //www.indiawteplan.com/NMPNote.pdf

MNES. (2004b). *Renewable Energy in India: Business Opportunities* [Report]. Ministry of Non-Conventional Energy Sources Retrieved June 8, 2007, from http: //mnes.nic.in/business%20oppertunity/index.htm

Modi, J. V., Ranjit S. Chauhan and Sneha Palnitkar (2002). *The Wealth of Waste: Waste Recyclers for SWM (A study of Mumbai)*. Mumbai: All India Institute of Local Self Government.

MOEF. (1997). *White Paper on Pollution in Delhi and an Action Plan*. New Delhi: Government of India, Ministry of Environment and Forests.

Mohanty, P. K. (2002). *Innovative Urban Development Financing Practices: A Case Study of Hyderabad City, India* [Working Paper], Building Municipal Capacity for Finance and Budgeting (pp. 1-32). Birmingham: University of Birmingham, International Development Department of the School of Public Policy.

Morris, S. (2002). 'The Challenge to Governance in India'. *India Infrastructure Report 2002*, 15-21.

Morris, S. (Ed.). (2004). *India Infrastructure Report 2004: Ensuring Value for Money*. New Delhi: Oxford University Press.

MPCB. (2005). *Model Tender Documents for Municipal Solid Waste Management* [Prepared by CRISIL Ltd.]. Maharashtra Pollution Control Board Retrieved May 14, 2007, from http: //mpcb.mah.nic.in/muncipal/msw.php

MPCB. (2007b). *Report on Assessment of Electronic Wastes in Mumbai-Pune Area*. Maharashtra Pollution Control Board Retrieved June 1, 2007, from http: //mpcb.mah.nic.in

MPCB. (2007c). *Role of Private Sector Partnership in Solid Waste Management*. Maharashtra Pollution Control Board Retrieved June 8, 2007, from http: //mpcb.mah.nic.in/muncipal/role.php

Mukhopadhyay, T. (2005). *'Decentralization of Urban Finances: A Case Study of Mumbai'*, a paper presented in a seminar on 'New Forms of Urban Governance in Indian Mega Cities'. Paper presented at the Seminar on New Forms of Urban Governance in Indian Mega-Cities, Jawaharlal Nehru University, New Delhi.

NAC. (2005). *Empowerment of Local Government, Stakeholders and Citizens*. National Advisory Council Retrieved May 6, 2007, from http: //nac.nic.in/concept.htm

Nair, J. (2005). *The Promise of the Metropolis—Bangalore's Twentieth Century*. New Delhi: Oxford University Press.

Narayan, L. (2007) 'Envisioning new perspectives to Solid Waste Management'. *Integrated Solid Waste Management for Pune* Retrieved June 8, 2007, from http: //iwm.emcentre.com/index.php?option=com_content&task=view&id=17&Itemid=26

Navtej, K. B. (2005). '*Ward Committees of Mumbai—Functions, Finances and Practices*'. Paper presented at the Seminar on New Forms of Urban Governance in Indian Mega-Cities, New Delhi.

NBJK. (2006). *Clean Jharkhand Project*. Nav Bharat Jagriti Kendra Retrieved April 23, 2007, from http: //www.nbjk.org/cjp/clean_jharkhand.htm

NEERI. (1996). *Strategy Paper on Solid Waste Management in India*. National Environmental Engineering Research Institute Retrieved June 15, 2005, from http: //www.neeri.res.in/swm.html#pastprj

NIUA. (2004). 'Urban Reform Incentive Fund (URIF)—A Status Report'. *Urban Finance—Quarterly Newsletter of the NIUA, 7*(4).

Niyogi, S. (2005, October 22). 'Disaster Waiting to Happen'. *The Times of India*.

ODC. (2003). *Tender Form/Bid Document for the Work of Cleaning, Sweeping, Collection & Transportation of Garbage and Debris*, Office of the Deputy Commissioner. Bangalore: Bangalore Mahanagara Palike.

Patel, K. (2004). *'A Strategy for Solid Waste Management in Areas within AUDA Governed by the Urban Local Bodies' [unpublished dissertation]*. CEPT University, Ahmedabad.

Patel, S. (2004). 'Cleaning up the Dirt!' *Citywatch* Retrieved May 16, 2007, from http: //www.citywatchnews.net/sep03mar04/sep03mar04.html#PRECE

Pedersen, J. (2001). 'India's Industrial Dilemmas in West Bengal'. *Asian Survey, 41*(4).

Pinto, M. R. (2005, January). '*Urban Governance in India-Spotlight on Mumbai*'. Paper presented at the Seminar on New Forms of Urban Governance in Indian Mega-Cities, Jawaharlal Nehru University, New Delhi.

Raju, B., Gupta Sourendu and Ogale Girish. (1997). *Slums: The Magnitude of the Problem*. Retrieved May 16, 2007, from http: //theory.tifr.res.in/bombay/index.html

Ramachandra, T.V. and S.K.V. (2003). 'Exploring Possibilities of Achieving Sustainability in Solid Waste Management'. *Indian Journal of Environmental Health, 45*(4), 255-264.

Rao, S. R. (2000). 'Municipal Affairs in India-A Critical Appraisal of Selected Issues' [Working Paper], *Andrew Young School of Policy Studies*. Atlanta: Georgia State University.

Rathi, S. (2007). 'Optimization model for integrated municipal solid waste Management in Mumbai, India'. *Environment* and *Development Economics, 12*, 105-121.

Redkar, S. (2005). *'New Management Tools for the Solid Waste of Mumbai: Some Concerns'*. Paper presented at the Seminar on New Forms of Urban Governance in Indian Mega-Cities, Jawaharlal Nehru University, New Delhi.

Rondinelli, D. A. (1983). *Secondary Cities in Developing Countries: Policies for Diffusing Urbanization*. Beverly Hills: Sage Publications.

Rosario, A. (1994). *'Decentralized Solid Waste Management Approach'*. Paper presented at the 20th Water, Engineering and Development Centre (WEDC) Conference, Colombo.

Rosario, A. and Anne Sheinberg (2004). 'Integrated Sustainable Waste Management in Bangalore'. In V. D. Vreede (Ed.), *UWEP City Series*. The Netherlands: Urban Waste Expertise Programme.

Sadhu, A. and Sandeep R. Bharatwaj (2003). 'Developing Municipal Bond Markets in India'. In S. Morris (Ed.), *India Infrastructure Report* (pp. 212-218). New Delhi: 3i Network, Oxford University Press.

Salvi, P. (1997). *Integrated Solid Waste Management—Citizens' Perceptions* [Report]. Save Bombay Committee/Hamburger Umwelt Institut Retrieved May 16, 2007, from http: //www.karmayog.com/images/waste_management.pdf

SCMC. (2005). *Fifth Quarterly Report*. Supreme Court Monitoring Committee on Hazardous Wastes Retrieved June 20, 2007, from http: //www.scmc.info/reports/quarterly_reports/fifth_quarterly_report.html

Sekher, M. (2001). 'Tackling Society's Detritus: Stakeholder's Partnerships and Urban Service Delivery'. *Asian Journal of Political Science*, 9(2).

Sekher, M. (2005). *'Ward Committees and Urban Governance: A Comparison of the Post 74th Constitutional Amendment Scenario in South India'*. Paper presented at the Seminar on New Forms of Urban Governance in Indian Mega-Cities, Jawaharlal Nehru University, New Delhi.

Sengupta, T. (2004). 'Suryapet: dustbin-free town'. *Toxics Link* Retrieved April 23, 2007, from http: //www.toxicslink.org/art-view.php?id=23

Seshu, G. (2002). *New Eco-Warriors*. Retrieved May 16, 2007, from http: //www.boloji.com/wfs/wfs051.htm

Singh, R. K. and R.C. Sinha (n.d.). 'Jharkhand Ek Parichaya' [Introduction]. In *Jharkhand*. Ranchi: Gyan Bharati Publication.

Singhal, S. and Suneel Pandey (2001). Solid Waste Management in India: Status and Future Directions. *TERI Information Monitor on Environmental Science*, 6(1), 1-4.

Sivaramakrishnan, K. C. (2004). *Municipal and Metropolitan Governance: Are they relevant to the Urban Poor?* Paper presented at the Forum on

Urban Infrastructure and Public Service Delivery for the Urban Poor, New Delhi.

Snel, M. (1998). 'NGOs/CBOs in SWM in Hyderabad (India)'. *24th Water, Engineering and Development Centre (WEDC) Conference* Retrieved May 14, 2007, from http: //wedc.lboro.ac.uk/conferences/pdfs/24/Snel.pdf

'Status Report on Municipal Solid Waste Management Konkan Division'. (2003). *SWM Department*. Mumbai: All India Institute of Local Self Government.

Sukthanker, D.M. (1995). 'The 74th Amendment of the Constitution: Initiating a Process of Government Reform in Maharashtra'. *Urban India, 15*(1), 13-26.

Sule, S. (2004). 'Municipalities overruling the SC'. *India Together* Retrieved August 28, 2006, from http: //www.indiatogether.org/2004/jul/env-muniswm.htm

Sule, S. (2005). 'Whose Garbage Is It, Anyway?' *India Together* Retrieved May 14, 2007, from http: //www.indiatogether.org/2005/jan/env-ragpick.htm

Supreme Court. (1999). *Report of the Committee Constituted by the Honorable Supreme Court of India, Solid Waste Management in Class I Cities in India.*

SWACHH. (2005). *SWACHH: Alliance of Wastepickers in India*. Retrieved June 20, 2007, from http: //www.wiego.org/ [search: 'SWACHH']

Tariq, M. (2006). 'Solid waste management programme towards self-sustainability at Defence Colony and Gautampuri'. *Toxics Link* Retrieved June 8, 2007, from http: //www.toxicslink.org/art-view.php?id=95

Thakur, S. (2006). *Evaluating the Financial Health of Indian Cities: A Diagnostic Report*. National Institute of Urban Affairs, New Delhi Retrieved June 18, 2007, from http: //www.niua.org/Publications/working_papers/financial_health_sandeep.pdf

ToxicsLink. (2005a). 'Upscaling People's Participation in Urban Solid Waste: Constraints and Prospects'. *Toxics Link* Retrieved May 14, 2007, from http: //www.toxicslink.org/pub-view.php?pubnum=55

Toxics Link. (2005b). 'Waste Management Company Gets a Taste of Democracy'. *Toxics Link* Retrieved June 20, 2007, from http: //www.thesouthasian.org/archives/2005/ waste_management_company_gets.html

Toxics Link. (2007). 'Mumbai Generating 19,000 Tonnes of E-waste Annually: Study'. *Toxics Link* Retrieved May 21, 2007, from http: //enews.toxicslink.org/news-view.php?id=19

UN-Habitat. (2000). 'The Global Campaign for Good Urban Governance'. *Environment* and *Urbanization, 12*(1), 197-203.

UN-Habitat. (2002). *The Global Campaign on Urban Governance*. Nairobi: UN-Habitat.

UNDP. (2006). *Democratic Governance*. United Nations Development Program Retrieved April 15, 2007, from http: //www.undp.org.in/index.php?option=com_content&task=view&id=20&Itemid=78

UNEP. (2007). *Municipal Solid Waste Management—Sound Practices in Composting*. United Nations Environment Program Retrieved May 14, 2007, from http: //www.unep.or.jp/ietc/ESTdir/Pub/MSW/SP/SP4/SP4_2.asp

USAid-India-FIRE. (2004). *Financial Institutions Reform and Expansion. Retrieved* April 14, 2007, from http: //www.usaid.gov/in/our_work/activities/Eco_growth/fir re_d.htm

Vaidya, C. and Hitesh Vaidya (2004). 'Recent Financial Initiatives Among Urban Local Bodies: A Report'. In S. Morris (Ed.), *India Infrastructure Report 2004: Ensuring Value for Money* (pp. 336-339). New Delhi: Oxford University Press.

Varkkey, B. (2003a). 'Ministry of Urban Development and Poverty Alleviation: a Critical Examination'. In S. Morris (Ed.), *India Infrastructure Report* (pp. 132-137). New Delhi: 3i Network, Oxford University Press.

Varkkey, B. (2003b). 'The Private Sector in Municipal Waste Management: The Case of POAB Green'. In S. Morris (Ed.), *India Infrastructure Report* (pp. 286-294). New Delhi: 3i Network, Oxford University Press.

Venkatachalam, P. (2005). *Innovative Approaches to Municipal Infrastructure Financing: A case study on Tamil Nadu, India*. London School of Economics and Political Science, Development Studies Institute Retrieved June 20, 2007, from www.lse.ac.uk/depts/destin

Verghese, G. (2005). *'Centralized and Decentralized SWM Practices in Bangalore'*. Paper presented at the FEDINA Workshop on Decentralized SWM.

Vyasulu, V. (2000). *'Decentralization from Above: Panchayati Raj in the 1990s'*. Paper presented at the International Budget Project, Bangalore, Centre for Budget and Policy Studies.

WBPCB. (2005). *A Status Report on Common Bio-Medical Waste Treatment & Disposal Facilities (CBMWTDF) at Howrah*. Kolkata: West Bengal Pollution Control Board.

WMC. (2007). *Waste Minimization Circle* Retrieved June 8, 2007, from http: //wmc.nic.in/index.asp

World Bank. (1992). *Governance and Development*. Washington, DC: World Bank.

World Bank. (1995). *Better Urban Services: Finding the Right Incentives*. Washington, DC: World Bank.

World Bank. (1997). *World Bank Development Report 1997: The State in a Changing World*. Washington, DC: World Bank.

Yashada. (2004a). *Community Participation in Solid Waste Management in Brihanmumbai Municipal Corporation*. Ford Foundation Project on Good Governance, Urban Local Bodies in Maharashtra Retrieved May 14, 2007, from http: //yashadacgg.mah.nic.in/yashada/cntFileVisit. asp?path=../upload/Uploads/ubp11.doc

Yashada. (2004b). *Solid Waste Management by Nasik Munipal Corporation*. Ford Foundation Project on Good Governance, Urban Local Bodies in Maharashtra Retrieved May 14, 2007, from http: // yashadacgg.mah.nic.in/yashada/cntFileVisit. asp?path=../upload/ Uploads/ubp777.doc

Zurbrügg, C., Silke Drescher, Almitra Patel and H.C. Sharatchandra. (2003). 'Taking a Closer Look at Decentralized Composting Schemes—Lessons from India' [Newsletter]. *Asian Society for Environmental Protection (ASEP)* Retrieved April 14, 2003, from http: //stweb.ait.ac.th/~tony/ asep/publication.php

Zurbrügg, C., Silke Drescher, Almitra Patel and H.C. Sharatchandra. (2004). 'Decentralized Composting of Urban Waste—An Overview of Community and Private Initiatives in Indian cities'. *Waste Management*, 24, 655-662.

Index

Active citizenry participation, 162
Additional Municipal Commissioner, 220, 228-29, 231
Administrative responsibilities, 327
Administrative Staff College of India, 136
Advanced Locality Management, 54, 228, 230
Ahmedabad Municipal Corporation, 312
Aid agencies, 35, 47
ALM
  groups, 231
  groups, 54, 250
  movements, 231, 246
  scheme, 230-31, 245
  system, 67, 231
  trusts, 246
Andhra Pradesh State Pollution Control Board, 142
Animal waste, 163
Assistant Community Coordinator, 264
Assistant Head Supervisor, 221
Assistant Project Manager, 264-65
Awareness building programmes, 277

Background and Project Genesis, 26
Bangalor Municipal Corporation, 119
Bangalore Agenda Task Force, 100
Bangalore Development Authority, 103
Bangalore District Composition, 88
Bangalore Mahanagar Palika, 50, 87-88, 103
Bangalore Municipal Corporation, 87
Bangalore's Urban Waste Composition, 93
Basic conservancy services, 234
Beneficiary participation, 281
Beukering, 25, 33, 47, 91-93, 99, 102, 127
Bhabha Atomic Research Centre, 34, 239
Bhagidari System, 43, 45
Bilateral agencies, 43
Bin-free city, 104
Bio-degradable waste, 34, 46, 61, 78, 101, 115, 117, 134, 179, 229, 292, 325
Bio-medical waste, 3, 18, 37, 41, 92-93, 97, 142-43, 176, 184-85, 189-90, 198, 203-04, 244, 284, 295, 314, 319
  treated, 179
  treatment norms, 143
Bio-methanation process, 243
Biogas plant, 41, 239-40
Biological processing, 20
Bombay Electricity Supply and Transport, 216, 239

Borough Committees, 172-73
Brihanmumbai Electric Supply and Transport, 215
Brihanmumbai Mahanagar Palika, 54
Budgetary allocation, 226-27
Bulk garbage, 143, 155, 284
Bulk waste, 284

Capacity-building
  initiative, 205
  programme, 175
  sessions organised, 269
  training programmes, 17
Census of India, 88, 90, 109
Central Finance Commission, 27, 70
Central Pollution Control Board, 30-31, 36-37, 81
Central Public Health Environment and Engineering Organisation, 19, 36
Central Road Research Institute and Bangalore University Center, 98
Centralised authorities, 328
Centralized Solid Waste Management, 103
Centre for Development Communication, 72, 77
Centre for Environment Education, 42, 53, 100-01, 103
Chuttu processing site, 277
Citizen
  anchor categories, 112
  awareness, 317
  committee, 189
  engagement, 54
  groups, 247
  interventions, 177
  local government initiative, 99
  monitor, 303
  participation, 99, 110
City garbage, 157, 219, 226
City and Industrial Development Corporation, 239
City Development Plans, 17, 74, 82
City Managers, 71
City Municipal Corporations, 103
City planning agencies, 213
Civic Centre, 313
Civil
  services, 5
  society groups, 186
Clean Bangalore, 100
Clean city, 24, 162, 249
Clean Hyderabad, 162
Clean Jharkhand Project, 28, 55-56, 258-59, 261, 270
Clean Mumbai Zones, 254
Clean sites, 325
Cleanest City Certification, 134
Cleaning services, 142
Clinical waste, 190, 198
Collection
  and Transportation of Waste, 94
  bins, 50, 144, 150, 273
  city-wide door-to-door waste, 192
  scheme, 193
  voluntary garbage, 140
Common Bio-Medical Waste Treatment & Disposal, 184
Communicable disease, 299, 300
Communist Party of India, 171
Community
  action network, 264
  awareness campaign, 114
  participation, 14, 44, 67, 109, 111, 113, 124, 128, 186, 268
  vats, 193-94, 199, 202, 204
  volunteer, 142
  waste bins, 112
  waste management, 73
Community-based
  model, 72
  organisations, 27, 43, 177, 233
  solid waste management, 258
  waste management scheme, 101

Community development
departments, 140
societies, 44, 198
Composition of Ward Committee, 208
Compostable waste, 115, 146
Composting
plants, 80, 97, 185, 276
sites, 113, 144, 278
and Waste to Energy, 62
Computerized administration system, 313
Conservancy Wing, 220, 222, 224
Constitutional Amendment Act, 12-14, 27, 36, 75, 90, 172, 174, 180
Container lifting systems, 318
Contract agencies, 157
Contracting systems, 134
Contractor vehicles, 224-25
Contractual
agreements, 105
collectors, 118
labour, 66, 81
obligations, 160
system, 119, 126-27
trucks, 107
workers, 103, 114, 116, 118-19, 198
Cow dung, 59, 62, 108, 144, 217, 243, 295, 297
Credit groups, 197
Credit Rating Services of India Limited, 71
Crow Bar, 163, 165-67

Daily Waste Generation, 291
Debt services, 74
Decentralised
community-based initiative, 72
governance, 285-86
Decentralization of Urban Governance in India, 12
Decentralized scheme, 86, 114
Decentralized Solid Waste Management, 85-86
Decentralized Waste Disposal, 243-44
Delivery and Collection of Municipal Solid Waste, 254
Deputy Municipal Commissioner, 220, 229
Development Communication, 72, 77
Dhapa
landfill site, 176, 194
waste disposal site, 185
Disinfectants, 137-38
Disposable plastics, 186
Disposal
costs, 199
of dead bodies, 217
efficient, 34
final, 20, 22, 104, 146
final garbage, 243
large-scale, 52
of Municipal Solid Wastes, 20, 225
site disputes, 127
of Solid Waste, 3, 119, 181, 192, 217, 227
sustainable, 228
unauthorized, 189
Donor agencies, 101, 186
Door-step garbage collection, 57
Door-to-door
services, 82
waste collection, 42, 126, 202, 228, 238, 244, 273-74, 285-86, 291, 319
Drainage
network, 258
system, 204
Drains
open, 3, 179, 183, 192, 198-99, 273, 284

roadside, 18, 258
storm water, 15, 252, 258
Dry waste, 46, 55, 58, 76, 103, 202, 223, 229, 240, 245, 296-97, 314, 319
Dumper placer
trucks, 146
vehicles, 145-46
Dumping
costs, 52, 58
daily waste, 104
ground, 104
illegal, 112, 193, 202
roadside, 69
sites, 41, 143, 146, 202, 224-26, 273-74, 278, 286, 297, 308
Dustbin Free and Zero Garbage Town, 289

E-based system, 317
E-complaint system, 317
E-governance, 312-13, 322-23
Emptying disposal bins, 115
Entrepreneurship development programmes, 298
Environmental decision-making processes, 174
Environmental Improvement Project, 175
Excel Industries, 67

Familiarisation period, 254
Fire services, 14
Foot paths, 95, 163, 165-67, 296
Formal contractual alliances, 54
Functional responsibilities, 36

Gaddiannarum fruit market, 34
Garbage, 3, 55-57, 75, 103-05, 109-10, 115, 136-37, 141-46, 152, 157, 221-25, 232-34, 273-74, 283-85, 319-20
bins, 307
collected, 134, 150, 222, 228, 243, 275, 291, 319, 322
collection, 134, 150, 222, 243, 291, 322
collection fee, 193
collection points, 163-64, 166-67
collection, primary, 220
commercial, 29
dry, 232
dumps, 249
dumpsites, 275
generated, 291
lifting, 275
mafia, 116
removal, 3
scattering, 194
throwing, 246
truck, 193, 230
wet, 232, 297
Garbage Treatment System, Choice of, 160
Geographical Information Systems, 34, 70, 117, 177
Geological Survey of India, 196
Golconda landfill site, 147
Good city management, 9
Green city, 72
Griridh and Medininagar municipalities, 282

Hand trolleys, 274
Hazardous wastes, 41, 80, 176, 179, 189
Health centers, 92
Heritage sites, 189
High-risk workers, 127
Hospital wastes, 30, 108
Household
affluent, 191
alert, 193
collection, 277-78

disposal, 233
garbage form, 143
waste, 33, 92, 116, 143, 281, 284, 308
Hyderabad Urban Development Authority, 133

ICEF, 258, 277, 279
funding, 268, 279, 283
project, 283
Improved Waste Management System, 247
India-Canada Environment Facility, 28
Indian Institute of Science, 109
Indian Institute of Technology, 34
Indian Medical Association, 295
Industrial
sites, 189
wastes, 29, 31, 36, 68, 112, 179, 190
Inert waste, 20, 78, 105-06, 115, 117-18, 297
Informal
agents, 47
sector waste pickers, 81
waste recovery system, 99
Infrastructure Leasing and Financing Services Ltd, 71, 79
Infrastructure projects, 47, 71, 101, 313
Insurance services, 231
Integrated Urban Environment Improvement Project, 116
International Agreements on Waste Management, 79
Iron Baskets, 163, 165-67

Jamadar, 263, 275

Kalyananagara Waste Committee Monthly Budget, 120
Karnataka Municipal Corporation Act, 87, 88
Karnataka Municipal Reforms Project, 97
Karnataka State Cooperative Marketing Federation, 63, 107
Khadi Gram Udyogh, 318
Kitchen waste, 240, 243, 259
Kolkata Environmental Improvement Programme, 29, 74, 176, 197
Kolkata Municipal Corporation, 170-72, 174, 201
Kolkata Municipal Corporation Act, 180-81
Kolkata Urban Agglomeration, 171
Kolkata's sewer system, 198

Labour contractors, 220, 317
Landfill sites, 21, 78, 82, 105, 145-47, 149-50, 152-53, 157, 160-61, 225-26, 262-63, 284, 297-98, 307-08, 325
engineered sanitary, 88
engineered, 105, 320
identifying potential, 147
integrated sanitary, 105
locating sanitary, 147
location, 161
Large-capacity trucks, efficient, 60
Leachate collection system, 60
Life Insurance Corporation of India, 70
Lifting operation, 165, 168
Litter-free Hyderabad, 157
Litter Free Small Town, 289
Lok Samitis, 264
Lok Satta, 301-02
Lottery system, 159

Maharashtra City Managers, 73
Maintenance Fund, 229
Male workers, 213

Management and Handling Rules, 19
Management Information System, 281
Management costs, 74
Management system, 176
Mandatory responsibilities, 158
Market waste, 29, 198
 vegetable, 243, 276, 319
Mass awareness, 259
Mayor-in-Council system, 180, 216
Mazdoors, 191-93
Micro finance initiative, 320
Ministry of Social Justice and Empowerment programmes, 67
Ministry of Urban Development, 16, 36, 37
Monitor garbage collection, 322
Mumbai Metropolitan Regional Development Authority, 213
Mumbai Municipal Corporation Act, 217
Mumbai slums, 2, 235
Mumbai Urban Agglomeration, 171, 213
Municipal
 management, 94, 170
 solid waste category, 179
 vehicles, 223-25
 waste, 76, 92, 176, 182-83, 194, 202
Municipal Corporation of Greater Mumbai, 67
Municipal Corporation of Hyderabad, 50, 54, 150, 168
Municipal solid wastes, 18-20, 65, 80, 164, 167, 179, 194, 201, 218, 223, 225, 250, 254
 management and handling rules, 18
 physical characteristics of, 218
Municipalities
 member, 13
 peri-urban, 325
 subsidizes, 64
 urban, 19, 22

Nagarpalika, 312-14, 317-25, 327
 officials, 320, 322-23, 326
 workers, 319
National Assistance Programme, 197
National authorities, 35, 36
National Environment Policy, 76, 79
Nav Bharat Jagriti Kendra, 28, 43, 57, 264
Neighbourhood
 associations, 43
 development programme, 236
Neighbourhood planning process, 101
NGO Council, 54, 73, 250
Notified sites, 166

Organic waste, 135, 196, 297, 303
Organising women rag-pickers, 42

Parisar Vikas Programme, 236-37, 240
Partnerships and Alliances, Examples of, 48
Pavement communities, 250
Pig-rearing communities, 284, 290, 295, 303, 309
Pilot project, 121, 233
Pits windrow system, 297
Planning Association of India, 239
Plastic bags, 3, 45, 186, 297, 319
Plastic wastes, 97, 284, 323
 management, 186
Pocket Development Committees, 55, 264, 266
Poverty alleviation programme, 189, 240
Private
 sanitation contract, 306

sector participation, 40, 42, 100, 138
Private contract system, 89, 137
Private-Public Partnership, 100
Processing of municipal solid wastes, 20
Project manager, 264
Property mapping data system, 177
Public Affairs Centre, 99
Public health criteria, 83
Public Private Peoples Partnership, 77, 184
Public services, 12, 40, 270
Public Works Department, 15
Publicity campaign, 189
Pune's City Development Plan, 71

Quality service delivery, 177

Rag-Pickers, 22, 42, 98-99, 115, 127, 145-46, 152, 159-60, 192, 194-96, 204-05, 220, 231, 236-37, 245-47
continuance of, 160
dump site, 236
education and development scheme, 101
engaged, 228
families, 108, 145
organisations, 42
scavenging, 199
trained local, 101
training, 42, 203
Ranchi Municipal Corporation, 28, 56, 258, 262
RCC bins, 164, 167
RDF Power Projects Ltd, 64, 148
Recyclable waste, 30, 42, 45, 72, 99, 115, 134, 141, 144-45, 152-53, 157, 190, 219-20, 229, 303
Constituents of, 219
Recycling
centres, 99
contracts, 238
system, 136, 160
workers, 236
Redressal system, 313
addressing grievance, 281
e-based complaint, 317
on-line complaint, 313
Reform of city management, 74
Regional agencies, 103
Remedial project, 74
Removal and Transportation of Waste, 224
Reserve Bank of India, 239
Resident Welfare Associations, 27, 40, 43, 72, 103, 134, 152, 162, 186, 323
Residential Solid Waste Processing, 38
Residents dump, 59
Revenue programmes, 177
Revolutionary Socialist Party of India, 171
Roads
arterial, 1, 223
bypass, 126
concrete, 197
four-lane, 139
inner, 273
Roadside dustbins, 292

Safai Mitras, 56, 264-67, 270, 273-75, 277-78, 280-81, 283
Safety equipment, 96, 108, 116
Salt Lake City, 194
Sanitary
services, 323
supervisor, 148-49
workers, 50, 148, 263, 314
Sanitation
contracts, 137
services, 252
Scientific Sanitary Landfill Site, 196

Secondary
  Collection and Disposal, 104
  Waste Transportation, 60
Segregation of municipal solid wastes, 20
Self Employed Women, 42, 314
Self Employment Programme, 43
Self Help Groups, 197
Service delivery, 15, 39, 98, 125, 173-74, 180, 186, 192, 201, 203
  decentralized, 173
  non-governmental, 6
  sustainable, 186
SEWA, 27, 42, 57, 314, 317-20, 322-24
Sewage system, 198
Sewer system, ancient, 204
Shriram Energy Systems Pvt, 65
Slaughterhouse waste, 295
Slum Adoption Scheme, 227, 233, 235
Slum development programme, 154
Slum dwellers, 235
Slum dwelling population, 233
Slum improvement initiative, 197
Slum Networking Programme, 45
Slum sanitation programmes, 246
Solid waste, 3-4, 18-19, 52-54, 78-79, 91, 136, 146-48, 164, 183, 191-93, 201-02, 204, 217-20, 225, 245-47
  bulk producers of, 123, 224
  composition of, 32, 92, 135, 219
  disposal site, 81
  dumping, 257
  problems, 106, 110, 251
Solid waste management, 3-4, 14, 18, 25, 27, 36-37, 47, 72, 74, 81, 98, 112, 117, 200-01, 247-48
  component, 47
  effective, 82
  rules, 27-28, 71, 77
  manual for, 36
  master plan on, 81
  new efforts in, 227
  special agencies, 15
Special Local Initiatives, 99
Special programme officers, 282
Sponsored initiative, 99
Stakeholder consultation processes, 175, 177
State pollution control boards, 37, 75, 77
State Road Development Corporation, 15
State Urban Development Department, 262
State Urban Development Funds, 70
Storm water disposal, 216
Stree Mukti Sanghathan, 42, 72, 227, 231, 236, 323
Street
  Beautifiers, 144
  Cleansing & Solid Waste Collection, 201
  lighting, 14, 324
  littering, 228
  sweepers, 127
  Sweeping and Collection of Waste, 223
Suryapet Municipality, 292, 298-300, 307-08
Sustainable
  city model, 157
  cleanliness programme, 101
  development system, 42
Swachchha Bangalore Programme, 100-01, 105, 109-10, 112, 114, 116-19, 121, 123, 125-26, 128
Swarna Jayanti Shahari Rozgar Yojana (SJSRY), 43, 67, 197, 238, 240, 308, 310
Systematic rotation, 310
Systemic problem, 158

Tata Consultancy Services, 239

Thermal gasification, 65, 67, 78
Three-way Partnerships, 56
Thrift and Credit groups, 197
Tipper trucks, 104, 146, 150
Toxics Link, 44, 61, 62, 72-73, 75
Train sanitation workers, 322
Transaction costs, 16
Transport
  waste, 183
  wing, 220, 222, 224
Transportation
  costs, 67, 104, 113, 156, 183
  of garbage, 95
  of waste, 20, 28, 41, 59, 78, 86, 88, 94-95, 126, 145, 156, 224, 237
Treatment unit, 290, 292, 297
Tricycle pullers, 140-41, 144-45, 158
Two-bin system, 292, 299

UNDP, 17, 68
Unhygienic work conditions, 236
Unified city system, 101
University of Agricultural Science in Bangalore, 115
Urban and Industrial wastes, 36, 68
Urban community development services, 154
Urban development departments, 15, 262, 274, 284
Urban employment programme, 308
Urban governance, 4-5, 12, 94, 100, 282, 286
Urban Local Bodies, 12, 14, 25, 37, 64
Urban management, 4-5, 25, 54, 74, 81, 99, 201
Urban planning, 13-14, 62, 100, 207
Urban policy initiative, 22
Urban service delivery, 156
Urban Solid Waste Management in India, 19
Urban Waste Expertise Programme, 44

Valmiki and Dalit communities, 318
Vegetable market associations, 299
Vermi-composting plant, 134, 143-44, 281
Volimory Garbage Disposal Scheme, 152
Voluntary agencies, 140
Voluntary Garbage Disposal Scheme, 54, 134

Ward adoption procedure, 270
Ward Committee Rules, 208-09
Ward Development Committee, 267-68
Ward Programme, 148
Ward Sabha, 110-11
Ward Vision Campaign, 101, 109-11
Waste bins
  industrial, 192
  near, 150
Waste collection/tender bids, 126
Waste collectors, 45, 98, 103, 115-18, 125, 127, 145, 190, 201, 223, 286, 297, 317
  formal, 192
  informal, 192
  itinerant, 98, 220
  private, 127
  regular, 69
Waste Disposed, Amount of, 226
Waste Generated and Composition, 30
Waste heaps, 205
Waste lifting, 274, 318
Waste littering, 302
Waste management, 19, 24-25, 28, 33, 36-37, 43, 57, 67, 73, 79, 103, 246-47, 273-74, 322, 324-27
  bio-medical, 97, 176, 182, 184-85, 190, 204

chain, 21, 45
decentralised, 158
hazardous, 18, 176
initiative, 313-14, 320
market, 180
mechanisms, 114
plan, 101
scheme, implementing, 239
services, Financial sustainability of, 76
tasks, 22, 310

Waste processing facilities, 21
Waste recoverable, 127
Waste recycling, 46, 52, 160, 284
Waste residential, 33, 179
Waste residual, 22
Waste segregate, 78, 152, 159, 223, 229
Waste segregation, 19, 58-59, 189, 228, 238, 240, 246, 285, 299, 310
Waste traders, 192
Waste transport, 42, 221, 267
Waste transportation system, 125
Waste treatment systems, 160
Waste un-segregated, 61, 98, 103, 106, 108, 127, 193
Water pumping costs, 204
Waterlogging, 326
Wet waste, 20, 58, 68, 103, 117, 201, 229, 239-40, 243, 245, 249, 296-97, 314
Wetlands, 171, 194, 196, 202, 204
Women's Rights Movement, 236
Worker training, 54
Works package, 162, 168

Yoga programme, 299

Zero
garbage, 66, 143, 239, 245
waste management, 290, 307

Solid Waste Management in Indian Cities
Status and Emerging Practices

ABOUT THE EDITORS

Prof. Darshini Mahadevia

Prof. Darshini Mahadevia teaches at Faculty of Planning and Public Policy, CEPT University, Ahmedabad. Educated at CEPT, Ahmedabad, School of Planning and Architecture, New Delhi, and PhD from Jawaharlal Nehru University, New Delhi. Has been visiting scholar at number of Universities abroad hand has also collaborated and collaborating with Universities abroad for reserach work. Has wide experience of working with NGOs in India and have number of publications in books and academic journals on urban issues, human and gender development, and environment.
Email: darshini@cept.ac.in

Prof. Jeanne Wolfe

Jeanne M. Wolfe is an Emeritus Professor at the School of Urban Planning, McGill University, Montreal. Has been educated at London University, the University of Western Ontario and McGill, she was Director of the School of Urban Planning from 1988 to 1999. She has worked extensively in the Caribbean region, helped in setting up a planning program at the University of the West Indies in Trinidad with CIDA aid, and has participated in projects in China and India. Her research interests include housing governance and institutional change, urban policy, and international development studies and have widely published in a variety of academic journals.
Email: jeanne.wolfe@mcgill.ca